Current Topics in Membranes, Volume 42

Chloride Channels

Current Topics in Membranes, Volume 42

Series Editors

Arnost Kleinzeller
Department of Physiology
University of Pennsylvania
School of Medicine
Philadelphia, Pennsylvania

Douglas M. Fambrough
Department of Biology
The Johns Hopkins University
Baltimore, Maryland

Current Topics in Membranes, Volume 42

Chloride Channels

Guest Editor
William B. Guggino
Departments of Physiology and Pediatrics
The Johns Hopkins University School of Medicine
Baltimore, Maryland

ACADEMIC PRESS
San Diego New York Boston London Sydney Tokyo Toronto

This book is printed on acid-free paper. ∞

Academic Press, Inc.
A Division of Harcourt Brace & Company
525 B Street, Suite 1900, San Diego, California 92101-4495

United Kingdom Edition published by
Academic Press Limited
24-28 Oval Road, London NW1 7DX

International Standard Serial Number: 0020-2161

International Standard Book Number: 0-12-153342-5

PRINTED IN THE UNITED STATES OF AMERICA
94 95 96 97 98 99 EB 9 8 7 6 5 4 3 2 1

Contents

CHAPTER 4 Anion Channels in the Mitochondrial Outer Membrane

Marco Colombini

CHAPTER 5 Regulation of Chloride Channels in Lymphocytes

Michael D. Cahalan and Richard S. Lewis

CHAPTER 6 Chloride Channels in Skeletal Muscle and Cerebral Cortical Neurons

Andrew L. Blatz

CHAPTER 7 The CFTR Chloride Channel

Michael J. Welsh, Matthew P. Anderson, Devra P. Rich, Herbert A. Berger, and David N. Sheppard

Contributors

Numbers in parentheses indicate the pages on which the authors' contributions begin.

Qais Al-Awqati (59), Departments of Medicine and Physiology, College of Physicians and Surgeons, Columbia University, New York, New York 10032

Matthew P. Anderson (153), Departments of Internal Medicine and Physiology and Biophysics, Howard Hughes Medical Institute, University of Iowa College of Medicine, Iowa City, Iowa 52242

Herbert A. Berger (153), Departments of Internal Medicine and Physiology and Biophysics, Howard Hughes Medical Institute, University of Iowa College of Medicine, Iowa City, Iowa 52242

Andrew L. Blatz (131), Department of Physiology, University of Texas Southwestern Medical Center at Dallas, Dallas, Texas 75235

David R. Burt (215), Department of Pharmacology and Experimental Therapeutics, University of Maryland School of Medicine, Baltimore, Maryland 21201

Michael D. Cahalan (103), Department of Physiology and Biophysics, University of California at Irvine, Irvine, California 92717

Marco Colombini (73), Laboratories of Cell Biology, Department of Zoology, University of Maryland, College Park, Maryland 20742

Raymond A. Frizzell (173), Department of Physiology and Biophysics, University of Alabama at Birmingham, Birmingham, Alabama 35294

David C. Gadsby (317), Laboratory of Cardiac/Membrane Physiology, The Rockefeller University, New York, New York 10021

William B. Guggino (265), Department of Physiology, The Johns Hopkins University School of Medicine, Baltimore, Maryland 21205

Rainer Hedrich (1), Institut für Biophysik, Universität Hannover, D-30419 Hannover 21, Germany

Tzyh-Chang Hwang (317), Laboratory of Cardiac/Membrane Physiology, The Rockefeller University, New York, New York 10021

Thomas J. Jentsch (35), Center for Molecular Neurobiology (ZMNH), Hamburg University, D-20246 Hamburg, Germany

Richard S. Lewis (103), Department of Molecular and Cellular Physiology, Medical Center, Stanford University, Stanford, California 94035

Anibal G. Lopes (265), Instituto de Biofisica Carlos Chagas Filho, Universidade do Rio de Janeiro, Rio de Janeiro, RJ, Brazil

Andrew P. Morris (173), Departments of Physiology and Cell Biology, and Gastroenterology, University of Texas Health Science Center at Houston, Houston, Texas 77030

Devra P. Rich (153), Departments of Internal Medicine and Physiology and Biophysics, Howard Hughes Medical Institute, University of Iowa College of Medicine, Iowa City, Iowa 52242

Erik M. Schwiebert (265), Department of Physiology, Johns Hopkins University School of Medicine, Baltimore, Maryland 21205

David N. Sheppard (153), Departments of Internal Medicine and Physiology and Biophysics, Howard Hughes Medical Institute, University of Iowa College of Medicine, 500 EMRB, Iowa City, Iowa 52242

Michael J. Welsh (153), Departments of Internal Medicine and Physiology and Biophysics, Howard Hughes Medical Institute, University of Iowa College of Medicine, Iowa City, Iowa 52242

Previous Volumes in Series

Current Topics in Membranes and Transport

Volume 19 Structure, Mechanism, and Function of the Na/K Pump* (1983)
Edited by Joseph F. Hoffman and Bliss Forbush III

Volume 20 Molecular Approaches to Epithelial Transport* (1984)
Edited by James B. Wade and Simon A. Lewis

Volume 21 Ion Channels: Molecular and Physiological Aspects (1984)
Edited by Wilfred D. Stein

Volume 22 The Squid Axon (1984)
Edited by Peter F. Baker

Volume 23 Genes and Membranes: Transport Proteins and Receptors* (1985)
Edited by Edward A. Adelberg and Carolyn W. Slayman

Volume 24 Membrane Protein Biosynthesis and Turnover (1985)
Edited by Philip A. Knauf and John S. Cook

Volume 25 Regulation of Calcium Transport across Muscle Membranes (1985)
Edited by Adil E. Shamoo

Volume 26 Na^+–H^+ Exchange, Intracellular pH, and Cell Function* (1986)
Edited by Peter S. Aronson and Walter F. Boron

Volume 27 The Role of Membranes in Cell Growth and Differentiation (1986)
Edited by Lazaro J. Mandel and Dale J. Benos

Volume 28 Potassium Transport: Physiology and Pathophysiology* (1987)
Edited by Gerhard Giebisch

* *Part of the series from the Yale Department of Cellular and Molecular Physiology*

Volume 29 Membrane Structure and Function (1987)
Edited by Richard D. Klausner, Christoph Kempf, and Jos van Renswoude

Volume 30 Cell Volume Control: Fundamental and Comparative Aspects in Animal Cells (1987)
Edited by R. Gilles, Arnost Kleinzeller, and L. Bolis

Volume 31 Molecular Neurobiology: Endocrine Approaches (1987)
Edited by Jerome F. Strauss, III and Donald W. Pfaff

Volume 32 Membrane Fusion in Fertilization, Cellular Transport, and Viral Infection (1988)
Edited by Nejat Düzgünes and Felix Bronner

Volume 33 Molecular Biology of Ionic Channels* (1988)
Edited by William S. Agnew, Toni Claudio, and Frederick J. Sigworth

Volume 34 Cellular and Molecular Biology of Sodium Transport* (1989)
Edited by Stanley G. Schultz

Volume 35 Mechanisms of Leukocyte Activation (1990)
Edited by Sergio Grinstein and Ori D. Rotstein

Volume 36 Protein–Membrane Interactions* (1990)
Edited by Toni Claudio

Volume 37 Channels and Noise in Epithelial Tissues (1990)
Edited by Sandy I. Helman and Willy Van Driessche

Current Topics in Membranes

Volume 38 Ordering the Membrane Cytoskeleton Tri-Layer* (1991)
Edited by Mark S. Mooseker and Jon S. Morrow

Volume 39 Developmental Biology of Membrane Transport Systems (1991)
Edited by Dale J. Benos

Volume 40 Cell Lipids (1994)
Edited by Dick Hoekstra

Volume 41 Cell Biology and Membrane Transport Processes* (1994)
Edited by Michael Caplan

CHAPTER 1

Voltage-Dependent Chloride Channels in Plant Cells: Identification, Characterization, and Regulation of a Guard Cell Anion Channel

Rainer Hedrich
Institut für Biophysik, Universität Hannover, D-30419 Hannover 21, Germany

I. INTRODUCTION

Since the early voltage-clamp experiments from Cole and Curtis (1938,1939) on giant algal cells, excitability has been recognized in membranes of animals and plants. The action potential in the plasma membrane of these algal cells, which were often called "green muscles," is generated by conductance changes with respect to calcium, chloride, and potassium ions (for review, see Tazawa *et al.*, 1987). From voltage-clamp studies we have now learned that this action potential is initiated by calcium ions entering the cytoplasm followed by an activation of a Ca^{2+}-dependent anion conductance which depolarizes the membrane.[1] This depolarization in turn induces a potassium efflux. Thus depolarization through calcium influx and chloride efflux is followed by a potassium-dependent repolarization phase. Even though these transient potential changes were detected simultaneously to those of the squid giant axon the function of the algal action potential is not yet fully understood, and a complete picture of the ion channels mediating the observed conductance changes is still missing. In order to provide a basis for comparison to animal counterparts within the frame of this issue I focus on ion channels of higher plant cells which have been approached at the molecular level by patch-clamp studies (for review, see Hedrich and Schroeder, 1989; Blatt, 1991; Tyerman, 1992) and by cloning of the related proteins (see Section VII, Outlook; Anderson *et al.*, 1992; Sentenac *et al.*, 1992).

In contrast to giant algal cells which allowed the application of conventional voltage-clamp and internal perfusion techniques, the elucidation of ion channels in smaller plant cells had to await the development of the patch-clamp technique (Neher and Sakmann, 1976; Hamill *et al.*, 1981; Schroeder *et al.*, 1984; Hedrich *et al.*, 1986). As in animal cells "the major biophysical advances of the last decade have depended heavily on the ability to isolate single sensory cells—often by using enzymes—and to patch clamp them" (Hille, 1992). The combined use of the patch-clamp technique and the development of new methods to isolate wall-free cells (protoplasts), which essentially were based on the availability of improved enzyme preparations, enabled direct studies on ion channels embedded in the plasma membrane, the vacuolar membrane, and the photosynthetic membrane (Hedrich and Neher, 1987; Schönknecht *et al.*, 1988; Hedrich and Schroeder, 1989).[2]

[1] Please note that anion gradients in plant cells are inverse when compared to animal cells (see Section III,B).

[2] For techniques to isolate vacuoles or inflate the stacked membrane continuum of thylakoids, see Keller and Hedrich (1992).

Even though the presence of ion channels has been shown in many plant cell types, I concentrate on plasma membrane ion transport in a specialized cell type, the guard cell, for three reasons: (i) they control the water status of the plant (for review see Raschke, 1979), (ii) isolated protoplasts of this cell type maintain their physiological function (see below and Zeiger and Hepler, 1977; Schnabl, 1981; Assmann *et al.*, 1985; Serrano *et al.*, 1988; Marten *et al.*, 1991), and (iii) since the first detection of potassium-selective channels, guard cells have become one of the electrophysiologically best-characterized cell type providing a broad background on the physiology and biophysics of ion channels and ion pumps in general (for review, see Schroeder and Hedrich, 1989; Blatt, 1991).

In the detailed presentation of a highly regulated *g*uard *c*ell *a*nion *c*hannel (GCAC1) I, however, compare its properties to anion channels from other plant cells and giant algae (for comparison to animal counterparts, see also Hedrich and Jeromin, 1992).

II. ION TRANSPORT AND STOMATAL MOVEMENT

Terrestrial plants exchange carbon dioxide and water with the atmosphere through turgor-operated cellular valves, the stomata, located in the outer epidermis of leaves and stems. Light, CO_2, and plant hormones have been shown to affect the stomatal aperture between the two pore-forming guard cells (Figs. 1 and 2, see also Raschke, 1979; Marten *et al.*, 1991).

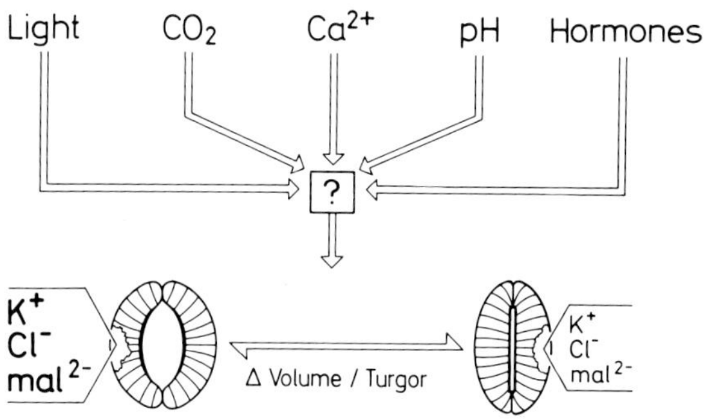

FIGURE 1 Control of stomatal aperture. External signals such as changes in light intensity; CO_2, Ca^{2+}, and H^+ concentration; or the hormone level result in changes in the salt content of guard cells and finally their volume and turgor.

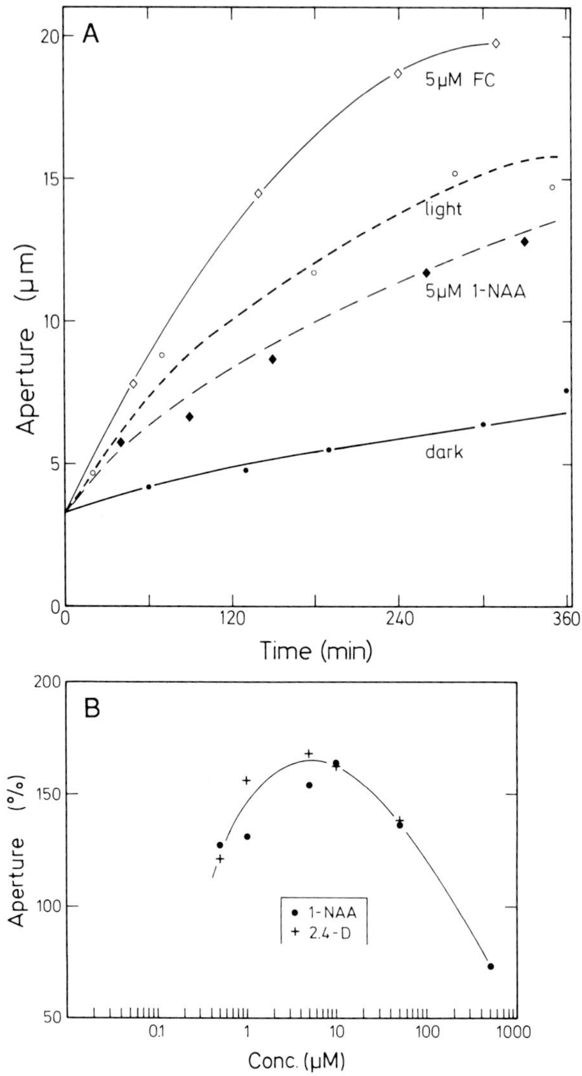

FIGURE 2 (A) Stomatal opening of broad bean epidermal strips in response to auxins (1-NAA, 2, 4-D), light, and the fungal toxin fusicoccin (FC) (Marten *et al.*, 1991, reprinted with permission from *Nature*, copyright 1991 Macmillan Magazines Limited). (B) Concentration dependence of auxin-induced stomatal opening (Lohse and Hedrich, 1992, with permission).

At the beginning of the century it was already possible to correlate stomatal aperture to the potassium content of the guard cells (Imamura, 1943). Microprobe analyses later unequivocally demonstrated that the potassium salts, KCl and K_2 malate, accumulate in guard cells during stomatal opening (Humble and Raschke, 1971).

Thus the increase in turgor and volume in a pair of guard cells, expressed through the characteristic structure of their cell wall, inflates the cell bodies along their axes to form a diffusion pore between them (Raschke, 1979). During stomatal closure potassium salts are released again and the stomatal pore closes. The basic function of guard cell ion transport is to initiate signal transduction pathways on the one hand (Assmann *et al.*, 1985; Serrano *et al.*, 1988; Schroeder and Hagiwara, 1990a; Marten *et al.*, 1991) and to regulate a net flow of ions and establish an electrical driving force that promotes the movement of other ions/solutes on the other (Hedrich and Schroeder, 1989).

Patch-clamp and conventional microelectrode measurements have demonstrated that inward- and outward-rectifying potassium channels represent the predominant potassium conductances in the plasma membrane of guard cells and higher plants in general (Fig. 3; Hedrich and Schroeder, 1989; Schroeder and Hedrich, 1989; for review, see Blatt, 1991). Activated by hyperpolarization and depolarization, these voltage-dependent potassium channels control potassium fluxes during stomatal movement. The resting potential can be decomposed into a K^+ conductance (E_{K^+} = −120 mV) achieved by the activity of inward-rectifying K^+ channels and an outward-directed H^+ ATPase pumping the membrane potential to −200 mV [or even more negative values (Fig. 4) see also Blatt, 1991; Lohse and Hedrich, 1992].

Apart from the resting level, sustained and transient depolarizations, called "action potentials," have been reported in guard cells and attributed to either stomatal closure or the transduction of external and internal signals (Thiel *et al.*, 1992).

In the following I further restrict myself to describing the major role of a voltage-dependent anion channel (GCAC1) in the control of salt release, excitability, and the transduction of extracellular and intracellular signals.

III. GCAC1, A VOLTAGE-DEPENDENT ANION CHANNEL

In order to study plasma membrane anion fluxes the patch-clamp technique was applied to guard cell protoplasts (Schroeder *et al.*, 1984; Raschke and Hedrich, 1990). Patch pipettes were sealed onto protoplasts isolated enzymatically from the epidermis of broad bean (*Vica faba*) leaves

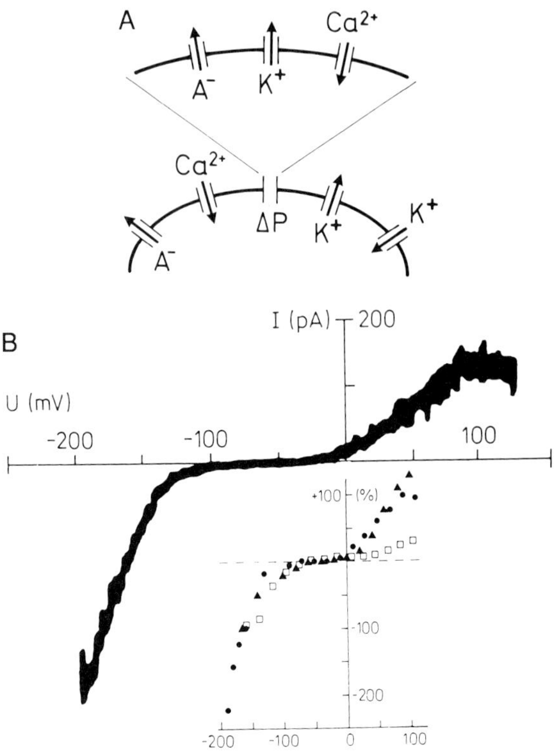

FIGURE 3 (A) Schematic diagram summarizing the ion channels found to date in the plasma membrane of guard cells (modified from Cosgrove and Hedrich, 1991, with permission). (B) Two types of voltage-dependent potassium channels in the plasma membrane of guard cells (current trace and ●), barley aleuron cells (□), and cells from a tobacco tumor cell line (▲). Inward- and outward-rectifying whole-cell K^+ currents following elicitation by a voltage ramp from −200 to 120 mV (R. Hedrich, unpublished).

(Raschke and Hedrich, 1990). Current and voltage recordings were performed in the whole-cell configuration and cell-free membrane patches.

A. Identification

In our initial investigations anion-selective channels appeared around the reversal potential of the potassium ion (−100 to −40 mV). Since these currents were small (25–50 pA) and the membrane resistance was high (> 10 GΩ), fluctuations of single anion channels could be resolved even in the whole-cell configuration. When the membrane voltage was depolarized

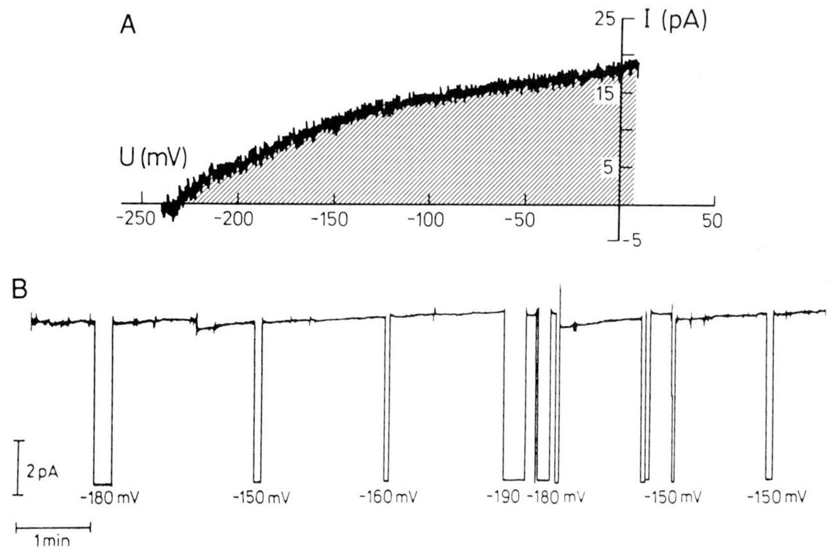

FIGURE 4 Electrogenic proton pumping hyperpolarizes the plasma membrane of guard cells. (A) Time course of whole-cell current and voltage generated by the outward-directed H^+-ATPase of the plasma membrane. (B) Current–voltage relation of the H^+-ATPase. (Lohse and Hedrich, 1992, with permission.)

beyond the potassium equilibrium potential, anion- and K^+-release channels could be recorded simultaneously (Keller *et al.*, 1989).

After replacement of potassium by the K^+ channel blockers Cs^+ or TEA^+ (tetraethylammonium) or impermeable NMG^+ (*N*-methylglucamine), K^+ channels were absent whereas anion channel activity was basically unchanged. In the presence of, e.g., 150 m*M* CsCl on the cytoplasmic side of the membrane and 30 m*M* CsCl in the bath, single 38-pS channels reversed direction close to E_{Cl^-}, indicating a higher permeability of this channel toward chloride than to potassium (Keller *et al.*, 1989).

B. Selectivity

In order to characterize the selectivity filter of the anion channel, patch-clamp experiments were performed with solutions containing salts of non-permeating cations in combination with permeant anions. When chloride in the extracellular solution was substituted by other halides as well as by the physiologically relevant anions, malate, and nitrate, the current reversal potential of GCAC1 was shifted in a way expected for an anion channel with a permeation path different from water (Robinson and Stokes, 1959). Our studies revealed a permeability sequence relative to chloride (Hedrich and Marten, 1993) of

$$\begin{array}{ccccccccc} NO_3^- & \geq & I^- & > & Br^- & > & Cl^- & >> & \text{malate}^{2-} \\ 4.2 & & 3.9 & & 1.9 & & 1 & & 0.1. \end{array}$$

This sequence resembles the Eisenman Series I, indicating "weak" electrostatic interaction(s) of anions with cationic binding site(s) where the change in free energy upon binding of the ion is mostly determined by the dehydration energy (Eisenman and Horn, 1983). When, on the other hand, chloride was replaced by glutamate or gluconate, anion currents were absent. Experiments using organic anions with molecule radii between those of malate and glutamate/gluconate will thus allow the determination of whether the pore in GCAC1 is in the same order of magnitude as that found for animal anion channels (3–5 Å; see Bormann *et al.*, 1987).

C. Single-Channel Conductance Depends on Extracellular Anions

The ionic composition of the extracellular milieu, the cell-wall space, is tissue dependent and varies from almost distilled water up to 50–60 mosmol potassium salts (Speer and Kaiser, 1991). The anion content of the medium around guard cells is dominated by chloride and malate which shuttle across the plasma membrane during stomatal movement (for review, see Raschke, 1979). Thus the cell-wall space of open stomata should be low in anions but should continuously increase during closure when the guard cells release a major fraction of their ionic content (from 500–800 to 100–200 m*M;* Raschke, 1979). Even though reaccumulation of K^+ salts by the mesophyll and/or neighboring cells may take place, the anion concentration changes manyfold.

Patch-clamp studies have revealed that the peak amplitude of anion currents increases as the extracellular chloride concentration is raised from 0 up to 300 m*M*. This reflects an inverse dependence of GCAC1 on the chloride gradient across the plasma membrane. In addition to whole-cell recordings it could be documented through single-channel experiments that the rise in the macroscopic current results from an increased single-channel conductance (Fig. 5, Hedrich and Marten, 1993). The chloride sensitivity of GCAC1 lies within the physiological range of the external anion content (Bowling, 1987; Speer and Kaiser, 1991).

This behavior is reminiscent of the dependence of the K^+ channel RCK4, a shaker K^+ channel subtype from rat brain, on external potassium (Pardo *et al.*, 1992). In contrast to GCAC1, extracellular ions increase

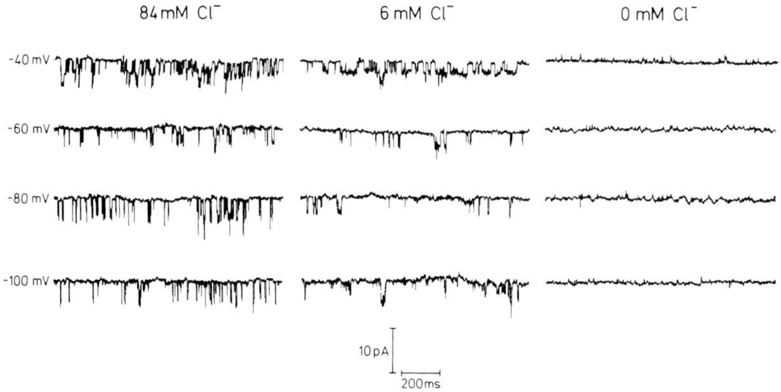

FIGURE 5 The single-channel conductance of GCAC1 reflects the extracellular chloride concentration. Decrease in chloride concentration suppresses anion release through the anion channel (Hedrich and Marten, 1993, with permission of Oxford University Press).

the mean open time or gating charge displacement of RCK4 rather than its single channel conductance. Whereas the role of a regulatory, extracellular ion binding site(s) in RCK4 expressing nerve cells is still unknown, one might suppose that the inverse relationship of the GCAC1 single-channel amplitude to the anion gradient allows anion efflux to occur and thus stomatal closure to proceed even at increasing extracellular anion levels.

D. Voltage Dependence and Kinetics

Current–voltage curve analysis of whole-cell ion fluxes has revealed that outward chloride fluxes are activated near −100 mV and peak at −30 to −40 mV. Chloride currents decline through the decrease in the electrochemical driving force when the membrane potential is stepped to values more depolarized than −30 mV, reversing direction at the Nernst potential for chloride (Fig. 6 and Hedrich *et al.*, 1990).

Activation kinetics of the anion channel have usually been determined by a series of voltage pulses where the membrane potential is clamped to the resting potential of the cell (negative to −100 mV; Hedrich *et al.*, 1990; Lohse and Hedrich, 1992). Upon stepping the voltage to values between −70 and −10 mV, currents activate with a half-time between 30

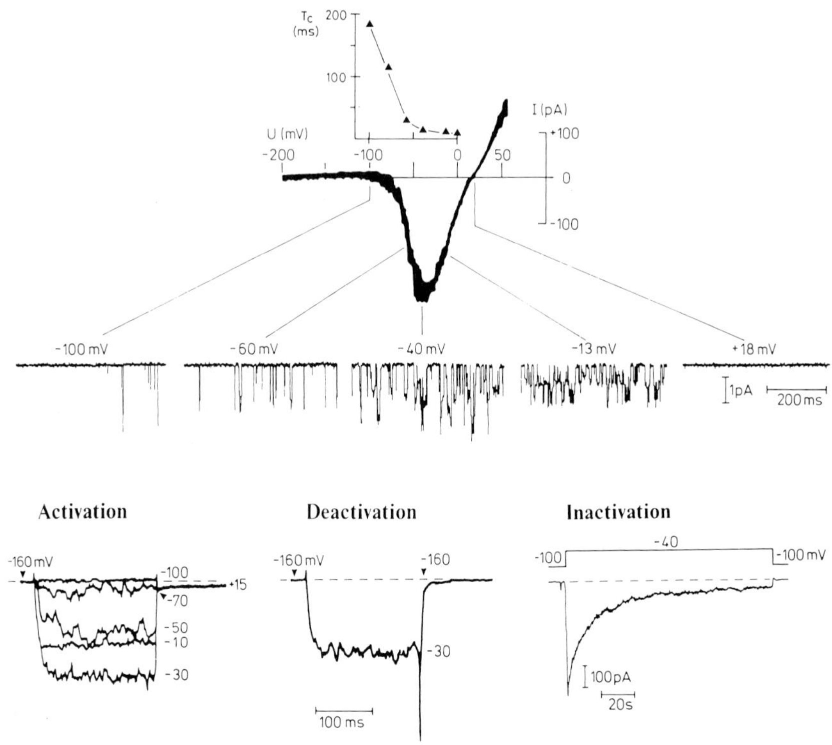

FIGURE 6 Voltage-and time-dependent activity of anion channels recorded in whole cells and outside-out patches. Current–voltage relation of whole-cell anion currents during a voltage ramp from −200 to 60 mV (upper traces). Closed times (T_c) of single anion channels (insets) recorded at voltages indicated. Anion currents generated by voltage pulses from a holding potential of −160 to −100, −70, −50, −30, and −10 mV each followed by a pulse to +15 mV (activation). Tail current resulting from a pulse to −30 mV followed by a pulse to −160 mV (deactivation). Inactivation resulted from prolonged voltage stimulation. (Hedrich *et al.*, 1990, with permission of Oxford University Press).

and 10 msec. During subsequent repolarization to the resting potential, large tail currents correspond to the instantaneous rise in the electrical driving force for anion efflux and the complete deactivation of GCAC1 takes place in less than 20 msec. Analysis of the single-channel activity under steady-state conditions have revealed that "deactivation" results from an increase in the mean closed-time of the anion channel with hyperpolarizing potentials.

During prolonged depolarization anion channels inactivate in a voltage-dependent manner (Hedrich *et al.*, 1990; Marten *et al.*, 1992, 1993). Inactivation is slow (half time 10–12sec) with respect to the inactivation of animal Na^+, K^+, and Ca^{2+} channels and activation kinetics of GCAC1.[3]

Anion channel activity could be restored from inactivation during repolarization.

Using the same experimental system and voltage protocol the kinetics of voltage-dependent activation and de- and inactivation of GCAC1 (Hedrich *et al.*, 1990) have been confirmed by Schroeder and Keller (1992).[4]

E. Calcium and Nucleotides

Anion currents in the presence of MgATP an 0.1–1 m*M* of the calcium buffer EGTA and in the absence of exogenous calcium in the cytoplasm (pipette solution) and 0.1 m*M* calcium in the extracellular medium did not exceed the background level until the external Ca^{2+} concentration was raised to ≥ 10 m*M* (Hedrich *et al.*, 1990). Upon increasing Ca^{2+} the current amplitude increased 5- to 20-fold. External Ca^{2+} failed, however, in activating GCAC1 when (i) cytoplasmic Ca^{2+} was buffered to nanomolar concentrations using 10 m*M* EGTA or (ii) MgATP was excluded from the pipette solution and ATP scavengers were present. These experiments documented the requirement of Ca^{2+} and MgATP for channel activation. Since Ca^{2+}- and nucleotide-dependent activation peaked 1–2 min after the whole-cell configuration was established, phosphorylation of the anion channel or regulator(s) may account for the delay. However, apart from MgATP (2–10 m*M*), ATPγs and GTPγs at 100 μ*M*, in the absence of ATP scavengers, were able to support Ca^{2+}-dependent activation. Thus it is unclear whether low levels of endogenous ATP suffice for phosphorylation or whether channel activation requires only the binding of nucleotides.

As mentioned in the Introduction, voltage-clamp experiments on giant algal cells have shown that a Ca^{2+}- and ATP-modulated anion conductance is involved in the action potential (Lunevsky *et al.*, 1983). A more detailed comparison to GCAC1 has, however, to wait until the single-channel equivalent(s) of the anion conductance is (are) identified.[5]

[3] It should be noted that the whole-cell anion currents did not decay completely but to a background level (25–50 pA) representing less than 20% of the current (Keller *et al.*, 1989; see also below).

[4] Their GCAC1 was named r-type for rapid activating current.

[5] Promising first attempts to answer this question through microdissection of the cell wall and attachment of patch pipettes have been made (Laver, 1991). Subsequent studies using intracellular perfusion or cell-free membrane patches might thus give new insights into the regulation of these channels by cytoplasmic factors.

F. Identification of the Putative Channel Protein

In a survey of compounds capable of interacting with the anion channel we have selected IAA-94 and IAA-23 in order to test their suitability as potential ligands for the isolation of the channel protein. We have selected these two compounds in particular since they have already been successfully employed in the isolation of anion channels from kidney epithelia (Landry *et al.*, 1989; Redhead *et al.*, 1992; see also Chapter 3, by Q. Al-Awqati).

IAA-23, a structural analogue of IAA-94, was used to enrich ligand-binding polypeptides from the plasma membrane of guard cells. From this protein fraction a 60-kDa polypeptide cross-reacted specifically with polyclonal antibodies raised against anion channels isolated from kidney membranes. In contrast to guard cells, mesophyll plasma membranes were deficient in voltage-dependent anion channels and lacked cross-reactivity with the antibody (Marten *et al.*, 1992).

When membrane fractions, other than plasma membranes, were analyzed immunologically cross-reactions with other polypeptides were observed as well. Whether this represents anion-permeable channels in vacuolar, photosynthetic, or mitochondrial membranes (for review, see Hedrich and Schroeder, 1989; Keller and Hedrich, 1992) needs further investigation.

IV. GATING MODIFIERS AND INHIBITORS

A. Growth Hormones

The opening of the stomatal aperture is elicited when guard cells are exposed to the growth hormones auxins (see Fig. 2A). Dose–response curves for auxins show an optimum around 5×10^{-6} *M*, followed by a decay at higher concentrations (see Fig. 2B; Marten *et al.*, 1991; Lohse and Hedrich, 1992). This indicates that guard cells possess recognition sites for auxins as well as mechanisms for translating the hormonal signal into ion fluxes which, in turn, cause volume increase and stomatal opening.

The early events in auxin signaling were characterized by patch-clamp studies on the hormone interaction with channels of guard cell protoplasts isolated from auxin-responsive epidermal layers. When auxins were applied to the external solution the activation potential and current amplitudes were altered in a dose-dependent manner. A comparison of the whole-cell currents and the single-channel activity ($N \cdot P_0$) in excised membrane patches (Fig. 7) demonstrated that macroscopic anion fluxes

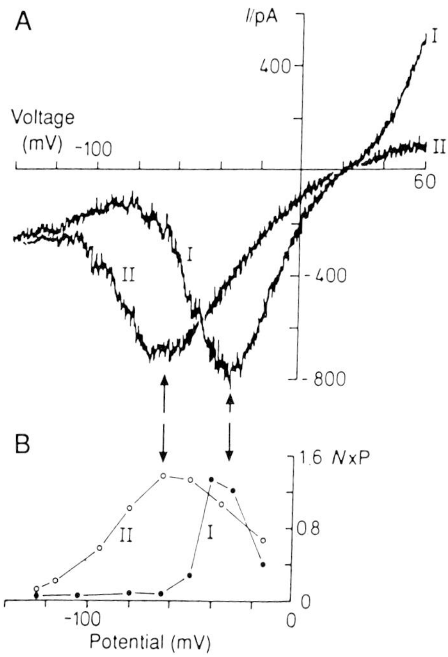

FIGURE 7 Modulation of the activation potential of GCAC1 by auxin. Comparison of whole-cell anion currents and single-channel activity (A) before (I) and after auxin stimulation (II). Single-channel activity (B) is expressed by the number (N) of single channels in the cell-free membrane patch and their open probability (P_0). (Marten *et al.*, 1991, reprinted from *Nature*, copyright 1991 Macmillan Magazines Limited.)

were generated by auxin-sensitive anion channels. The slope of the current–voltage curve results from a shift in the threshold potential of activation toward more negative voltages superimposed by a voltage-dependent block (Marten *et al.*, 1991,1993).

The specificity of the auxin interaction with GCAC1 is further characterized by:

(i) reversibility of channel modulation upon removal of auxins from the medium;
(ii) side-specific recognition site(s) for auxins located only at the external face of the channel;
(iii) no requirement of cytoplasmic factors for hormone targeting through the anion channel;
(iv) discrimination against other phytohormones;
(v) channel-specific interaction since the inward-rectifying K^+ channel is not affected.

Furthermore, preliminary experiments have shown that the level of expression of a putative auxin receptor (Palme, 1991) in transgenic tobacco alters the dose dependence of the auxin-induced stomatal opening (G. Lohse and R. Hedrich, unpublished) and that voltage-dependent anion currents exist in cells of an auxin autotroph tobacco culture (Fig. 8, P. Dietrich and R. Hedrich, unpublished).[6]

In the absence of auxin the voltage gate of these cells is located at more hyperpolarized potentials compared to guard cells, reminiscent of GCAC1 in the presence of high hormone levels. Future studies will show whether "receptor" channel density or the location of its voltage gate is crucial for hormone sensitivity.

B. Malate

Is malate a possible CO_2 sensor; how do guard cells sense the intercellular CO_2 concentration?

Stomata close upon an increase in intercellular CO_2, one of the conditions reflecting the inability of the photosynthetic apparatus to take advantage of excess CO_2. Low CO_2 levels, on the other hand, cause stomatal opening (for review, see Raschke, 1979).

Sudden changes in CO_2 concentration and light intensity cause oscillations in stomatal aperture, a feedback mechanism which might involve activation–inactivation cycles of GCAC1.

Photosynthetic subtypes are characterized by the pattern of their assimilation products, where malate content, transport, and its cellular and subcellular compartmentation reflect the metabolic level of different cell types.[7]

When the extracellular side of the plasma membrane is exposed to malate the amplitude of whole-cell anion currents increases while the voltage gate shifts toward hyperpolarized potentials (Fig. 9A), a behavior we have already recognized during auxin action (see Fig. 7; Marten *et al.*, 1991). The half-saturation constant for malate action was 0.4 m*M*, a concentration which is well within the physiological range of this dicarboxylate (Speer and Kaiser, 1991). Malate levels above 1 m*M* shifted the voltage gate by almost −80 mV (Hedrich and Marten, 1993).

The increase in current amplitude in the presence of this extracellular effector is a consequence of channel openings at a voltage range of higher

[6] Similar observations on another tobacco cell line have also been reported at the 9th International Workshop on Plant Membrane Transport by Barbier-Brygoo, 1992.

[7] One exception are the Crassulaceae (CAM plants) to which most of the desert plants belong and where stomata open at night, but close during the day, opposite to the other plant types.

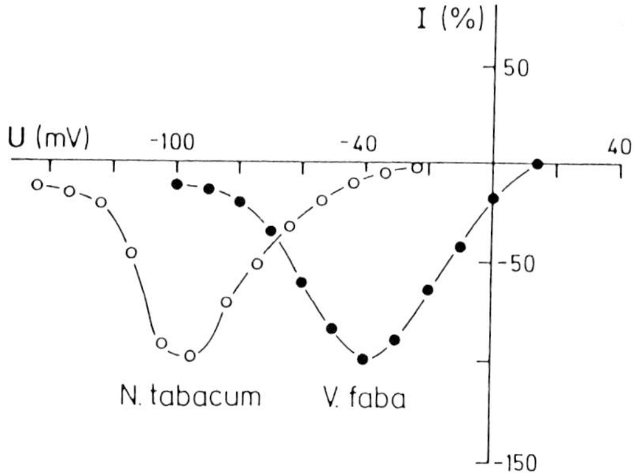

FIGURE 8 Two GCAC1-like anion channels with different working ranges of activity, respectively, with the location of their voltage gate in guard cells (●) and in an auxin-autotrophic tobacco cell line (○). (P. Dietrich and R. Hedrich, unpublished.)

driving force for anion release. On the single-channel level malate causes a shift in voltage dependence rather than a change in unit conductance (cf. Section III,C.).

External malate released from the photosynthetic tissue, neighboring cells, or guard cells during stomatal closure, therefore, causes the following chain of events (cf. Fig. 13B): (i) a shift of the voltage gate of GCAC1 toward the resting potential; (ii) induction of anion release, a fraction of which is malate (Van Kirk and Raschke, 1978); (iii) elevated malate levels in turn feedforward anion release. After inactivation to the background conductance (Fig. 9B, R. Hedrich and I. Marten, 1993) release of anions proceeds until malate dissociates from GCAC1 and the binding site is occupied by excess chloride (cf. Section III,C.), which in turn causes resetting of the voltage gate.

Therefore malate-sensitive anion channels may provide a mechanism whereby stomata adapt to the current photosynthetic capacity of the leaf. Upon changes in the environmental conditions cross-talk between guard cells and the photosynthetic tissue should enable the control of CO_2 uptake and water loss.

C. Anion Channel Blockers

When anion channel inhibitors are applied to the bathing solution both the voltage dependence and current amplitude are altered. The shift and block of the anion channel saturate as a function of the effector concentra-

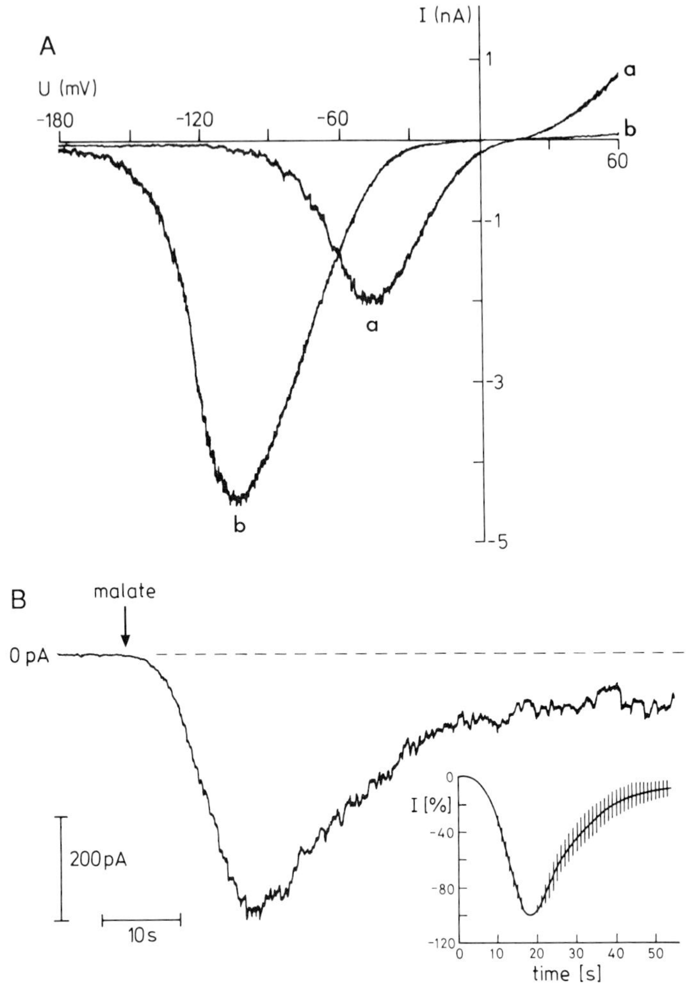

FIGURE 9 Modulation of the plasma membrane anion channel of guard cells by malate. (A) Malate-induced shift in the activation potential and peak anion fluxes. Current–voltage relation of whole-cell anion currents before (a) and during (b) bath perfusion with malate resulting from voltage ramp stimulation. (B) Transient activation of GCAC1 following malate application. Time course of malate-induced whole-cell anion currents in guard cells clamped to −90 mV. Anion currents of a single guard cell (inset) and malate-induced ion fluxes of three different cells normalized with respect to the peak current. (Hedrich and Marten, 1993, with permission).

tion. However, the shape of the curve and thus the kind of interaction with a putative "binding site" differs strongly (Marten *et al.*, 1993). Whereas the channel block can be explained by an interaction of the ligand with the open mouth of the channel, the "gate-shifting" action may result from adsorption of the ligand(s) at sites which directly or indirectly alter the intramembraneous electrical field (cf. Armstrong and Cota, 1991).

The effects on GCAC1 elicited by anion channel blockers are characterized by reversibility of modulation when effectors are removed from the extracellular medium,[8] and side specificity since cytoplasmic inhibitor levels up to 100 μM are not effective. The sensitivity of the channel to inhibitors[9] is expressed by the sequence of the inhibitor constants (K_d in micromolars, Marten *et al.*, 1992,1993): PA (>100) < A−9−C (100) < NA (20) < IAA−94 (7) < NPPB (4) = SITS (4) < DNDS (0.5) < DIDS (0.2).

Interaction with the inhibitor (Fig. 10, right panel) is characterized by a flickering block (Gögelein, 1988). Whereas DIDS only affected ion permeation inhibition by e.g., SITS and DNDS was superimposed with a shift in the voltage gate.[10] As expected from voltage (depolarization)-dependent blockers the decay times of the anion current are strongly reduced at positive potentials (Fig. 11; Marten, 1993; Marten *et al.*, 1993).

In sections IV, A–C I have shown that extracellular growth hormones, the key metabolite malate, and anion channel blockers are able to alter the voltage sensor (Marten *et al.*, 1991,1993; Hedrich and Marten, 1993). All gating modifiers studied so far were anionic under our experimental conditions. It is thus tempting to speculate that the nature of the ligand occupying an "anion binding site" at the external face of GCAC1 controls the location of the voltage gate. On the other hand residues such as aryl, alkyl, or sulfonyl groups may determine whether and how the compound occludes the channel pore.

[8] Prolonged exposure of guard cells to NPPB and DIDS irreversibly increased the leak conductance.

[9] PA = *p*-(dipropylaminosulfonyl) benzoic acid; A-9-C = anthracene-9-carboxylic acid; NA = 2-(α,α,α-trifluoro-*m*-toluidino)-pyridine-3-carboxylic acid; IAA-94 = [(6,7-dichloro-2-cyclopentyl-2,3-dihydro-2-methyl-1-oxo-1H-inden-5-yl)oxy]acetic acid; NPPB = 5=nitro-2-(3-phenylpropylamino)benzoic acid; SITS = 4-acetamido-4-isothiocyanostilbene-2,2′-disulfonic acid; DNDS = 4,4′-dinitrostilbene-2,2′-disulfonic acid; DIDS = 4,4′-diisothiocyanostilbene-2,2′-disulfonic acid.

[10] The possibility that DIDS shifts the voltage sensor out of the potential range tolerated by guard cell protoplasts cannot be excluded.

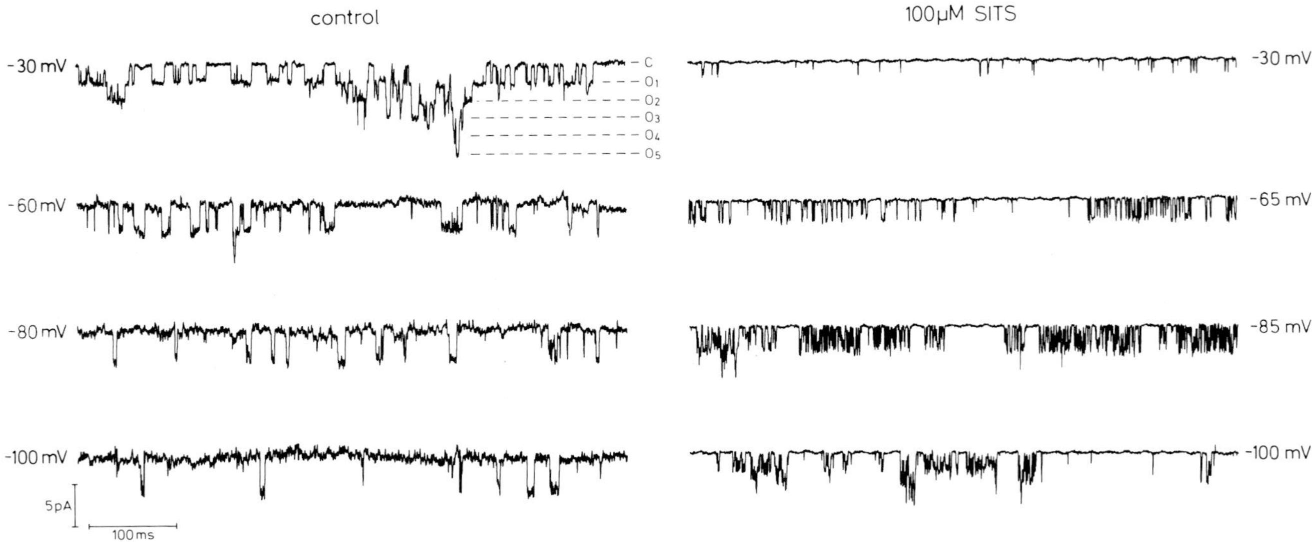

FIGURE 10 Modulation of the single-channel properties of GCAC1 by the stilbene derivative SITS. Single-channel activity recorded from a cell-free membrane patch in the absence (left) and presence of SITS (right). (Hedrich *et al.*, 1990, with permission).

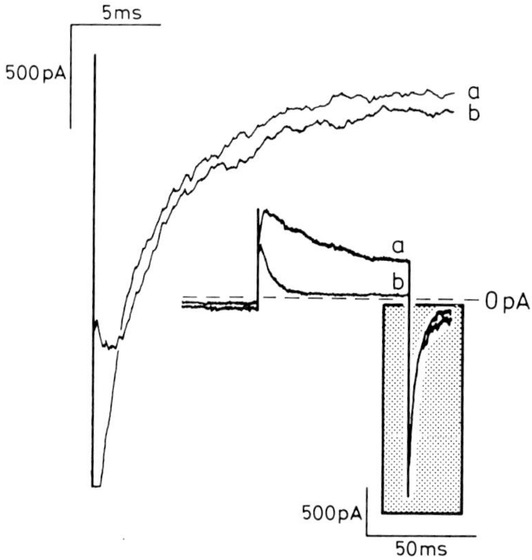

FIGURE 11 Voltage-dependent block of the anion current by the stilbene derivative SITS. Depolarization-(+56 mV) induced anion currents in the absence (a) and presence (b) of 10 μM SITS. Repolarization to −204 mV causes tail currents (framed and enlarged on the left). Note that tail currents in the presence of SITS could be induced by repolarization after a prepulse where currents were absent. The increase of the tail current represents unblock (Marten *et al.*, 1993, with permission).

V. DIVERSITY OF PLANT ANION CHANNELS

A. Anion Channels in Higher Plants and Algae

Anion channels comprise the most diverse class of ion channels in animal cells, with respect to both the single-channel conductance and their regulation (Frizzell and Halm, 1991; Greger, 1992; and chapters of this issue). Table I summarizes related features of plant and algal anion channels.[11] The conductance spectrum ranges from 1 to 200 pS. Here, three conductance levels can be distinguished: 1-pS channels, which could only be detected by noise analysis (Schroeder and Keller, 1992); 20- to 50-pS channels, e.g., GCAC1; and large conductance anion channels. With respect to their voltage dependence, anion channels fall into two major classes at least: inward- and outward-rectifiers. A third class is activated mechanically by stretching the membrane (for animal counter-

[11] For the sake of comparison studies which lack single-channel or noise analysis have not been included.

TABLE I

Comparison of the Basic Characteristics of Plant and Algal Anion Channels[a]

	Membrane	Conductance (pS)	Selectivity	Activation	Reference(s)
Plant					
Suspension cells	PM	100	$Cl^- > K^+$	Hyperpolarization	Schauf and Wilson (1987)
Asclepias tuberosa					
Suspension cells	PM	200	$NO_3^- > Cl^- > K^+ > ASP^-$	Hyperpolarization	Terry *et al.* (1991)
Amaranthus tricolor					
Suspension cells	PM	15		Depolarization	Zimmermann (pers. comm.)
Daucus carota					
Suspension cells	PM	100		Hyperpolarization ATP	Barbara *et al.* (1994)
Nicotiana tabacum					
Roots					
Triticum aestivum	PM	4	$NO_3^- \geq Cl^- > I^- >> PO_4^{3-}, ClO_4^-$		Skerrett and Tyerman (1994)
Triticum turgidum					
Mesophyll cells	TM	65–150	$NO_3^- > Cl^-$	Depolarization	Schönknecht *et al.* (1988)
Peperomia metallica					
Cotyledons	PM	5–40		Voltage independent	Lew (1991)

Arabidopsis thaliana					
Stem cells	PM	86 ; 146	$Cl^- > K^+$	Stretch	Falke *et al.* (1988)
Pisum sativum					
Epidermal cells	PM	300	$NO_3^- > Cl^- = Br^- > I^- > F^- > mal^{2-}$	Hyperpolarization	Theo *et al.* (1994)
Nicotiana tabacum					
Guard cells	PM	34 ; 59	$A^- > K^+$	Stretch	Schroeder (1987)
Commelina communis					
Guard cells				Depolarization	Keller *et al.* (1989); Hedrich *et al.* (1990);
Vicia faba	PM	24–39	$NO_3^- > Cl^- > mal^{2-}$	Ca^{2+}, ATP	Hedrich and Marten (1993); Linder and Raschke (1992)
Xanthium strumarium					
Guard cells	PM	1/33		Depolariation	Schroeder and Keller (1992);
V. faba				Ca^{2+}	Schroeder *et al.* (1993)
Guard cells	PM	27 ; 13	$Cl^- > K^+$	Stretch	Cosgrove and Hedrich (1991)
V. faba				Depolarization	
Algae					
Chara australis	PM	10–55			Laver (1991)
C. australis	PM	7–44		Hyperpolarization	Coleman (1986)
C. corallina	PM	9–25		Hyperpolarization Ca^{2+}	Okihara *et al.* (1991)
C. corallina	PM	16 ; 52		Depolarization (transient) Ca^{2+}, Ba^{2+}	G. Thiel (personal communication)
C. australis	VM	21	$NO_3^- > Cl^- > Asp^-$		Tyerman and Findlay (1989)

[a] PM, Plasma membrane; TM, Thylakoid membrane; VM, Vacuolar membrane; A^-, anion.

parts, see Sachs, 1986; Morris, 1990). The role of Ca^{2+} and nucleotides is unclear or has not been studied in cells other than guard cells (see Section IV); a detailed inhibitor analysis is not yet available.[12]

B. Different Anion Channels or Different Modes of GCAC1?

Just before Keller *et al.* (1989) identified GCAC1,[13] Schroeder and Hagiwara (1989) detected an anion conductance in whole cells with a voltage dependence, Ca^{2+} activation, and kinetics remarkably different from those of GCAC1. Ionic currents were elicited by membrane polarization only when cytoplasmic Ca^{2+} was buffered to or above 1.5 μM, with the membrane clamped to a depolarized potential (+40 mV) close to the anion equilibrium. These currents activated almost instantaneously and did not deactivate (when compared to those in Fig. 6), during 1 sec pulses to potentials negative and positive of +40 mV indicating that channels activate fast (Schroeder and Hagiwara, 1989) or were already open at the holding potential. The instantaneous current–voltage curve revealed an almost ohmic behavior whereas in the steady-state anion fluxes were supposed to exhibit the same current–voltage relation as GCAC1 (Schroeder and Hagiwara, 1990b).

However, in recent experiments of Schroeder and Keller (1992), where cytoplasmic Ca^{2+} was buffered to 150 nM and the membrane held at −100 mV, anion currents could be elicited by extremely long pulses to depolarized potentials as well. Under these conditions anion currents activated very slowly. Since 10 sec voltage steps did not create steady-state currents a 15-min ramp from +40 to −100 mV was used to obtain a quasi-steady-state current–voltage curve. In contrast to the steady-state current–voltage curve of anion currents in high cytoplasmic calcium which was identical to GCAC1-like currents the current–voltage curve derived from a 15 min ramp revealed (i) a less steep voltage dependence, (ii) a peak current at a potential of about 0 mV (−30 to −40 mV for GCAC1), (iii) only partial deactivation at hyperpolarized potentials, and (iv) a unitary conductance of about 1 pS (30–40 pS for GCAC1).

In order to distinguish this slow-activating current from GCAC1, which was often superimposed on recordings, GCAC1 was named r-type (r for rapid) and the former Ca^{2+}-dependent type (Schroeder and Hagiwara, 1989) was named s-type (s for slow; Schroeder and Keller, 1992).

[12] In contrast to Keller *et al.* (1989), studies by Marten *et al.* (1993) did not observe a Zn^{2+} block on GCAC1.

[13] At that time, the voltage-dependent anion channel was not yet named GCAC1. This terminology was introduced by Hedrich and Jeromin (1992) and Marten *et al.* (1992).

Linder and Raschke (1992), have also studied the slow-activating anion conductances in guard cells of broad bean and *Xanthium strumarium* (here called slow anion channel, SLAC) and have shown s-type-like kinetics. However, the single-channel conductance (33 pS)[14] and steady-state current–voltage curve characterized by a peak current potential of −30 to −40 mV, were very similar to those of GCAC1. In addition, potent inhibitors GCAC1 such as NPPB and IAA-94 block s-type channels, too (Schroeder *et al.*, 1993).

Because of the striking similarity of GCAC1 and SLAC on the one hand with respect to voltage dependence and unitary conductance and, on the other, due to their differences in kinetics, one might suggest that GCAC1 and SLAC/s-type represent different gating modes of the same channel. Thus the peculiar properties of GCAC1 reviewed here serve as diagnostic tools for answering the question of relationship.

VI. VOLUME REGULATION, EXCITATION, AND SIGNAL TRANSDUCTION

A. Downregulation of Volume and Turgor

Above we have discussed how the voltage window negative to the reversal potential of K^+ and positive to those of anions could provide the working range for anion and K^+ channels to mediate salt release during stomatal closure (Fig. 12, shaded areas).

Under experimental conditions the activity of the anion channel can be triggered by a rise in cytoplasmic Ca^{2+} (through sudden application of steep Ca^{2+} gradients; extracellular high). However, a spontaneous increase in anion channel activity emerged sometimes during repetitive voltage ramps in low external Ca^{2+}. This observation allows one to speculate that under physiological conditions the increase in Ca^{2+} levels is mediated by voltage-dependent or ligand-dependent gates in the plasma membrane (Schroeder and Hagiwara, 1990a; Cosgrove and Hedrich, 1991) or release from intracellular stores (Blatt *et al.*, 1990).

In the activated state the anion permeability was dominating. During inactivation of the anion channel K^+ permeability increased and generally dominated the whole-cell conductance (Hedrich *et al.*, 1990; Thiel *et al.*, 1992). The shift in the state of activation of the anion channel was accompanied by a change in the reversal potential of the whole-cell cur-

[14] Later, Schroeder *et al.* (1993) related a 36-pS conductance to the s-type current as well.

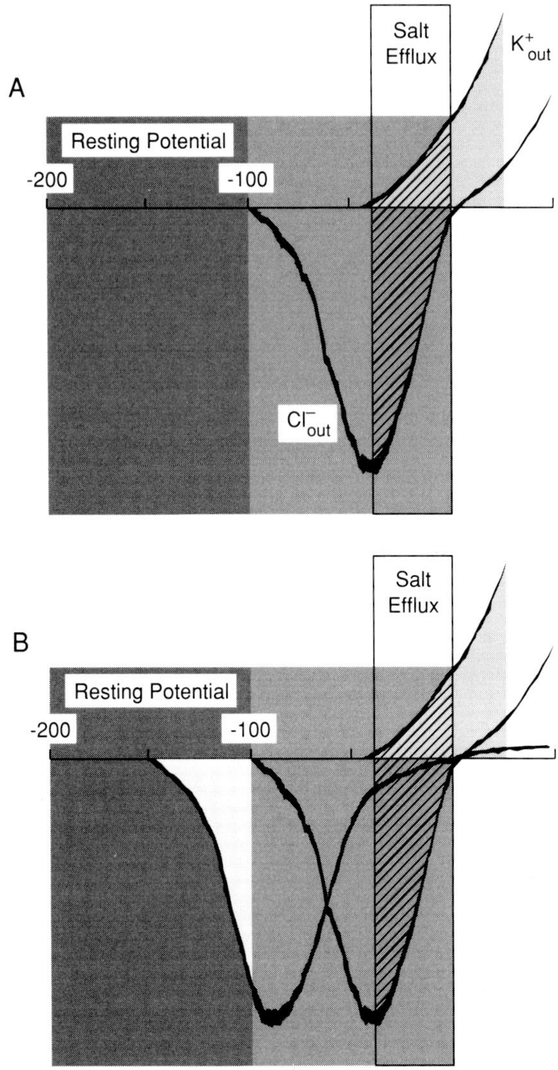

FIGURE 12 The working range of anion and potassium channels to mediate salt efflux (A) and signal transduction and excitability (B).

rent. Since the residual current (maximal inactivation) was in the same order of magnitude as the anion currents in the inactivated (background) state it is tempting to suggest that Ca^{2+} and nucleotides modulate the number of active channels or the inactivation gate. The activation state and gating mode (GCAC1 or SLAC see Section IV,B) of the anion channel depending on Ca^{2+} and nucleotides may thus determine both the magnitude and time course of salt release.

In addition to voltage-dependent potassium channels and GCAC1, three types of stretch-activated channels selectively permeable to Ca^{2+}, K^+, and Cl^- (Cosgrove and Hedrich, 1991)[15] may serve as pressure valves in the plasma membrane to adjust the steady-state volume with respect to the settings given by the various effectors and the biochemical machinery of the guard cell.

B. Do Voltage Dependence and Kinetics of GCAC1 Ask for Excitability?

The voltage dependence and kinetics of GCAC1 described above are reminiscent of Na^+, K^+, and Ca^{2+} channels in excitable membranes (for review, see Hille, 1992). In analogy to the Sodium Hypothesis of Hodgkin and Huxley (1952), one can describe the voltage- and time-dependent cycles of activation and inactivation by

$$I_{(Cl-)} = m{\cdot}\exp X \cdot h{\cdot}\exp Y \cdot G_{(Cl-)}\max \cdot (E - E_{Cl-}),$$

where G_{max} represents the maximal anion conductance of the membrane; m is the activation gate, closed at rest and opening with polynome X, describing the time course of current increase; and h is the inactivation gate which is open at the resting potential and closes exponentially with Y during depolarization (Hedrich *et al.*, 1990). As illustrated in the Hodgkin–Huxley cycle of GCAC1 (Fig. 13) initial depolarization to slightly negative potentials to -100 mV activates a few anion channels which in turn further depolarize the plasma membrane toward E_{Cl-}. Due to the voltage- and time-dependent inactivation of GCAC1 (see Fig. 6), the membrane repolarizes again. Calcium and K^+ channels may also contribute to the shape of this transient change in membrane potential. Wilting hormone (absicisic acid)-induced or CRAC-like calcium conduc-

[15] Please note, that it seems to be unlikely that mechano-sensitive (MS) channels represent stretch-activated voltage-sensitive channels, since e.g. MS chloride channels have a smaller conductance and different voltage dependence than GCAC1).

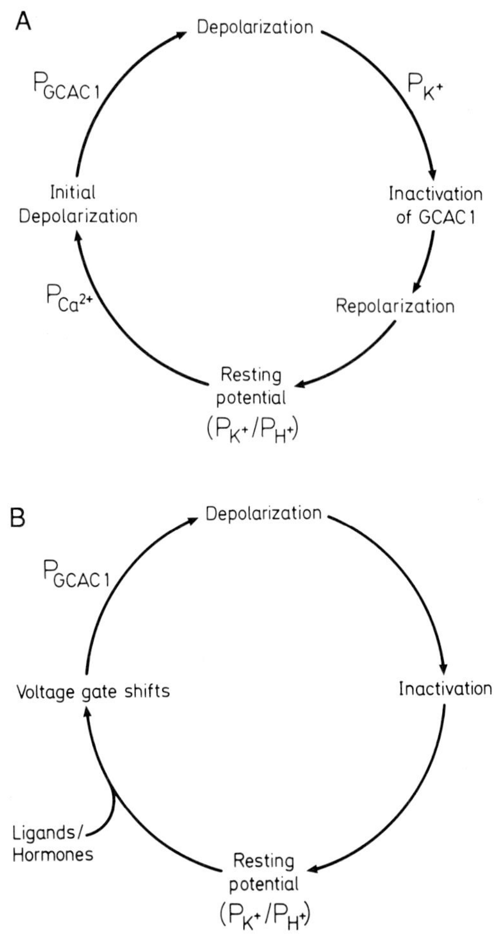

FIGURE 13 Hodgkin-Huxley cycle of transient conductance changes in the plasma membrane of guard cells, during excitation, when the initial "input" was a depolarization (A, *voltage* change) caused by a Ca^{2+} permeability (upper cycle) and signal transduction, when (B) *ligands* modify GCAC1 directly (lower cycle).

tances (Schroeder and Hagiwara, 1989; Cosgrove and Hedrich, 1991; Hoth and Penner, 1992) may account for the initial depolarization (Fig. 13, upper cycle) and outward-rectifying K^+ channels for enhanced repolarization, under conditions where the voltage gate of GCAC1 is in its resting position (unchanged by gating modifiers such as auxins or malate).

C. How Do Plant Hormones Work?[16]

When auxins, malate, or other stimuli reach their target sites on the anion channel the activation potential of the channel is shifted toward the resting potential of the cell (usually −100 to −160 mV). This shift in activation and peak potential of the anion channel toward hyperpolarized values, a voltage range where K^+_{out} channels are not active (see Fig. 12B), results in the opening of anion channels causing membrane depolarization rather than salt efflux. Transient activation and inactivation of anion channels may provide a potential mechanism for excitability (Fig. 13, lower cycle) in response to sudden changes in water relations (Raschke, 1970), hormone level, and other stimuli (Johnsson *et al.*, 1976).

VII. OUTLOOK

In this review describing the current status of anion channel research in plants I have raised many more questions about the properties and function of this channel type than I could answer. In order to answer the queries and to relate plant anion channels to animal counterparts at least three different fields of research might emerge or develop:

(i) Future inhibitor-structure channel-function analysis may give further insights into the nature and location of channel sites responsible for gate shifting and block (Fig. 14). In Section IV,C I have shown that structurally nonrelated compounds like growth hormones and anion channel blockers alter the voltage-dependent activity of GCAC1 in a similar fashion. Thus colabeling of the putative channel protein with azido-auxins and anion channel blockers or antibodies as well as the functional reconstitution of the receptor and channel properties (c.f., see Landry *et al.*, 1989; Palme, 1991) may help to identify the primary sequence of GCAC1.

(ii) Functional expression of ion channels in heterologous expression systems following injection of poly(A) mRNA or cRNA into frog oocytes or animal cell lines may shed light on the topic. This strategy developed to express or/and clone ion channels from animal tissues (for review, see Rudy and Iverson, 1992) has already been successfully used to express plant outward-rectifying K^+ channels

[16] For the dilemma in explaining the mode of hormone action in general, see Trewavas and Gilroy (1981, 1991) and Guern *et al.* (1989).

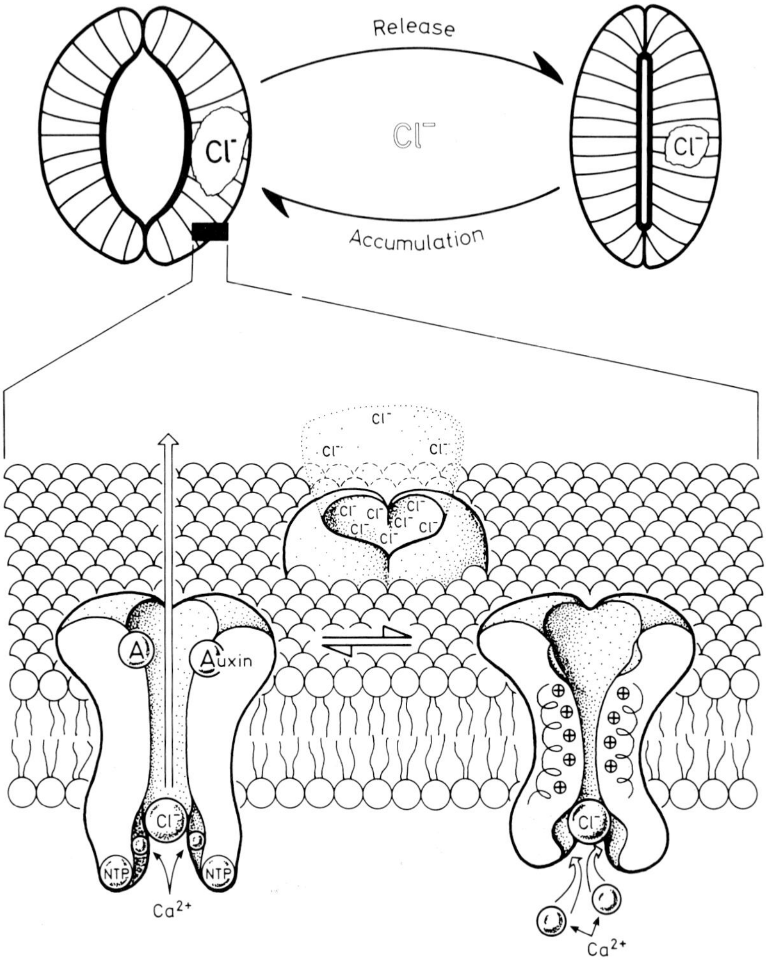

FIGURE 14 Schematic drawing of the regulation of GCAC1 by calcium and nucleotides and voltage. In the upper part a pair of guard cells and the dynamics of chloride movements are illustrated. GCAC1 is assumed to be a homodimer (due to symmetry considerations) embedded into the 2D lipid matrix. The front view of a half GCAC1 dimer represents the open state on the left and closed state on the right side. Binding of Ca^{2+} and nucleotides (NTP) to sites on the cytoplasmic surface opens the pore leading to the release of anions. The binding pockets of gating modifiers (A) like auxin and malate and inhibitors are located on the external surface. For clearity the putative selectivity filter and pore of the channel outlined by the positive charged α-helix have only been drawn in the left closed state.

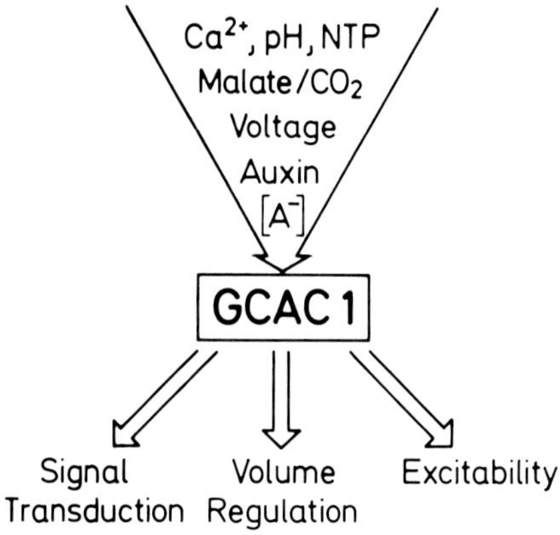

FIGURE 15 GCAC1, a channel with multiple regulation sites. Schematic diagram of the various inputs/signals and supposed outputs/physiological responses integrated by the cell through GCAC1 activity (cf. Fig. 1).

(Cao *et al.*, 1992). It should be noted that inward-rectifying plant K^+ channels have been cloned by complementation of K^+ uptake-deficient yeast mutants (Sentenac *et al.*, 1992; Anderson *et al.*, 1992).[17] Thus protein isolation, molecular genetics, and functional expression (Schachtman *et al.*, 1992) will give clues to protein structure and channel function.

(iii) Auxin- or hormone-sensitive mutants may allow signal transduction pathways to be decomposed. Analysis of the early events should help to identify the receptor and transduction elements linked to it. These studies will reveal whether ion channels represent key elements of the transduction machinery or only one of many inputs the cell integrates into a physiological response (Fig. 15).

Acknowledgments

I thank D. G. Robinson and K. Raschke for comments on the manuscript as well as P. Dietrich and I. Marten for critical discussion. Help from B. Raufeisen should also be acknowledged.

Supported by DFG Grants He 1640 1/1-2.

[17] Sequence information might be used now to clone the structurally unknown animal counterpart. Complementation of yeast with anion-efflux channels requires a physiological state, in other words a screening procedure, where anion release is crucial for viability, mutants which in contrast to uptake mutants are to my knowledge not yet described.

References

Anderson, J. A., Huplikar, S. S., Kochian, L. V., Lucas, W. J., and Gaber, R. F. (1992). Functional expression of a probable *Arabidopsis thaliana* potassium channel in *Saccharomyces cerevisiae*. *Proc. Natl. Acad. Sci. U.S.A.* **89,** 3736–3740.

Armstrong, C. M., and Cota, G. (1991). Calcium ion as a cofactor in Na^+ channel gating. *Proc. Natl. Sci. U.S.A.* **88,** 6528–6531.

Assmann, S. M., Simoncini, L., and Schroeder, J. I. (1985). Blue light activates electrogenic ion pumping in guard cell protoplasts of *Vicia faba*. *Nature (London)* **318,** 285–287.

Barbara, J.-G., Stoeckel, H., and Takeda, K. (1994). Hyperpolarization-activated inward chloride current in protoplasts from suspension-cultured carrot cells. *Protoplasma*, in press.

Blatt, M. R. (1991). Ion channel gating in plants: Physiological implications and integration for stomatal function. *J. Membr. Biol.* **124,** 95–112.

Blatt, M. R., Thiel, G., and Trentham, D. R. (1990). Reversible inactivation of K^+ channels of *Vicia* stomatal guard cells following the photolyses of caged inositol-1,4,5-triphosphate. *Nature (London)* **346,** 766–769.

Bormann, J., Hamill, O. P., and Sakmann, B. (1987). Mechanism of anion permeation through channels gated by glycine and γ-aminobutyric acid in mouse cultured spinal neurones. *J. Physiol. (London)* **385,** 243–286.

Bowling, D. (1987). Measurement of the apoplastic activity of K^+ and Cl^- in the leaf epidermis of *Commelina communis* in relation to stomatal activity. *J. Exp. Bot.* **38,** 1351–1355.

Cao, Y., Anderova, M., Crawford, N. M., and Schroeder, J. (1992). Expression of an outward-recifiying potassium channel from maize mRNA and complementary RNA in *Xenopus oocytes*. *Plant Cell* **4,** 961–969.

Cole, K. S., and Curtis, H. J. (1938). Electric impedance of *Nitella* during activity. *J. Gen. Physiol.* **22,** 37–64.

Cole, K. S., and Curtis, H. J. (1939). Electrical impedance of the squid giant axon during activity. *J. Gen. Physiol.* **22,** 649–670.

Coleman, H. A. (1986). Chloride currents in Chara—A patch clamp study. *J. Membr. Biol.* **93,** 55–61.

Cosgrove, D. J., and Hedrich, R. (1991). Stretch-activated chloride, potassium, and calcium channels coexisting in the plasma membranes of guard cells of *Vicia faba* L. *Planta* **186,** 143–153.

Eisenman, G., and Horn, R. (1983). Ionic selectivity revisited: The role of kinetic and equilibrium process in ion permeation through channels. *J. Membr. Biol.* **76,** 197–225.

Falke, L., Edwards, K. L., Pickard, B. G., and Misler, S. A. (1988). A stretch-activated anion channel in tobacco protoplasts. *FEBS Lett.* **237,** 141–144.

Frizzell, R. A., and Halm, D. R. (1991). Chloride channels in epithelial cells. *Curr. Top. Membr. Transp.* **37,** 247–282.

Gögelein, H. (1988). Chloride channels in epithelia. *Biochim. Biophys. Acta* **947,** 521–547.

Greger, R. (1992). Cl channels. *In* "New Comprehensive Biochemistry: Molecular Aspects of Transport Proteins" (J. J. H. M. de Pont and E. M. Wright, eds.). Elsevier, Amsterdam (in press).

Guern, J., Mathieu, Y., Kurkdjian, A., Manigault, P., Manigault, J., Gillet, B., Beloeil, C., and Lallemand, J.-Y. (1989). Regulation from within: The hormone dilemma. *Ann. Bot. (London)* **4,** 75–102.

Hamill, O. P., Marty, A., Neher, E., Sakmann, B., and Sigworth, F. J. (1981). Improved patch-clamp techniques for high-resolution current recording from cells and cell-free membrane patches. *Pfluegers Arch.* **391,** 85–100.

Hedrich, R., and Jeromin, A. (1992). A new scheme of symbiosis: Ligand- and voltage-gated anion channels in plants and animals. *Proc. R. Soc. London, Ser. B* **338,** 31–38.
Hedrich, R., and Neher, E. (1987). Cytoplasmic calcium regulates voltage-dependent ion channels in plant vacuoles. *Nature (London)* **329,** 833–836.
Hedrich, R., and Schroeder, J. I. (1989). The physiology of ion channels and electrogenic pumps in higher plants. *Annu. Rev. Plant Physiol.* **40,** 539–569.
Hedrich, R., Schroeder, J. I., and Fernandez, J. M. (1986). Patch-clamp studies on higher plants: A perspective. *Trends Biochem. Sci.* **12,** 49–52.
Hedrich, R., Busch, H., and Raschke, K. (1990). Ca^{2+} and nucleotide dependent regulation of voltage dependent anion channels in the plasma membrane of guard cells. *EMBO J.* **9,** 3889–3892.
Hedrich, R., and Marten, I. (1993). Malate-induced feedback regulation of plasma membrane anion channels could provide a CO_2 sensor to guard cells. *EMBO J.* **12,** 897–901.
Hille, B. (1992). "Ionic Channels of Excitable Membranes." Sinauer Assoc., Sunderland, MA.
Hodgkin, A. L., and Huxley, A. F. (1952). A quantitative description of membrane current and its application to conduction and excitation in nerve. *J. Physiol. (London)* **117,** 500–544.
Hoth, M., and Penner, R. (1992). Depletion of intracellular calcium stores activates a calcium current in mast cells. *Nature (London)* **355,** 353–356.
Humble, D. G., and Raschke, K. (1971). Stomatal opening quantitatively related to potassium transport. Evidence from microprobe analysis. *Plant Physiol.* **48,** 447–453.
Imamura, S. (1943). Untersuchungen über den Mechanismus der Turgorschwankung der Spaltöffnungsschliesszellen. *Jpn. J. Bot.* **12,** 251–346.
Johnsson, M., Issaias, S., Brogårdh, T., and Johnsson, A. (1976). Rapid, blue-light-induced transpiration response restricted to plants with grass-like stomata. *Physiol. Plant.* **36,** 229–232.
Keller, B. U., and Hedrich, R. (1992). Patch clamp techniques to study ion channels from organelles. *In* "Methods in Enzymology" (B. Rudy and L. E. Iverson, eds.), Vol. 207, pp. 673–681. Academic Press, London.
Keller, B. U., Hedrich, R., and Raschke, K. (1989). Voltage-dependent anion channels in the plasma membrane of guard cells. *Nature (London)* **341,** 450–453.
Landry, D. W., Akabas, M. H., Redhead, C., Edelman, A., Cragoe, E. J., Jr., and Al-Awqati, Q. (1989). Purification and reconstitution of chloride channels from kidney and trachea. *Science* **244,** 1469–1472.
Laver, D. R. (1991). A surgical method for accessing the plasma membrane of *Chara australis*. *Protoplasma* **161,** 79–84.
Lew, R. R. (1991). Substrate regulation of single potassium and chloride ion channels in *Arabidopsis* plasma membrane. *Plant Physiol.* **95,** 642–647.
Linder, B., and Raschke, K. (1992). A slow anion channel in guard cells, activating at large hyperpolarization, may be principal for stomatal closing. *FEBS Lett.* **313,** 27–30.
Lohse, G., and Hedrich, R. (1992). Characterization of the plasma membrane H+-ATPase from *Vicia faba* guard cells. Modulation by extracellular factors and seasonal changes. *Planta* **188,** 206–214.
Lunevsky, V. Z., Zherelova, O. M., Vostrikov, I. Y., and Berestovsky, G. N. (1983). Excitation of *Characeae* cell membranes as a result of acitivation of calcium and chloride channel. *J. Membr. Biol.* **72,** 43–58.
Marten, I. (1993). Untersuchung der Struktur-Funktions-Beziehung eines Anionenkanals in der Plasmamembran von Schlieβzellen aus *Vicia faba* L. Dissertation, University of Hannover.

Marten, I., Lohse, G., and Hedrich, R. (1991). Plant growth hormones control voltage-dependent activity of anion channels in plasma membrane of guard cells. *Nature (London)* **353,** 758–762.

Marten, I., Zeilinger, C., Redhead, C., Landry, D. W., Al-Awqati, Q., and Hedrich, R. (1992). Identification and modulation of a voltage-dependent anion channel in the plasma membrane of guard cells by high-affinity ligands. *EMBO J.* **11,** 3569–3575.

Marten, I., Busch, H., Raschke, K., and Hedrich, R. (1993). Modulation and block of plasma membrane anion channels of guard cells by stilbene derivatives. *Eur. J. Biophys.* **21,** 403–408.

Morris, C. E. (1990). Mechanosensitive ion channels. *J. Membr. Biol.* **113,** 93–107.

Neher, E., and Sakmann, B. (1976). Single-channel currents recorded from membrane of denervated frog muscle fibers. *Nature (London)* **260,** 779–802.

Okihara, K., Ohkawa, T., Tsutsui, I., and Kasai, M. (1991). A Ca^{2+}- and voltage-dependent Cl^--sensitive anion channel in the *Chara plasmalemma:* A patch-clamp study. *Plant Cell Physiol.* **32,** 593–601.

Palme, K. (1991). Molecular analysis of plant signalling elements: The relevance of eucaryotic signal transduction models. *Int. Rev. Cytol.* **132,** 223–283.

Pardo, L. A., Heinemann, S. H., Terlau, H., Ludewig, U., Lorra, C., Pongs, O., and Stühmer, W. (1992). Extracellular K^+ specifically modulates a rat brain K^+ channel. *Proc. Natl. Acad. Sci. U.S.A.* **89,** 2466–2470.

Raschke, K. (1970). Stomatal responses to pressure changes and interruptions in the water supply of detached leaves of *Zea mays* L. *Plant Physiol.* **45,** 415–423.

Raschke, K. (1979). Movements using turgor mechanisms. *Encycl. Plant Physiol., New Ser.* **7,** pp. 383–441.

Raschke, K., and Hedrich, R. (1990). Patch-clamp measurements on isolated guard-cell protoplasts and vacuoles. *In* "Methods in Enzymology" (S. Fleischer and B. Fleischer, eds.), Vol. 174, pp. 312–330. Academic Press, San Diego.

Redhead, C. R., Edelman, A. E., Brown, D., Landry, D. W., and Al-Awqati, Q. (1992). A ubiquitous 64 kDa protein is a component of a chloride channel of plasma and intracellular membranes. *Proc. Natl. Acad. Sci. U.S.A.* **89,** 3716–3720.

Robinson, R. A., and Stokes, R. H. (1959). "Electrolyte Solutions." Butterworth, London.

Rudy, B., and Iverson, L. E., eds. (1992). "Methods in Enzymology," Vol. 207. Academic Press, London.

Sachs, F. (1986). Mechanotransducing ion channels. *In* "Ionic Channels in Cells and Model Systems" (R. Latorre, ed.). Plenum, New York, pp. 181–193.

Schauf, C. L., and Wilson, K. J. (1987). Properties of single K^+ and Cl^- channels in *Asclepias tuberosa* protoplasts. *Plant Physiol.* **85,** 413–418.

Schnabl, H. (1981). The compartmentation of carboxylating and decarboxylating enzymes in guard cells. *Planta* **152,** 307–313.

Schönknecht, G., Hedrich, R., Junge, W., and Raschke, K. (1988). A voltage-dependent chloride channel in the photosynthetic membrane of a higher plant. *Nature (London)* **336,** 589–592.

Schroeder, J. I. (1987). K^+-Kanäle in der Plasmamembran von Schliesszellen. Eine Patch-Clamp Untersuchung molekularer Mechanismen des K^+-Transports in höheren Pflanzenzellen. Dissertation, University of Göttingen.

Schroeder, J. I., and Hagiwara, S. (1989). Cytosolic calcium regulates ion channels in the plasma membrane of *Vicia faba* guard cells. *Nature (London)* **338,** 427–430.

Schroeder, J. I., and Hagiwara, S. (1990a). Repetitive increases in cytosolic Ca^{2+} of guard cells by abscisic acid activation of nonselective Ca^{2+} permeable channels. *Proc. Nat. Acad. Sci. U.S.A.* **87,** 9305–9309.

Schroeder, J. I., and Hagiwara, S. (1990b). Voltage-dependent activation of Ca^{2+}-regulated anion channels and K^+ uptake channels in *Vicia faba* guard cells. *In* "Calcium in Plant Growth and Development" (R. T. Leonard and P. K. Hepler, eds.), Am. Soc. Plant Physiolo. Symp. Ser., Vol. 4, pp. 144–150.

Schroeder, J. I., and Hedrich, R. (1989). Involvement of ion channels and active transport in osmoregulation and signaling of higher plant cells. *Trends Biochem. Sci.* **5,** 187–192.

Schroeder, J. I., Schmidt, C., and Sheaffer, J. (1993). Identification of high-affinity slow anion channel blockers and evidence for stomatal regulation by slow anion channels in guard cells. *Plant Cell* **5,** 1831–1841.

Schroeder, J. I., and Keller, B. (1992). Two types of anion channel currents in guard cells with distinct voltage regulation. *Proc. Natl. Acad. Sci. U.S.A.* **89,** 5025–5029.

Schroeder, J. I., Hedrich, R., and Fernandez, J. M. (1984). Potassium-selective single channels in guard cell protoplasts of *Vicia faba*. *Nature (London)* **312,** 361–362.

Schroeder, J. I., Raschke, K., and Neher, E. (1987). Voltage-dependence of K^+ channels in guard-cell protoplasts. *Proc. Natl. Acad. Sci. U.S.A.* **84,** 4108–4112.

Sentenac, H., Bonneaud, N., Minet, M., Lacroute, F., Salmon, J.-M., Gaymard, F., and Grignon, C. (1992). Cloning and expression in yeast of a plant potassium ion transport system. *Science* **256,** 663–665.

Serrano, E. E., Zeiger, E., and Hagiwara, S. (1988). Red light stimulates an electrogenic proton pump in *Vicia* guard cell protoplasts. *Proc. Natl. Acad. Sci. U.S.A.* **85,** 436–440.

Skerrett, M., and Tyerman, S. D. (1994). A channel that allows inwardly directed fluxes of anions in protoplasts derived from wheat roots. *Planta* **192,** 295–305.

Speer, M., and Kaiser, W. M. (1991). Ion relations of symplastic and apoplastic space in leaves from *Spiniacia oleracia* L. and *Pisum sativum* L. under salinity. *Plant Physiol.* **97,** 990–997.

Tazawa, M., Shimmen, T., and Mimura, T. (1987). Membrane control in the *Characeae*. *Annu. Rev. Plant Physiol.* **38,** 95–117.

Terry, B. R., Tyerman, S. D., and Findlay, G. P. (1991). Ion channels in the plasma membrane of *Amaranthus* protoplasts: One cation and one anion channel dominate the conductance. *J. Membr. Biol.* **121,** 223–236.

Theo, J., Elzenga, M., and Van Volkenburgh, E. (1994). Characterization of ion channels in the plasma membrane of epidermal cells of expanding pea (*Pisum sativum* arg) leaves. *J. Membr. Biol.* **137,** in press.

Thiel, G., MacRobbie, E. A. C., and Blatt, M. R. (1992). Membrane transport in stomatal guard cells: The importance of voltage control. *J. Membr. Biol.* **126,** 1–18.

Trewavas, A., and Gilroy, S. (1981). How do plant hormones work? *Plant Cell Environ.* **4,** 203–228.

Trewavas, A., and Gilroy, S. (1991). How do plant hormones work? *Plant Cell Environ.* **14,** 1–12.

Tyerman, S. D. (1992). Anion channels in plants. *Annu. Rev. Plant Physiol. Plant Mol. Biol.* **43,** 351–373.

Tyerman, S. D., and Findlay, G. P. (1989). Current-voltage curves of single Cl^- channels which coexist with two types of K^+ channel in the tonoplast of *Chara corallina*. *J. Exp. Bot.* **40,** 105–117.

Van Kirk, C. A., and Raschke, K. (1978). Presence of chloride reduces malate production in epidermis during stomatal opening. *Plant Physiol.* **61,** 361–364.

Zeiger, E., and Hepler, P. K. (1977). Light and stomatal function: Blue light stimulates swelling of guard cell protoplasts. *Science* **196,** 887–889.

CHAPTER 2

Molecular Biology of Voltage-Gated Chloride Channels

Thomas J. Jentsch
Center for Molecular Neurobiology, Hamburg University, D-20246 Hamburg, Germany

I. INTRODUCTION

While the molecular characterization of voltage-gated cation channels has made extraordinary progress since the first cloning of a sodium channel by Numa's group (Noda *et al.*, 1984), molecular biology of voltage-gated chloride channels was nonexistent until late 1990. This may reflect in part a greater interest in cation channels at that time, but is undoubtedly also due to the specific difficulties encountered with chloride channels. The availability of high-affinity ligands for cation channels (e.g., tetrodotoxin for the sodium channel) has greatly facilitated their identification and purification. Unfortunately, there are no inhibitors of similar specificity for voltage-gated chloride channels. Attempts to use less than ideal inhibitors to identify and clone such channels either have failed (Jentsch *et al.*,

1989) or are at a stage where functional expression from a cloned cDNA has not yet been reported (Landry *et al.*, 1989; Redhead *et al.*, 1992; Landry *et al.*, 1993; Dermietzel, *et al.* 1994).

Because of the lack of specific inhibitors and of the difficulties of biochemical purification, the first voltage-gated chloride channel family was obtained by expression cloning (Jentsch *et al.*, 1990). This approach obviates the need for identification and purification of the corresponding protein. Homology screening using the first cloned voltage-gated chloride channel, ClC-0 (Jentsch *et al.*, 1990) as a starting point has led to a growing family of Cl^- channels (Steinmeyer *et al.*, 1991a; Thiemann *et al.*, 1992; Uchida *et al.*, 1993; Kieferle *et al.*, 1994).

Surprisingly, it was reported that expression of phospholemman, a single transmembrane span protein not suspected of being chloride channel, also induces chloride currents (Moorman *et al.*, 1992). However, it seems likely that phospholemman is not a chloride channel, but rather activates endogenous *Xenopus* oocyte chloride channels upon overexpression (Attali *et al.*, 1993). This may also apply for pI_{cln} (Krapivinsky *et al.*, 1994), a putative chloride channel isolated from MDCK cells by expression cloning (Paulmichl *et al.*, 1992).

This review focuses exclusively on cloned voltage-gated chloride channels. It therefore excludes other cloned Cl^- channels such as GABA and glycine receptors (Schofield *et al.*, 1987; Grenningloh *et al.*, 1987), which belong to the ligand-gated channel superfamily, and CFTR (Riordan *et al.*, 1989), the protein defective in cystic fibrosis. This review will not cover mdr1, a protein involved in the active transport of hydrophobic substances which enhances volume-activated chloride currents upon overexpression (Valverde *et al.*, 1992).

II. THE ClC FAMILY OF Cl^- CHANNELS

A. ClC-0, the Torpedo Electric Organ Cl^- Channel

The electric organ of *Torpedo* and related species is an exceptionally rich source for voltage-gated chloride channels. In the pioneering studies of Chris Miller and colleagues this channel was extensively studied after reconstitution into lipid bilayers (White and Miller, 1979; Miller, 1982; Richard and Miller, 1990; for review, see Miller and Richard, 1990). It is probably one of the best-characterized chloride channels and was the first voltage-gated Cl^- channel to be cloned (Jentsch *et al.*, 1990).

The *Torpedo* channel is sensitive to micromolar concentrations of the inhibitor DIDS, which probably binds covalently to the channel protein.

Early work therefore tried to identify the channel protein by DIDS or SITS binding (Taguchi and Kasai, 1980; Jentsch *et al.*, 1989). However, this approach failed due to the low binding specificities of these inhibitors (Jentsch *et al.*, 1989).

The cDNA for the *Torpedo marmorata* channel, now dubbed ClC-0, was finally isolated (Jentsch *et al.*, 1990) using a negative expression cloning approach (Lübbert *et al.*, 1987). It predicts a protein of 805 amino acids (calculated molecular mass ≈ 89 kDa) which was unrelated to any other known protein. Hydropathy analysis suggests 12 to 13 membrane spanning domains (Fig. 1). It lacks a cleavable signal peptide for ER translocation. Based on that observation and on the charge distribution in the neighborhood of the first transmembrane domain (Hartmann *et al.*, 1989) of different members of the ClC chloride channel family, we therefore propose a cytoplasmic localization of the amino-terminus. It should be stressed that the proposed transmembrane topology is hypothetical and will have to be confirmed, e.g., by site-specific antibodies. In fact, recent glycosylation studies (Kieferle *et al.*, 1994) suggest that the D8–D9 linker is extracellular, which requires a revision of the model.

Several of the putative transmembrane domains contain charges. These may be involved in the conduction process or may provide a voltage sensor akin to the charged S4 domains of voltage-gated cation channels (Stühmer *et al.*, 1989). ClC chloride channels do not have a similar segment.

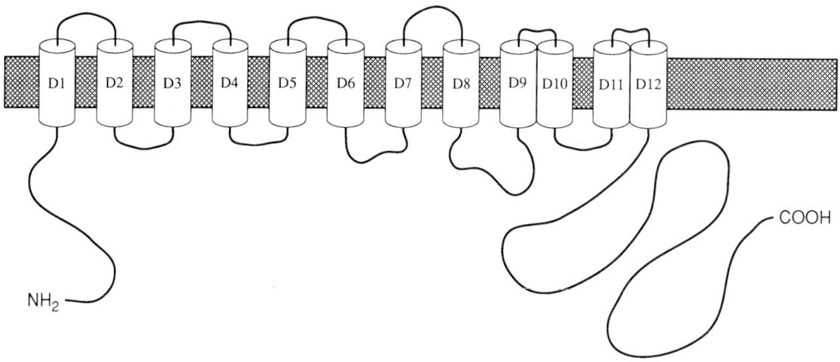

FIGURE 1 Proposed transmembrane topology of voltage-gated chloride channels of the ClC family. Hydropathy analysis suggests a maximum of 13 transmembrane domains. Some of these domains (D3, D4, D6, D8, and D13) are only weak candidates for transmembrane segments (see, e.g., top of Fig. 3), and the broad hydrophobic regions D9/D10 and D11/D12 are assumed to cross the lipid bilayer twice. This model is thus hypothetical and requires experimental confirmation. There is no cleavable signal peptide and the amino-terminus is assumed to be cytoplasmic.

Using sequence information from the *T. marmorata* channel (Jentsch *et al.*, 1990), the *Torpedo californica* channel was subsequently cloned and sequenced (O'Neill *et al.*, 1991). Not surprisingly, it is 97% identical in sequence. Of the 24 amino acid differences (most of them conservative exchanges) only one occurs in a putative transmembrane span (D2). Most differences are located in the amino-terminus or in the large segment between D12 and D13. These parts are also poorly conserved between other members of the ClC family.

Expression of ClC-0 in *Xenopus* oocytes (Jentsch *et al.*, 1990) generates chloride currents which clamp the resting voltage close to the oocyte Cl^- equilibrium potential (−25 to −30 mV). While partially open at resting voltages, the current slowly activates further upon hyperpolarization. Depending on the degree of hyperpolarization, 10 to 40 sec are needed to reach steady-state activation. Deactivation upon stepping back to resting potentials is much slower and needs several minutes to be complete. Once activated by hyperpolarization, the channel is strongly outward rectifying in the negative voltage range (Fig. 2a). This leads to maximum steady-state currents in the range between −80 and −120 mV. This is similar to results obtained earlier with the *T. californica* channel expressed in oocytes from total electric organ mRNA (Sumikawa *et al.*, 1984). It is also consistent with the studies of Miller and colleagues (Richard and Miller, 1990) on the reconstituted *T. californica* channel. It slowly activates within seconds with negative voltage on the *cis* side of the bilayer; once activated, the current increases (fast) with positive voltages. Comparison of these voltage dependences strongly suggests that the *cis* side of the bilayer corresponds to the cytoplasmic side of ClC-0.

The ion selectivity of ClC-0 was measured in the oocyte system by partially exchanging extracellular chloride with other anions in two-electrode voltage-clamp experiments (Bauer *et al.*, 1991). These experiments revealed a high selectivity of ClC-0 for chloride. Among the anions tested, only bromide was appreciably permeable, while iodide and SCN actually inhibited the channel. These observations are in agreement with data obtained in the lipid bilayer system.

ClC-0 currents in *Xenopus* oocytes can be inhibited (>80%) using rather high concentrations (1 m*M*) of extracellular DIDS or diphenylamine carboxylate (DPC). In the lipid bilayer system, the channel is quite sensitive to DIDS (in the μM range) from the *cis* side. This corresponds, however, to the cytoplasm, which explains the poor effect of this inhibitor to the oocyte studies.

Reconstituted in lipid bilayers, the *Torpedo* channel has two conductance states of 9 and 18 pS which appear together in bursts. The single-channel conductance is linear over a broad voltage range. Miller and

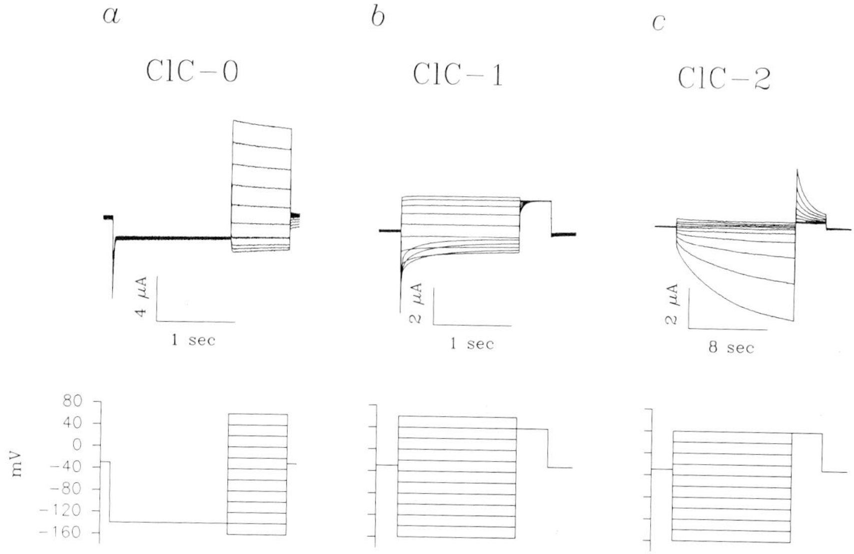

FIGURE 2 Representative voltage-clamp traces showing electrophysiological properties of ClC-0, ClC-1, and ClC-2. Channels were expressed in *Xenopus laevis* oocytes and measured after 1–3 days using conventional two-electrode voltage-clamp technique. Top panels show the current necessary to clamp the oocyte to the voltages indicated below. Note that different voltage-clamp programs were used. (a) Properties of ClC-0. The program was chosen to display the properties of the fast gate operating on the "protochannels." The oocyte was first hyperpolarized to −140 mV to open the slow gate. Then tail currents were measured by clamping to voltages between +60 and −160 mV. This demonstrates the typical outward rectification of activated ClC-0 in the negative voltage range. The current peak observed after the initial hyperpolarization to −160 mV reflects the closure of the fast gate by hyperpolarization. (b) Properties of ClC-1. Typical features of ClC-1 are the macroscopic inward rectification in the positive voltage range, and the deactivation in the negative voltage range (above ≈−100 mV). (c) Properties of ClC-2. This channel is closed in the physiological voltage range (+20 to −100 mV), but opens slowly by stronger hyperpolarization. Deactivation by stepping to positive voltages is faster than the initial activation.

colleagues proposed a "double-barrel" structure of the channel (Miller, 1982; Miller and Richard, 1990). In this model, the channel is a functional homodimer with two identical pores ("protochannels") of 9 pS which can open and close independently of each other. The gate operating on a "protochannel" is fast and opens with depolarization (or positive voltage on the *cis* side). This leads to the macroscopic outward rectification of ClC-0 currents after activation. A different common gate operates on both protochannels simultaneously. This gate is slow and opens with hyperpolarization.

Patch-clamp studies of ClC-0 expressed in *Xenopus* oocytes (Bauer *et al.*, 1991) show that the single polypeptide encoded by the cloned cDNA is sufficient to generate this double-barreled channel. The single-channel conductance, the voltage dependence, and the binomial distribution of open states were remarkably similar to those of the *T. californica* channel studied in reconstitution. Thus, all information for generating this complex channel is contained in a single polypeptide. It will be fascinating to investigate these mechanisms by site-directed mutagenesis.

In *Torpedo,* ClC-0 is highly expressed in the electric organ and to a lesser degree in skeletal muscle and in brain (C. Ortland, K. Steinmeyer, and T. J. Jentsch, unpublished). This has been determined both by Northern blotting and by immunofluorescence using several different anti-ClC-0 antibodies. Its function in the electric organ is to ensure a high conductivity of the noninnervated membrane during its discharge. Upon receiving a stimulus from the electric lobe, the membrane potential at the innervated surface of an electrocyte will break down due to sodium influx through the nicotinic acetylcholine receptors which are present in very high density. To generate a transcellular voltage, the potential difference across the noninnervated membrane has to remain constant, while a high conductivity is needed to pass large currents. This is accomplished by ClC-0, which is open appreciably under resting conditions and which will increase further its open probability upon depolarization. Since chloride equilibrium potential in the electric organ is probably close to resting potential, the membrane will be clamped to near constant voltages. The voltages across single cells add up to 100 V or more, as the electrocytes are arranged in large stacks on top of each other—like batteries connected in series. In *Torpedo* skeletal muscle it may play a role identical to ClC-1 in mammalian muscle: it ensures electrical membrane stability by clamping the resting potential to the chloride equilibrium potential.

B. ClC-1, the Major Skeletal Muscle Cl^- Channel

ClC-1 was cloned from rat skeletal muscle by homology screening with ClC-0 (Steinmeyer *et al.,* 1991a). Mammalian skeletal muscle was an obvious choice for searching for a homolog of the *Torpedo* channel, since the electric organ is derived from muscle. Also, ClC-0 is expressed in *Torpedo* skeletal muscle.

ClC-1 has an overall homology of about 55% to ClC-0, with a much higher degree of identity in some transmembrane domains. Considerable homology is also present in D13. D13 is a weak candidate for a transmembrane helix based on hydropathy analysis, but the homology found tends to support this view (or suggests another important role). ClC-1 is a slightly larger protein than ClC-0 due to N- and C-terminal extensions and due to a larger segment between D12 and D13 (predicted to be 110 kDa). Homology is very weak or nonexistent in the NH_2-terminus, the COOH-terminus, and the large putative cytoplasmic loop between D12 and D13.

ClC-1 is expressed very predominantly in skeletal muscle, with minor expression in smooth muscle and heart. In rat muscle, its mRNA levels increase steeply during the first 3 weeks after birth. This parallels an increase of chloride conductance observed during that time in rat skeletal muscle (Conte Camerino *et al.*, 1989). This suggests a transcriptional regulation of this channel during development which may be analogous to that of sodium channel isoforms (Trimmer *et al.*, 1990) and subunits of the muscle nicotinic acetylcholine receptor (Mishina *et al.*, 1986; Witzemann *et al.*, 1989). In fact, the ClC-1 upstream region has typical motifs found in muscle-specific enhancers (Lorenz *et al.*, 1994).

When expressed in *Xenopus* oocytes, ClC-1 induces chloride currents which are inwardly rectifying at positive voltages, and nearly linear in the physiological voltage range, and deactivate (with time constants in the range of 100 msec, depending on the voltage) upon hyperpolarization in excess of ≈ -100 mV (Fig. 2b). This leads to steady-state chloride currents which are maximal at about -100 mV and decrease with more negative voltages. Chloride conductance in intact muscle fibers from rat diaphragm was reported for the negative voltage range and shows very similar properties (Palade and Barchi, 1977).

ClC-1 currents can be blocked by comparatively low concentrations of 9-anthracene carboxylic acid (9-AC) (0.1 m*M* leads to >80% block). This agrees well with the sensitivity of macroscopic muscle Cl^- conductance to that drug (for review, see Bretag, 1987). The effect of 9-AC on *Xenopus*-expressed ClC-1 was very slow and dependent on drug concentration. At 0.1 m*M*, more than 10 min were needed to achieve maximal inhibition, which may mean that this (rather hydrophobic) inhibitor has to enter or cross the lipid bilayer to exert its effect. 9-AC can serve to pharmacologically differentiate between different members of the ClC Cl^- channel family, since ClC-0 and ClC-2 are only slightly affected even by concentrations of 9-AC as high as 1 m*M*.

Recent studies using noise analysis of ClC-1 transfected HEK293 cells have revealed a low single-channel conductance (~1pS) of ClC-1 (Pusch *et al.*, 1994). This makes classical single-channel analysis of ClC-1 impossible.

Only in rare cases have chloride channels been observed in patch-clamp studies of intact differentiated mammalian skeletal muscle, although other types of channels were easily found (Chua and Betz, 1991). The properties of these chloride channels, however, are inconsistent with ClC-1. In contrast, several different types of chloride channels have been identified in cultured muscle (Blatz and Magleby, 1985; Fahlke *et al.*, 1992; 1993), but their characteristics are also difficult to reconcile with ClC-1 currents. Moreover, characteristics of whole-cell chloride currents in cultured skeletal muscle (Zachar *et al.*, 1992) are incompatible with ClC-1 currents. This may not be surprising, since ClC-1 expression is developmentally regulated (Steinmeyer *et al.*, 1991a), and because denervation of muscle leads to a large decrease in Cl^- conductance (Camerino and Bryant, 1976; Conte Camerino *et al.*, 1989). Thus, ClC-1 may not be expressed in cultured muscle cells at significant levels.

Several results summarized above strongly indicate that ClC-1 is the major skeletal muscle chloride channel: First, it is preferentially expressed in skeletal muscle; moreover, its developmental regulation, electrophysiology, and pharmacology fit with the known characteristics of macroscopic muscle Cl^- conductance.

Skeletal muscle is unusual in that chloride conductance accounts for roughly 70–80% of resting conductance (Bretag, 1987). Since chloride equilibrium potential is close to resting potential, this conductance stabilizes membrane voltage in a manner similar to the K^+ conductance in most other cells. Its pharmacological block by 9-AC slows the rate of repolarization after action potentials. During that time sodium channels can recover from inactivation and fire again, leading to a train of action potentials. This leads to an impairment of muscle relaxation and is similar to findings made in myotonic diseases.

Myotonia (muscle stiffness) is a symptom of several human inheritable diseases (Rüdel and Lehmann-Horn, 1985). Myotonic dystrophy is a dominant systemic disorder, which in addition to skeletal muscle also affects, e.g., the eye, the CNS, and the heart. The gene responsible for that disease predicts a protein with typical structural features of protein kinases (Fu *et al.*, 1992; Brook *et al.*, 1992). Phosphorylation of different targets in different tissues could explain the multitude of symptoms in myotonic dystrophy. In contrast, autosomal recessive generalized myotonia (GM, Becker's myotonia) (Becker, 1977) and autosomal dominant myotonia congenita (MC, Thomsen's disease) (Thomsen, 1876) are purely myotonic diseases. Here symptoms are restricted to skeletal muscles.

There are several animal models for (pure) myotonia. A colony of goats with a form of myotonia which probably shows dominant inheritance has been investigated by Bryant and co-workers (Lipicky and Bryant, 1966; Adrian and Bryant, 1974; Bryant and Conte-Camerino, 1991). Several independent mouse mutants displaying a recessive pattern of inheritance have also been described, one of which (ADR) has been thoroughly investigated by Jockusch and collaborators (Stuhlfauth *et al.*, 1984; Reininghaus *et al.*, 1988; Mehrke *et al.*, 1988; Jockusch *et al.*, 1988). A significant decrease in muscle Cl^- conductance was described in both animal models, although many other parameters were also changed. These included reduced levels of parvalbumin (Stuhlfauth *et al.*, 1984) and altered myosin light-chain phosphorylation (Jockusch *et al.*, 1988). Many of these parameters, however, could be restored back to near normal values by symptomatic treatment of myotonic mice with tocainide (Jockusch *et al.*, 1988) which acts on the sodium channel. This demonstrates that many of the observed changes are secondary to muscle hyperactivity. Thus, alteration of muscle chloride channels, either by mutation or by defective regulation (Brinkmeier and Jockusch, 1987), remained an attractive hypothesis to explain myotonia.

The cloning of ClC-1 (Steinmeyer *et al.*, 1991a) allowed a direct test of this hypothesis. It was found that *ETn*, a known mouse transposon, had inserted into an intron interrupting exons coding for the putative transmembrane region D9 (Steinmeyer *et al.*, 1991b). On the mRNA level, this leads to several different splice products, none of which can encode functional channels. In other myotonic mouse strains, no abnormalities were found by Southern or Northern analysis. But since these mutants are known to be allelic, it is clear that myotonia in these strains will also be due to mutations in the ClC-1 locus. In fact, point mutations in ClC-1 were recently described in these mice (Gronemeier *et al.*, 1994).

It is interesting to note that myotonia in mice develops only some weeks after birth (Füchtbauer *et al.*, 1988). This coincides with the developmental expression of ClC-1 (Steinmeyer *et al.*, 1991a) which dominates resting conductance only after 2–3 weeks. Before that time, the contribution of potassium conductance is prevalent (Conte Camerino *et al.*, 1989).

In human myotonia, there has been conflicting evidence about the role of chloride conductance. Early work reported a decrease in Cl^- conductance both in dominant (Lipicky *et al.*, 1971) and in recessive (Rüdel *et al.*, 1988) human myotonia, while more recent reports rather stressed the role of Na^+ conductance (Franke *et al.*, 1991), especially for dominant Thomsen's disease (Iaizzo *et al.*, 1991). In a recent study, we localized the human ClC-1 gene on human chromosome 7q32-qter in the vicinity of the T-cell receptor β (TCRB) locus (Koch *et al.*, 1992). Restriction fragment length polymorphisms (RFLPs) in the CLC-1 locus and in the TCRB locus were used to link CLC-1 to both the recessive and the dominant form of myotonia in several German families (Koch *et al.*, 1992). Confirming these results, several Canadian families with dominant myotonia were linked to the TCRB locus (Abdalla *et al.*, 1992). One of the restriction endonucleases used to detect RFLPs fortuitously recognized a novel restriction site in the ClC-1 locus in two families with Becker's myotonia, but not in >100 control individuals (Koch *et al.*, 1992). This novel site was then used to search for the underlying mutation, which very likely causes myotonia in these Becker families. A point mutation was found which leads to a phenylalanine to cysteine exchange toward the end of putative transmembrane domain D8 (Koch *et al.*, 1992). This phenylalanine is highly conserved in all known members of the ClC family, suggesting an important role. In addition, cysteine may lead to the formation of improper disulfide bonds, which may interfere with the correct folding of the protein. This mutation was found in several, but not all, GM families, suggesting the presence of different ClC-1 mutations in GM which is a genetically homogeneous disorder (Koch *et al.*, 1993). ClC-1 mutations leading to dominant Thomsen-type myotonia were also described (George *et al.*, 1993; Steinmeyer *et al.*, 1994), including the mutation present in Thomsen's own family (Steinmeyer *et al.*, 1994).

How can different mutations in the same channel gene lead to recessive and dominant patterns of inheritance? A dominant effect can be easily explained by a gain-of-function mutation, e.g., a mutation leading to constitutively open channels. To explain a dominant negative mutation as in myotonia, it seems necessary to postulate a (homo)multimeric structure of the channel. A dominant mutation would lead to a "dead" channel subunit which could still associate with the normal subunits encoded by the normal allele and lead to their inactivation. This mechanism was previously proposed to explain the dominant effects of *Shaker* potassium channel mutants expressed in transgenic *Drosophila* (Gisselmann *et al.*, 1989). Detailed functional analysis of two Thomsen-type ClC-1 mutations lend strong support for this model (Steinmeyer *et al.*, 1994). It was concluded that the most likely stoichiometry of the homomultimeric channel is four. A recessive mutation could be explained by two different models: (a) the dead channel subunits can no longer associate with the normal one, or (b) they can associate, but the resulting complex is functional.

This model is consistent with the clinical observation (Becker, 1977) that myotonia is more severe in the recessive than in the dominant form: probably in dominant myotonia some normal subunits can escape inactivation by the mutated ones; this would also explain nicely the fact that reduction of muscle chloride conductance was more evident in Becker's than in Thomsen's myotonia.

C. *ClC-2, a Ubiquitously Expressed Mammalian Cl^-* Channel

ClC-2 was originally cloned from heart and brain cDNA libraries, but it soon became clear that it is ubiquitously expressed (Thiemann *et al.*,

1992). The complete cDNA was then obtained from rat brain, where expression is especially high.

Overall sequence homology is again about 50% to both ClC-0 and ClC-1, with highest degree of identity in transmembrane regions (Fig. 3). Expression in *Xenopus* oocytes gave currents which activated very slowly (within seconds) upon strong, unphysiological hyperpolarization (in excess of −90 mV) (Fig. 2c). More than 20 sec were needed to reach steady-state activation. In contrast, deactivation upon stepping back to holding potentials of −30 mV was several times faster. Once activated by hyperpolarization, it has a nearly linear, slightly inward-rectifying current–voltage relationship. Its conductivity sequence is Cl≥Br>I. Extracellular DIDS (1 m*M*) was nearly ineffective in blocking it, while 1 m*M* DPC or 9-AC were able to block it by roughly 50%.

By Northern analysis, ClC-2 expression was demonstrated in rat skeletal muscle, heart, brain, lung, kidney, pancreas, stomach, intestine, and liver. Since this could have meant that ClC-2 is expressed in a cell type common to all these tissues (e.g., fibroblasts or vascular smooth muscle cells), Northern analysis was repeated with several different cell lines and was again found to be ubiquitous. This included fibroblasts (NIH 3T3), epithelial cells (e.g., T84, CFPAC-I, LLC-PK1), and neuronal-like cells (Neuro-2a, PC-12). In some tissues (heart, fetal skeletal muscle, and T84 human colon epithelial cells) expression was confirmed by additional cDNA cloning. Expression in cells affected by cystic fibrosis is especially interesting as one could speculate that it could be used to bypass the chloride transport defect in that disease. A prerequisite, however, is expression of ClC-2 in apical membranes. This has not yet been examined.

It is very unlikely that strong hyperpolarization represents the physiological means of activating ClC-2, since these voltages are not reached *in vivo*. On the other hand, the ubiquitous expression suggests a function important for every cell. In many cells regulatory volume decrease is accomplished by the parallel opening of independent Cl^- and K^+ channels (Hoffmann and Simonsen, 1989). We thus hypothesized that ClC-2 could have a role in cell volume regulation. In fact, when expressed in oocytes, ClC-2 is activated by hypotonicity (Gründer *et al.*, 1992). Thus ClC-2 has a basic function for cell biology by mediating regulatory volume decrease. Extensive site-directed mutagenesis studies identified a N-terminal domain in ClC-2 which is essential for activation by cell swelling (Gründer *et al.*, 1992). This domain is position-independent, as it can be transplanted to the C-terminus without loss of function. A model was proposed in which this domain binds to a putative receptor on the cytoplasmic side of the channel; hypotonicity and hyperpolarization were postulated to reduce the affinity of this receptor, resulting in channel activation (Gründer *et al.*, 1992).

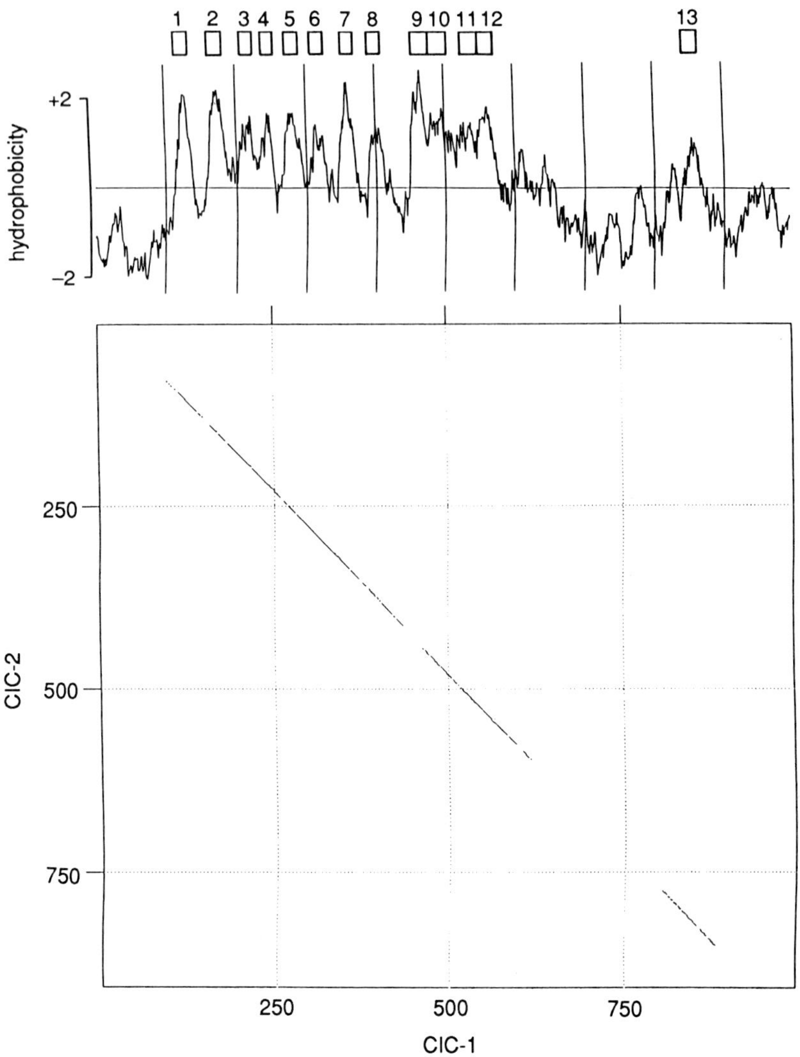

FIGURE 3 Homology between ClC-1 and ClC-2 chloride channel proteins. A dot-plot matrix is shown where homology is shown by dots at the appropriate positions. The top panel shows the hydropathy analysis of ClC-1 for comparison. It is evident that homology is found mainly in putative transmembrane regions. This includes D13, which on the basis of hydrophobicity analysis is only a poor candidate for a transmembrane span. No homology exists at both ends of the protein nor in the putative loop between D12 and D13.

D. ClC-K1 and ClC-K2, Kidney-Specific Chloride Channels

Using primers against regions conserved between different members of ClC chloride channels and the polymerase chain reaction (PCR), Uchida *et al.* (1993) isolated a new, kidney-specific member (ClC-K1) of this gene family from the rat. Using a similar approach and homology screening, Kieferle *et al.* (1994) obtained a different, highly homologous member of this gene family (rClC-K2), and again two highly homologous cDNAs (hClC-Ka and hClC-Kb) from human kidney. The high degree of identity within a species does not allow correlation between human proteins and their rat counterparts.

Using RT-PCR on microdissected tubule segments, the intrarenal distribution of these putative channels was investigated. While rClC-K1 is most highly expressed in the thin ascending limb of Henle's loop (Uchida *et al.*, 1993), the thick ascending limb and the distal convoluted tubule (Kieferle *et al.*, 1994), rClC-K2 shows a broader distribution (Kieferle *et al.*, 1994). Interestingly, rClC-K1 message could be increased by dehydration of the animals (Uchida *et al.*, 1993). While Uchida *et al.* (1993) reported that expression of ClC-K1 in *Xenopus* oocytes leads to time-independent outwardly rectifying chloride currents, no expression could be obtained in the second study (Kieferle *et al.*, 1994). Thus, though it is tempting to speculate that these proteins are involved in the reabsorption of chloride along the nephron, a defined physiological role for these ClC members has yet to be demonstrated.

III. CHANNELS OR CHANNEL ACTIVATORS?

A. pI_{Cln}

Paulmichl *et al.* (1992) have identified a totally different protein inducing Cl^- currents by positive expression cloning using *Xenopus* oocytes. The source of the messenger RNA was MDCK cells, a well-characterized epithelial cell line derived from dog kidney. A clone was finally isolated, which, when functionally expressed in oocytes, induced strongly outward-rectifying chloride currents. Only very small currents were present at physiological voltages (−50 to −70 mV). At positive voltages, (+60 mV), currents inactivated slowly with a time constant of ≈250 msec. Currents could be blocked by extracellular DIDS (IC_{50} = 20 μM) and NPPB (5-nitro-2-(3-phenylpropylamino)-benzoic acid, IC_{50} = 1.5 μM), known inhibitors of chloride channels. They could not be influenced by intracellular calcium, suggesting that I_{Cln} is not due to an activation of the endogenous

Ca-activated oocyte chloride channel. I_{Cln} is inhibited by extracellular nucleotides, including ATP, ADP, GTP, cAMP, and cGMP in rather high concentrations, with 70% inhibition reached by millimolar levels of cAMP and cGMP.

It came as a surprise that the rather small protein encoded by the cDNA (235 amino acids, predicted molecular mass 26 kDa) did not have any hydrophobic domain predicted to cross the lipid bilayer as an α-helix. Paulmichl *et al.* therefore proposed a model in which four amphipathic β-strands cross the membrane. Assuming that the functional unit of I_{Cln} is a dimer (which is not known), the pore of the putative channel might be formed by an eight-stranded β-barrel. A similar model has been proposed previously for the voltage-gated VDAC channel from yeast mitochondria (porin) (Blachly-Dyson *et al.*, 1990). However, there is no sequence homology to VDAC, nor to any other known protein.

Site-directed mutagenesis suggested that I_{Cln} is a Cl^- channel, and not just an activator of preexistent channels. The sequence of I_{Cln} predicts a nucleotide binding domain (GXGXG), which according to the model is located near the extracellular region. In principle, it could mediate the inhibitory effect of extracellular nucleotides. Mutating the first glycine of the consensus sequence (to alanine) indeed largely reduced the sensitivity to extracellular cAMP, while mutating all three glycines totally abolished its inhibitory effect (Paulmichl *et al.*, 1992). The triple mutation also slightly changed channel kinetics and made currents dependent on extracellular calcium.

Thus, I_{Cln} may represent a novel chloride channel with an unusual structure. Its function, including the role of extracellular nucleotides in its regulation, is unclear, especially since currents are very small in the physiological voltage range. It is expressed in many different tissues (Paulmichl *et al.*, 1992; Ishibashi *et al.*, 1993) and may have some basic role for the cell.

Paulmichl and co-workers have shown that anti-sense oligonucleotides directed against I_{Cln} suppress the endogenous volume-activated chloride current of 3T3 fibroblasts (Gschwentner *et al.*, 1994). In addition, injection of antibodies directed against pI_{Cln} into *Xenopus* oocytes suppressed their endogenous swelling-activated chloride current (Krapivinsky *et al.*, 1994). However, serious doubts about the role of pI_{Cln} as a chloride channel have been raised after it was found to be a soluble, cytoplasmic protein which interacts with actin and other proteins (Krapivinsky *et al.*, 1994). Therefore, it was suggested that it plays a role in mediating channel activation induced by cell swelling by interacting with the cytoskeleton. However, a specific physiological role for pI_{Cln} in this regulation has still to be demonstrated as gross alteration of the cytoskeleton (e.g., by cytochalasins) also can lead to changes in volume regulation. Moreover, the results obtained by mutagenesis (Paulmichl *et al.*, 1992) need to be explained, and the possibility remains that a small proportion of pI_{Cln} inserts into the plasma membrane to form a channel.

B. Phospholemman

Surprisingly phospholemman, a small polypeptide of 72 amino acids with a single predicted transmembrane span, also induces chloride currents when overexpressed in *Xenopus laevis* oocytes (Moorman *et al.*, 1992). These currents activate very slowly upon hyperpolarization with a threshold potential near −80 mV and are intriguingly similar to those observed with the entirely different ClC-2 protein (Thiemann *et al.*, 1992). Extracellular acidification increased phospholemman-induced chloride currents, which were still dependent on hyperpolarization. Currents could be blocked by injection of 9-AC into the oocytes. Currents were also inhibited by extracellular barium ($IC_{50} \approx 0.36$ mM), which is normally used to block K^+ channels. There was no dependence on intracellular calcium, since injection of EGTA into the oocyte had no effect on the expressed current. This observation also excludes the possibility that phospholemman acts by activating endogenous oocyte Ca-dependent Cl^- channels, though it may activate other *Xenopus* oocyte chloride channels.

Moorman *et al.* (1992) also mutated phospholemman in an attempt to ascertain whether it is a chloride channel by itself. In fact, mutating two amino acids in the transmembrane span (G25P and F28Y) changed kinetics of channel activation. However, in contrast to changes in ion selectivity, changes in kinetics are not particularly strong arguments in proving that the mutated protein is an ion channel.

Phospholemman (first called "15-kDa protein") has been originally identified as the major target for phosphorylation by cAMP-dependent kinase and protein kinase C in the sarcolemma of the myocardium (hence its name). It is also phosphorylated *in vivo* in response to isoproterenol (Presti *et al.*, 1985). Phospholemman is expressed in several muscular tissues and in liver, but not in brain or kidney. Its sequence (Palmer *et al.*, 1991) predicts a protein with a cleavable signal peptide for ER translocation and a single hydrophobic transmembrane span. Significant sequence homology was found to a partial amino acid sequence of the so-called γ-subunit of the (Na,K)-ATPase (Collins and Leczyk, 1987; Mercer *et al.*, 1993), a small protein copurifying with the ATPase, but whose function is unknown. This homology includes the putative transmembrane region. It will be interesting to see whether these proteins belong to the same gene family and whether the "γ-subunit" is also able to induce currents when expressed in oocytes. Limited homology was also found to phospholamban (Simmermann *et al.*, 1986), a major substrate for phosphorylation in cardiac sarcoplasmic reticulum believed to regulate Ca-ATPase activity. This homology, however, is restricted to the recognition sites for phosphorylation of phospholamban.

Although interest in phospholemann originally stemmed from phosphorylation studies, Moorman *et al.* (1992) unfortunately do not report experiments aimed at studying the effect of phosphorylation on chloride currents. Thus it is unclear whether it has any relationship to cardiac chloride currents stimulated by adrenergic substances (Bahinski *et al.*, 1989; Harvey and Hume, 1989; Ehara and Ishihara, 1990), which have a very different *I–V* relationship and are probably mediated by CFTR (Nagel *et al.*, 1992). The physiological role of phospholemman remains to be determined.

As in the case of pI_{Cln}, there are now doubts that phospholemman is a chloride channel. First, hyperpolarization-induced chloride currents are observed in some batches of noninjected oocytes (Kowdley *et al.*, 1994). These currents display an I > Cl halide selectivity which differs from the Cl > I selectivity of ClC-2 (which is also activated by hyperpolarization), but coincides with that observed with phospholemman overexpression. A common feature of both currents is also their sensitivity to external barium. Second, overexpression of I_{sK} (minK), a protein having also just a single transmembrane span and postulated to be a potassium channel (Takumi *et al.*, 1988), also induces currents resembling phospholemman-induced currents (Attali *et al.*, 1993). Further, overexpression of a dead ClC-1 mutant also induces similar currents (Steinmeyer *et al.*, 1994). Thus, it seems likely that phospholemman activates chloride channels endogenous to the cells in which it is overexpressed. Whether this activation plays a physiological role or is just an artifact of overexpression remains to be determined.

IV. SUMMARY AND OUTLOOK

Due to the impact of molecular biological techniques, an increasingly complex structural picture of chloride channels has emerged during the past few years (Table I). While for the ClC family of chloride channels there seems to be no reasonable doubt that at least some of their members (ClC-0, −1, and −2) represent chloride channels, the role of pI_{Cln} and phospholemman as chloride channels has been questioned recently. There is good reason to believe that these proteins activate chloride channels

TABLE I

Cloned Voltage-Gated Chloride Channels (or Channel Activators)

	Tissue distribution	Function	Voltage dependence	Pore	Structure
CIC-0	*Torpedo* Electric organ, skeletal muscle, brain	Stabilization of V in electric organ and muscle	Slow gate opened by hyperpolarization, fast gate opened by depolarization	Linear 10 pS, double-barrel CI > Br, I block	~12 hydrophobic transmembrane domains (TMD)
CIC-1	Mammals Skeletal muscle (smooth muscle, heart)	Stabilization of V in skeletal muscle defect: myotonia	Deactivates with hyperpolarization, inward rectifier in positive V range	~1 pS CI > Br > I I block	~12 TMDs ≈tetramer
CIC-2	Mammals ubiquitous	Cell volume regulation	Slowly activates with hyperpolarization, closed at resting V linear once activated	~3–5 pS CI ≥ Br > I I block	~12 TMDs
CIC-K1	Mammalian kidney (more restricted than −K2)	Cl reabsorption ?	? or outward rectifier	? pS (Br > CI > I ?)	~12 TMDs
CIC-K2	Mammalian kidney	Cl reabsorption?	?	?	~ 12 TMDs
pI_{Cln} (regulator ?)	Mammals Broad distribution (e.g., kidney, lung, leukocytes)	Cell volume regulation (?)	Outward rectifier deactivates with depolarization, inhibited by nucleotides	??	β-barrels ?
Phospholemman (regulator ?)	Mammals Muscle, heart, liver, not brain, kidney	??	Slowly activates with hyperpolarization	??	1 single TMD

endogenous to the cell used for expression, although a direct role as a chloride channel cannot be ruled out at present. Interpretation of expression data should be very cautious, especially if currents resemble those observed in control cells. Rigid criteria should be met before a claim that a cloned protein represents a chloride channel is justified.

The complexity of the picture is likely to increase dramatically in the next few years as new members of cloned channel families are relatively easy to identify by methods based on homology. In addition, new unrelated families are likely to be identified by sophisticated expression cloning strategies.

Challenges for the future will be the molecular identification of new channels, the elucidation of their physiological roles, and the dissection of their structure–function relationships by site-directed mutagenesis. The coming years promise to be very exciting.

Acknowledgments

I thank the co-workers in my lab for their enthusiasm, hard work, and stimulating discussions. The work in this laboratory is supported by the Bundesministerium für Forschung und Technologie, the Deutsche Forschungsgemeinschaft (Grants Je164/1-1, Je164/2-1), the Cystic Fibrosis Foundation, and the Muscular Dystrophy Association.

References

Abdalla, J. A., Casley, W. L., Cousin, H. K., Hudson, A. J., Murphy, A. J., Cornélis, F. C., Hashimoto, L., and Ebers, G. C. (1992). Linkage of Thomsen disease to the T-cell receptor beta (TCRB) locus on chromosome 7q35. *Am. J. Hum. Genet.* **51,** 579–584.

Adrian, R. H., and Bryant, S. H. (1974). On the repetitive discharge in myotonic muscle fibres. *J. Physiol. (London)* **240,** 505–515.

Attali, B., Guillemare, E., Lesage, F., Honoré, E., Romey, G., Lazdunski, M., and Barhanin, J. (1993). The protein Isk is a dual activator of K^+ and Cl^- channels. *Nature (London)* **365,** 850–852.

Bahinski, A., Nairn, A. C., Greengard, P., and Gadsby, D. C. (1989). Chloride conductance regulated by cyclic AMP-dependent protein kinase in cardiac myocytes. *Nature (London)* **340,** 718–721.

Bauer, C. K., Steinmeyer, K., Schwarz, J. R., and Jentsch, T. J. (1991). Completely functional double-barreled chloride channel expressed from a single *Torpedo* cDNA. *Proc. Natl. Acad. Sci. U.S.A.* **88,** 11052–11056.

Becker, P. E. (1977). "Myotonia Congenita and Syndromes Associated with Myotonia. Clinical-Genetic Studies of the Non-dystrophic Myotonias." Thieme, Stuttgart.

Blachly-Dyson, E., Peng, S., Colombini, M., and Forte, M. (1990). Selectivity changes in site-directed mutants of the VDAC ion channel: Structural implications. *Science* **247,** 1233–1236.

Blatz, A. L., and Magleby, K. L. (1985). Single chloride-selective channels at resting membrane potentials in cultured rat skeletal muscle. *Biophys. J.* **47,** 119–123.

Bretag, A. H. (1987). Muscle chloride channels. *Physiol. Rev.* **67,** 618–724.

Brinkmeier, H., and Jockusch, H. (1987). Activators of protein kinase C induce myotonia by lowering chloride conductance in muscle. *Biochem. Biophys. Res. Commun.* **148,** 1384–1389.

Brook, J. D., McCurrach, M. E., Harley, H. G., Buckler, A. J. *et al.* (1992). Molecular basis of myotonic dystrophy: Expansion of a trinucleotide (CTG) repeat at the 3′ end of a transcript encoding a protein kinase family member. *Cell (Cambridge, Mass.)* **68,** 799–808.

Bryant, S. H., and Conte-Camerino, D. (1991). Chloride channel regulation in the skeletal muscle of normal and myotonic goats. *Pfluegers Arch.* **417,** 605–610.

Camerino, D., and Bryant, S. H. (1976). Effects of denervation and colchicine treatment on the chloride conductance of rat skeletal muscle fibers. *J. Neurobiol.* **7,** 221–228.

Chua, M., and Betz, W. J. (1991). Characterization of ion channels on the surface membrane of adult rat skeletal muscle. *Biophys. J.* **59,** 1251–1260.

Collins, J. H., and Leszyk, J. (1987). The "γ-subunit" of Na-K-ATPase: A small, amphiphilic protein with a unique amino acid sequence. *Biochemistry*, **26,** 8665–8668.

Conte Camerino, D., De Luca, A., Mambrini, M., and Vrbovà, G. (1989). Membrane ionic conductances in normal and denervated skeletal muscle of the rat during development. *Pfluegers Arch.* **413,** 568–570.

Dermietzel, R., Hwang, T.-K., Buettner, R., Hofer, A., Dotzler, E., Kremer, M., Deutzmann, R., Thinnes, F. P., Fishman, G. I., Spray, D. C., and Siemen, D. (1994). Cloning and in situ localization of a brain-derived porin that constitutes a large-conductance anion channel in astrocytic plasma membranes. *Proc. Natl. Acad. Sci. U.S.A.* **91,** 499–503.

Dulhunty, A. (1982). Effect of chloride withdrawal on the geometry of the T-tubules in amphibian and mammalian muscle. *J. Membr. Biol.* **67,** 81–90.

Ehara, T., and Ishihara, K. (1990). Anion channels activated by adrenaline in cardiac myocytes. *Nature (London)* **347,** 284–286.

Fahlke, C., Zachar, E., and Rüdel, R. (1992). Single-channel recordings of chloride currents in cultured human skeletal muscle. *Pfluegers Arch.* **421,** 108–116.

Fahlke, C., Zachar, E., and Rüdel, R. (1993). Chloride channels with reduced single-channel conductance in recessive myotonia congenita. *Neuron* **10,** 225–232.

Franke, C., Iaizzo, P. A., Hatt, H., Spittelmeister, W., Ricker, K., and Lehmann-Horn, F. (1991). Altered Na^+ channel activity and reduced Cl^- conductance cause hyperexcitability in recessive generalized myotonia (Becker). *Muscle Nerve* **14,** 762–770.

Fu, Y.-H., Pizzuti, A., Fenwick, R. G., King, J., Rajnarayan, S., Dunne, P. W., Dubel, J., Nasser, G. A., Ashizawa, T., de Jong, P., Wieringa, B., Korneluk, R., Perryman, M. B., Epstein, H. F., and Caskey, C. T. (1992). An unstable triplet repeat in a gene related to myotonic muscular dystrophy. *Science* **255,** 1256–1258.

Füchtbauer, E.-M., Reininghaus, J., and Jockusch, H. (1988). Developmental control of the excitability of muscle: Transplantation experiments on a myotonic mouse mutant. *Proc. Natl. Acad. Sci. U.S.A.* **85,** 3880–3884.

George Jr., A. L., Crackower, M. A., Abdalla, J. A., Hudson, A. J., and Ebers, G. C. (1993). Molecular basis of Thomsen's disease (autosomal dominant myotonia congenita). *Nature Genet.* **3,** 305–310.

Gisselmann, G., Sewing, S., Madsen, B. W., Mallart, A., Angaut-Petit, D., Müller-Holtkamp, F., Ferrús, A., and Pongs, O. (1989). The interference of truncated with normal potassium channel subunits leads to abnormal behaviour in transgenic *Drosophila melanogaster. EMBO J.* **8,** 2359–2364.

Grenningloh, G., Rienitz, A., Schmitt, B., Methfessel, C., Zensen, M., Beyreuther, K., Gundelfinger, E. D., and Betz, H. (1987). The strychnine-binding subunit of the glycine receptor shows homology with nicotinic acetylcholine receptors. *Nature (London)* **328,** 215–220.

Gronemeier, M., Condie, A., Prosser, J., Steinmeyer, K., Jentsch, T. J., and Jockusch, H. (1994). Nonsense and missense mutations in the muscular chloride channel gene Clc-1 of myotonic mice. *J. Biol. Chem.* **269,** 5963–5967.

Gründer, S., Thiemann, A., Pusch, M., and Jentsch, T. J. (1992). Regions involved in the opening of ClC-2 chloride channel by voltage and cell volume. *Nature (London)* **360,** 759–762.

Gschwentner, M., Nagl, U. O., Wöll, E., Schmarda, A., Ritter, M., and Paulmichl, M. (1994). Antisense oligonucleotides suppress cell volume-induced activation of chloride channels. (Submitted for publication).

Hartmann, E., Rapoport, T. A., and Lodish, H. F. (1989). Predicting the orientation of eukaryotic membrane-spanning proteins. *Proc. Natl. Acad. Sci. U.S.A.* **86,** 5786–5790.

Harvey, R. D., and Hume, J. R. (1989). Autonomic regulation of a chloride current in heart. *Science* **244,** 983–985.

Heiny, J. A., Valle, J. R., and Bryant, S. H. (1990). Optical evidence for a chloride conductance in the T-system of frog skeletal muscle. *Pflüegers Arch.* **416,** 288–295.

Hoffmann, E. K., and Simonsen, L. O. (1989). Membrane mechanisms in volume and pH regulation in vertebrate cells. *Physiol. Rev.* **69,** 315–382.

Iaizzo, P. A., Franke, C., Hatt, H., Spittelmeister, W., Ricker, K., Rüdel, R., and Lehmann-Horn, F. (1991). Altered sodium channel behaviour causes myotonia in dominantly inherited myotonia congenita. *Neuromuscular Disord.* **1,** 47–53.

Ishibashi, K., Sasaki, S., Uchida, S., Imai, T., and Marumo, F. (1993). Tissue expression of mRNA of chloride channel from MDCK cells and its regulation by protein kinases. *Biochem. Biophys. Res. Comm.* **192,** 561–567.

Jentsch, T. J., Garcia, A. M., and Lodish, H. F. (1989). Primary structure of a novel 4-acetamido-4′-isothiocyanostilbene-2,2′-disulphonic acid (SITS)-binding membrane protein highly expressed in *Torpedo californica* electroplax. *Biochem. J.* **261,** 155–166.

Jentsch, T. J., Steinmeyer, K., and Schwarz, G. (1990). Primary structure of *Torpedo marmorata* chloride channel isolated by expression cloning in *Xenopus* oocytes. *Nature (London)* **348,** 510–514.

Jockusch, H., Reininghaus, J., Stuhlfauth, I., and Zippel, M. (1988). Reduction of myosin-light-chain phosphorylation and of parvalbumin content in myotonic mouse muscle and its reversal by tocainide. *Eur. J. Biochem.* **171,** 101–105.

Kieferle, S., Fong, P., Bens, M., Vandevalle, A., and Jentsch, T. J. (1994). Two highly homologous members of the ClC chloride channel family in both rat and human kidney. *Proc. Natl. Acad. Sci. U.S.A.* (in press).

Koch, M. C., Steinmeyer, K., Lorenz, C., Ricker, K., Wolf, F., Otto, M., Zoll, B., Lehmann-Horn, F., Grzeschik, K.-H., and Jentsch, T. J. (1992). The skeletal muscle chloride channel in dominant and recessive human myotonia. *Science* **257,** 797–800.

Koch, M. C., Ricker, K., Otto, M., Wolf, F., Zoll, B., Lorenz, C., Steinmeyer, K., and Jentsch, T. J. (1993). Evidence for genetic homogeneity in autosomal recessive generalised myotonia (Becker). *J. Med. Genet.* **30,** 914–917.

Kowdley, G. C., Ackerman, S. J., John, III, J. E., Jones, L. R., and Moorman, J. R. (1994). Hyperpolarization-activated chloride currents in *Xenopus* oocytes. *J. Gen. Physiol.*, (in press).

Krapivinsky, G. B., Ackerman, M. J., Gordon, E. A., Krapivinsky, L. D., and Clapham, D. E. (1994). Molecular characterization of a swelling-induced chloride conductance regulatory protein, pI_{Cln}. *Cell* **76,** 439–448.

Landry, D. W., Akabas, M. H., Redhead, C., Edelman, A., Cragoe, E. J., and Al-Awqati,

Q. (1989). Purification and reconstitution of chloride channels from kidney and trachea. *Science* **244,** 1469–1472.

Landry, D. W., Sullivan, S., Nicolaides, M., Redhead, C., Edelman, A., Field, M., Al-Awqati, Q., and Edwards, J. (1993). Molecular cloning and characterization of p64, a chloride channel protein from kidney microsomes. *J. Biol. Chem.* **268,** 14948–14955.

Lipicky, R. J., and Bryant, S. H. (1966). Sodium, potassium, and chloride fluxes in intercostal muscle from normal goats and goats with hereditary myotonia. *J. Gen. Physiol.* **50,** 89–111.

Lipicky, R. J., Bryant, S. H., and Salmon, J. H. (1971). Cable parameters, sodium, potassium, chloride, and water content, and potassium efflux in isolated external intercostal muscle of normal volunteers and patients with myotonia congenita. *J. Clin. Invest.* **50,** 2091–2103.

Lorenz, C., Meyer-Kleine, C., Steinmeyer, K., Koch, M. C., and Jentsch, T. J. (1994). Genomic organization of the human muscle chloride channel ClC-1 and analysis of novel mutations leading to Becker type myotonia. *Hum. Mol. Genet.* **3,** in press.

Lübbert, H., Hoffman, B. J., Snutch, T. P., van Dyke, T., Levine, A. J., Hartig, P. R., Lester, H. A., and Davidson, N. (1987). cDNA cloning of a serotonin 5-HT_{1C} receptor by electrophysiological assays of mRNA-injected *Xenopus* oocytes. *Proc. Natl. Acad. Sci. U.S.A.* **84,** 4332–4336.

Mehrke, G., Brinkmeier, H., and Jockusch, H. (1988). The myotonic mouse mutant adr: Electrophysiology of the muscle fiber. *Muscle Nerve* **11,** 440–446.

Mercer, R. W., Biemesderfer, D., Bliss, Jr., D. P., Collins, J. H., and Forbush III, B. (1993). Molecular cloning and immunological characterization of the γpolypeptide, a small protein associated with the Na,K-ATPase. *J. Cell Biol.* **121,** 579–586.

Miller, C. (1982). Open-state substructure of single chloride channels from *Torpedo* electroplax. *Philos. Trans. R. Soc. London, Ser. B* **299,** 401–411.

Miller, C., and Richard, E. A. (1990). The voltage-dependent chloride channel of *Torpedo* electroplax. Intimations of molecular structure from quirks of single-channel function. *In* "Chloride Channels and Carriers in Nerve, Muscle, and Glial Cells" (F. J. Alvarez-Leefmans and J. M. Russell, eds.), pp. 383–405. Plenum, New York.

Mishina, M., Takai, T., Imoto, K., Noda, M., Takahashi, T., Numa, S., Methfessel, C., and Sakmann, B. (1986). Molecular distinction between fetal and adult forms of muscle nicotinic acetylcholine receptor. *Nature (London)* **321,** 406–411.

Moorman, J. R., Palmer, C. J., John, J. E., III, Durieux, M. E., and Jones, L. R. (1992). Phospholemman expression induces a hyperpolarization-activated chloride current in *Xenopus* oocytes. *J. Biol. Chem.* **267,** 14551–14554.

Nagel, G., Hwang, T.-C., Nastiuk, K. L., Nairn, A. C., and Gadsby, D. C. (1992). The protein kinase A-regulated cardiac Cl^- channel resembles the cystic fibrosis transmembrane conductance regulator. *Nature (London)* **360,** 81–86.

Noda, M., Shimizu, S., Tanabe, T., Takai, T., *et al.* (1984). Primary structure of *Electrophorus electricus* sodium channel deduced from cDNA sequence. *Nature (London)* **312,** 121–127.

O'Neill, G. P., Grygorczyk, R., Adam, M., and Ford-Hutchinson, A. W. (1991). The nucleotide sequence of a voltage-gated chloride channel from the electric organ of *Torpedo californica*. *Biochim. Biophys. Acta* **1129,** 131–134.

Palade, P. T., and Barchi, R. L. (1977). Characteristics of the chloride conductance in muscle fibers of the rat diaphragm. *J. Gen. Physiol.* **69,** 325–342.

Palmer, C. J., Scott, B. T., and Jones, L. R. (1991). Purification and complete sequence determination of the major plasma membrane substrate for cAMP-dependent protein kinase and protein kinase C in myocardium. *J. Biol. Chem.* **266,** 11126–11130.

Paulmichl, M., Li, Y., Wickman, K., Ackermann, M., Peralta, E., and Clapham, D. (1992). New mammalian chloride channel identified by expression cloning. *Nature (London)* **356,** 238–241.

Presti, C. F., Jones, L. R., and Lindemann, J. P. (1985). Isoproterenol-induced phosphorylation of a 15-kilodalton sarcolemmal protein in intact myocardium. *J. Biol. Chem.* **260,** 3860–3867.

Pusch, M., Steinmeyer, K., and Jentsch, T. J. (1994). Low single channel conductance of the major skeletal muscle chloride channel, ClC-1. *Biophys. J.* **66,** 149–152.

Redhead, C. R., Edelman, A. E., Brown, D., Landry, D. W., and Al-Awqati, Q. (1992). A ubiquitous 64-kDa protein is a component of a chloride channel of plasma and intracellular membranes. *Proc. Natl. Acad. Sci. U.S.A.* **89,** 3716–3720.

Reininghaus, J., Füchtbauer, E. M., Bertram, K., and Jockusch, H. (1988). The myotonic mouse mutant ADR: Physiological and histochemical properties of the muscle. *Muscle Nerve* **11,** 433–439.

Richard, E. A., and Miller, C. (1990). Steady-state coupling of ion-channel conformations to a transmembrane ion gradient. *Science* **247,** 1208–1210.

Riordan, J. R., Rommens, J. M., Kerem, B.-S., Alon, N., Rozmahel, R., Grzelczak, Z., Zielenski, J., Lok, S., Plavsic, N., Chou, J.-L., Drumm, M. L., Iannuzzi, M. C., Collins, F. S., and Tsui, L.-C. (1989). Identification of the cystic fibrosis gene: Cloning and characterization of complementary DNA. *Science* **245,** 1066–1073.

Rüdel, R., and Lehmann-Horn, F. (1985). Membrane changes in cells from myotonia patients. *Physiol. Rev.* **65,** 310–356.

Rüdel, R., Ricker, K., and Lehmann-Horn, F. (1988). Transient weakness and altered membrane characteristics in recessive generalized myotonia (Becker). *Muscle Nerve* **11,** 202–211.

Schofield, P. R., Darlison, M. G., Fujita, N., Burt, D. R., Stephenson, F. A., Rodriguez, H., Rhee, L. M., Ramachandran, J., Reale, V., Glecorse, T. A., Seeburg, P., and Barnard, E. A. (1987). Sequence and functional expression of the $GABA_A$ receptor shows a ligand-gated receptor super-family. *Nature (London)* **328,** 221–227.

Simmermann, H. K. B., Collins, J. H., Theibert, J. L., Wegener, A. D., and Jones, L. R. (1986). Sequence analysis of phospholamban. *J. Biol. Chem.* **261,** 13333–13341.

Steinmeyer, K., Ortland, C., and Jentsch, T. J. (1991a). Primary structure and functional expression of a developmentally regulated skeletal muscle chloride channel. *Nature (London)* **354,** 301–304.

Steinmeyer, K., Klocke, R., Ortland, C., Gronemeier, M., Jockusch, H., Gründer, S., and Jentsch, T. J. (1991b). Inactivation of muscle chloride channel by transposon insertion in myotonic mice. *Nature (London)* **354,** 304–308.

Steinmeyer, K., Lorenz, C., Pusch, M., Koch, M. C., and Jentsch, T. J. (1994). Multimeric structure of ClC-1 chloride channel revealed by mutations in dominant myotonia congenita (Thomsen). *EMBO J.* **13,** 737–743.

Stuhlfauth, I., Reininghaus, J., Jockusch, H., and Heizmann, C. W. (1984). Calcium-binding protein, parvalbumin, is reduced in mutant mammalian muscle with abnormal contractile properties. *Proc. Natl. Acad. Sci. U.S.A.* **81,** 4814–4818.

Stühmer, W., Conti, F., Suzuki, H., Wang, X., Noda, M., Yahagi, N., Kubo, H., and Numa, S. (1989). Structural parts involved in activation and inactivation of the sodium channel. *Nature (London)* **339,** 597–603.

Sumikawa, K., Parker, I., Amano, T., and Miledi, R. (1984). Separate fractions of mRNA from *Torpedo* electric organ induce chloride channels and acetylcholine receptors in *Xenopus* oocytes. *EMBO J.* **3,** 2291–2294.

Taguchi, T., and Kasai, M. (1980). Identification of an anion channel protein from electric organ of *Narke japonica*. *Biochem. Biophys. Res. Commun.* **96,** 1088–1094.

Takumi, T., Ohkubo, H., and Nakanishi, S. (1988). Cloning of a membrane protein that induces a slow voltage-gated potassium current. *Science* **242,** 1042–1045.

Thiemann, A., Gründer, S., Pusch, M., and Jentsch, T. J. (1992). A chloride channel widely expressed in epithelial and non-epithelial cells. *Nature (London)* **356,** 57–60.

Thomsen, J. (1876). Tonische Krämpfe in willkürlich beweglichen Muskeln in Folge von ererbter psychischer Disposition. *Arch. Psychiatr. Nervenkr.* **6,** 702–718.

Trimmer, J. S., Cooperman, S. S., Agnew, W. S., and Mandel, G. (1990). Regulation of muscle sodium channel transcripts during development and in response to denervation. *Dev. Biol.* **142,** 360–367.

Valverde, M. A., Díaz, M., Sepúlveda, F. V., Gill, D. R., Hyde, S. C., and Higgins, C. F. (1992). Volume-regulated chloride channels associated with the human multidrug-resistance P-glycoprotein. *Nature (London)* **355,** 830–833.

Uchida, S., Sasaki, S., Furukawa, T., Hiraoka, M., Imai, T., Hirata, Y., and Marumo, F. (1993). Molecular cloning of a chloride channel that is regulated by dehydration and expressed predominantly in kidney medulla. *J. Biol. Chem.* **268,** 3821–3824.

White, M. M., and Miller, C. (1979). A voltage-gated anion channel from the electric organ of *Torpedo californica*. *J. Biol. Chem.* **254,** 10161–10166.

Witzemann, V., Barg, B., Criado, M., Stein, E., and Sakmann, B. (1989). Developmental regulation of five subunit specific mRNAs encoding acetylcholine receptor subtypes in rat muscle. *FEBS Lett.* **242,** 419–424.

Zachar, E., Fahlke, C., and Rüdel, R. (1992). Whole-cell recordings of chloride currents in cultured human skeletal muscle. *Pfluegers Arch.* **421,** 101–107.

CHAPTER 3

An IAA-Sensitive Vacuolar Chloride Channel

Qais Al-Awqati

Departments of Medicine and Physiology, College of Physicians and Surgeons, Columbia University, New York, New York 10032

I. INTRODUCTION

All eukaryotic cells tested have one or more types of chloride channel. In some cells they are present in the plasma membrane where they participate in the control of cell volume (probably in most cells), membrane potential (in muscle and nerve), or the secretion of NaCl and water (in epithelia). In addition, chloride channels have been found in intracellular organelles such as endosomes, lysosomes, and Golgi where they are present in parallel to an electrogenic proton translocating ATPase and hence can control the pH gradient and membrane potential of these organelles. Based on single-channel behavior, chloride channels exhibit marked diversity in conductance, current–voltage relation, and regulation by modulators. Using these functional criteria, it was found that a single cell might contain as many as three types of plasma membrane Cl channels, in addition to the one or more types of intracellular channels.

Studies have identified a number of molecules that code for Cl channels which do not belong to a single family. The GABA and glycine receptors resemble the nicotinic receptor, a cation channel (Olsen and Tobin, 1990). The cystic fibrosis gene product (CFTR) and the multidrug resistance gene bear similarities to ATP-dependent solute transporters (Riordan, 1992; Welsh *et al.*, 1992). A family of voltage-sensitive chloride channels found in *Torpedo* electric organ, mammalian muscle, and epithelial cells do not resemble any known sequences (Steinmeyer *et al.*, 1991). Porins in outer membranes of bacteria and mitochondria do not even contain "canonical" transmembrane hydrophobic helices (Weiss and Schulz, 1992). Neither does a channel cloned from MDCK epithelia (Paulmichl *et al.*, 1992). This diversity suggests that there may be even more types of chloride channels to be discovered. Indeed, using the biochemical approach of solubilization and reconstitution, we (Landry *et al.*, 1989) and others (Ran and Benos, 1991) have found two additional Cl channels in epithelial cells.

A. p64, a Vacuolar Cl Channel

Most of the studies referred to above were directed toward the identification of plasma membrane Cl channels. We had discovered that Golgi, endosomes, and secretory granules contain chloride channels which are intimately involved in the control of the pH gradient generated by the H^+-ATPase (Glickman *et al.*, 1983). That this Cl channel can control vacuolar pH *in vivo* was shown when we found that secretagogues induce acidification of secretory granules by opening Cl channels (Barasch *et al.*, 1988). Since we had demonstrated that Golgi vesicles also contain Cl channels and were not maximally acidified, we speculated that control of Golgi pH might profoundly affect the function of Golgi enzymes many of which are pH sensitive (Al-Awqati, 1986).

To provide molecular reagents for further analyses of these processes we decided to purify the Cl channel of intracellular organelles which we term here as a vacuolar chloride channel. We do not at present know whether there is a single molecular species that codes for vacuolar Cl channels or whether there are many. Given the level of molecular diversity in Cl channels already identified, it is probable that there are several kinds of intracellular Cl channels. As discussed in detail below, we used an inhibitor derived from ethacrynic acid, IAA-94, to purify the drug binding proteins from bovine kidney cortex intracellular vesicles and showed by reconstitution that the affinity-purified proteins contain a Cl channel

(Landry *et al.*, 1987, 1989). Two proteins were eliminated as drug binding proteins and one of the remaining, a 64-kDa protein (p64), generated a specific antibody. This antibody was able to deplete all Cl channel activity from bovine kidney cortex suggesting that it is a necessary component of the vacuolar channel (Redhead *et al.*, 1992). We cloned and sequenced the cDNA for p64 and obtained a novel sequence of a membrane protein. Using antibodies we found that p64 is located in the perinuclear region (probably Golgi) and in the apical membrane of some epithelial cells.

B. Cystic Fibrosis

The discovery that the phenotypic defect in cystic fibrosis is an abnormality of activation of an epithelial Cl channel has galvanized the field of study of Cl channels and awarded it a high visibility (Schoumacher *et al.*, 1987; Welsh and Liedtke, 1986). The gene, CFTR, was rapidly identified (Welsh and Liedtke, 1986) and its function was analyzed in detail by transfection into heterologous cells where it causes the appearance of a new cyclic AMP and an ATP-activated small Cl channel with a linear *I–V* relationship (Riordan, 1992; Welsh *et al.*, 1992). The major mutation, ΔF508, appears to be a temperature-sensitive mutation that causes an abnormality in proper folding of the protein (Chen *et al.*, 1990; Denning *et al.*, 1992). Such proteins get rapidly identified by a "quality control" mechanism that leads to their degradation from the endoplasmic reticulum preventing their progress through the Golgi and on to the apical plasma membrane (Hurtley and Helenius, 1989). Reduction of temperature results in the appearance of cyclic AMP-regulated chloride channels in the plasma membrane that have the same conductance and *I–V* relation as the wild-type protein but with a lower open probability.

These findings suggest that the mutant CFTR is not present in the Golgi. This might explain why the Golgi is abnormally acidified in cystic fibrosis. We found that this defective acidification is accompanied by dramatic changes in Golgi function (Barasch *et al.*, 1991). There is a widespread decrease in sialylation of membrane and secreted proteins. Because sialylation and sulfation are competitive in the Golgi, it is likely that this defect in sialylation provides an explanation for the increased sulfation. Because there is also a defect in glycosphingolipid sialylation, we had proposed that the greater amount of asialogangliosides could act as receptors for *Pseudomonas* organisms, a major cause of morbidity and mortality in CF.

One puzzling finding is that CF cells demonstrate abnormalities in two other channels. The amiloride-sensitive Na channel is tonically open (Willumsen and Boucher, 1991). An outwardly rectifying Cl channel (ORCC) is present in CF cells but it cannot be regulated by protein kinase A (PKA) (Schoumacher *et al.*, 1987; Welsh and Liedtke, 1986); transfection with the wild-type CFTR corrects this defect (Egan *et al.*, 1992). Because the mutant and wild-type CFTR have essentially the same electrophysiological characteristics (except for open probability), it is likely that the ORCC is a different protein. Since the outward rectifier exhibits defective regulation in CF cells, we can conclude that when CFTR does not leave the endoplasmic reticulum abnormalities in other proteins can result.

As far as we know, p64 is not directly involved in CF. The protein exists in CF epithelial cells and does not seem to have either an abnormal location or molecular weight, in that none of these parameters are changed when the cells are transfected with the wild-type CFTR. However, it remains possible that p64 is the outward rectifier channel and we are now attempting to reconstitute the overexpressed protein in order to test this question.

II. THE IAA-SENSITIVE CHLORIDE CHANNEL

A. *Purification and Reconstitution*

Using bovine kidney cortex we isolated membranes that were depleted of brush borders but contained a variety of light intracellular organelles and some basolateral membranes. Using this preparation we developed an assay based on loading the vesicles with KCl and passing them down an anion exchange resin. Vesicles that had Cl but no significant K conductance will develop a membrane potential, positive inside. Addition of extracellular ^{36}Cl leads to the accumulation of the tracer in only the vesicles that have a Cl channel. Using this assay we screened a large number of potential inhibitors and found that the indanyl oxyacetic acids are reasonable inhibitors with an IAA-94 being the most potent in this system. The K_i in the bovine kidney cortex vesicles was in the range of 1–2 μM (Landry *et al.*, 1987). Subsequent studies by others showed that the potency in other systems was weaker.

We constructed an IAA-23 affinity resin and developed a simple procedure for solubilization and reconstitution of Cl channel activity from bovine kidney cortex membranes. Solubilized material was adsorbed to the affinity resin and washed extensively. Exposure of the column to 100 μM IAA-94 resulted in the elution of four proteins of 29, 40, 64, and 96 kDa. As seen in Fig. 1A, the 29- and 40-kDa proteins were the most abundant. Excess phospholipids were added and the protein:lipid mixture was dialyzed extensively to remove the IAA-94 and detergent. The resultant

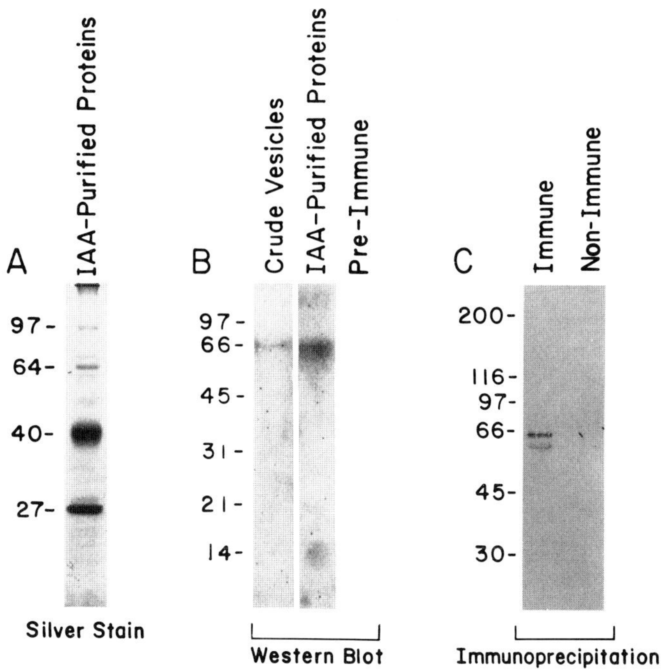

FIGURE 1 (A) Silver-stained gels of proteins purified from solubilized bovine kidney cortex vesicles using the IAA-23 column. (B) Immunoblots using antibodies generated against 64-kDa proteins cut from gels similar to those in panel A. (C) Immunoprecipitation using the antibodies described in panel B; after elution of the proteins from the antibody beads they were biotinylated and probed with avidin.

proteoliposomes were used for reconstitution experiments. It became clear that the phospholipid vesicles had a ^{36}Cl uptake that was independent of protein content. Toyoshima and Thompson (1975) had previously found that phospholipids act as a Cl:Cl exchange system. To overcome this problem we co-reconstituted the purified proteins with bacteriorhodopsin, a light-driven proton pump. When these vesicles were exposed to light they developed a membrane potential, positive inside. This energized the uptake of ^{36}Cl into these vesicles by a potential-driven, i.e., channel-mediated, mechanism. In the presence of potassium, addition of the electrogenic ionophore, valinomycin, led to the collapse of the membrane potential and reduction of the uptake of ^{36}Cl to the basal level. Quantitative analysis suggests that the affinity column has resulted in a minimum of a 1000-fold purification. Which of the four protein was the Cl channel was not clear. Indeed, it could be that the purified Cl channel was in such low abundance that it was not visible in the silver-stained gels. We purified larger amounts of the proteins and were able to obtain N-terminal sequences on the 29- and 96-kDa protein. They turned out to be glutathione-*S*-transferase and Na,K-ATPase, respectively. Both of these proteins are known to bind to ethacrynic acid, the parent compound of IAA-94; hence, this explains why they were purified with an IAA-23 column.

B. p64, a Component of the IAA-Sensitive Cl Channel in Kidney and Other Cells

To identify which of the remaining proteins were components of the Cl channel we "scaled up" the preparation and separated the purified proteins on SDS–PAGE. The 64- and 40-kDa bands were cut and injected into guinea pigs. Only the 64-kDa bands generated useful antibodies. These antisera recognized a 64-kDa band in kidney vesicles (Fig. 1B). Immunoblots of membranes from a variety of cell types in a variety of species, including plants, also yielded bands in the 60- to 70-kDa range. It was surprising to see that the antibody recognizes the protein across such a wide species range. All cells examined had the protein expressed in them. However, the protein was expressed in low abundance such that positive immunoblots were seen frequently only after membranes were used rather than cell lysates.

Immunocytochemistry did not reveal any staining except in CF-PAC, a pancreatic adenocarcinoma cell line established from a patient with cystic fibrosis. In that cell, there was staining in perinuclear (Golgi?) vesicles and some of the cells showed staining in the apical cell membrane (Fig. 2).

To demonstrate that the antigen for this antibody is a component of a Cl channel we performed immunodepletion experiments. We solubilized kidney cortex vesicles and incubated them with preimmune, immune, and no immunoglobulins. The antigen–antibody complexes were precipitated with protein A beads and the supernatants were reconstituted and assayed for voltge-sensitive ^{36}Cl uptake. The material incubated with no Ig or with preimmune Ig exhibited vigorous valinomycin-sensitive uptake. Preincubation with anti-p64 sera led to a complete removal of any Cl channel activity in the reconstituted proteoliposomes. To provide further evidence that p64 was a necessary component of the IAA-sensitive Cl channel we partially purified Cl channel activity by separating solubilized kidney cortex vesicles on a gel filtration column. The column effluent fractions were reconstituted. Valinomycin-sensitive ^{36}Cl uptake was present only in the 400- to 550-kDa fraction. When these fractions were stained by anti-p64

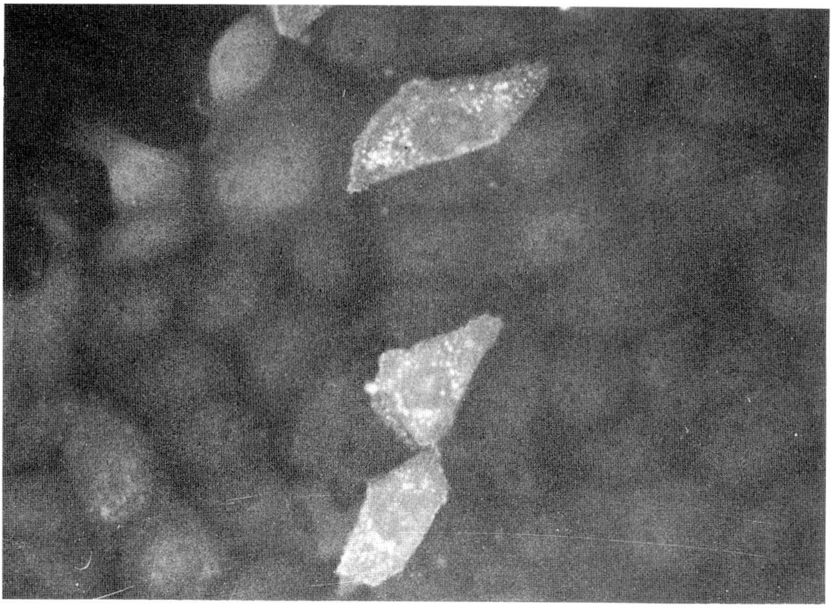

FIGURE 2 Immunocytochemistry of permeabilized CF-PAC cells using the anti-p64 antibody.

antibodies, only those fractions that contained p64 showed reconstitutable ^{36}Cl uptake (Redhead *et al.*, 1992). These results provide strong evidence that the antigen recognized by anti-p64 antibody is an important component of a Cl channel. That reconstitutable activity traveled at such a high M_r suggests that this Cl channel complex is either composed of several copies of p64 or composed of a heteromultimer (Fig. 3 demonstrates these results using an antibody against the cloned p64 described below).

C. Cloning of the p64 cDNA

We initially searched several cell lines and tissues to identify one that overexpressed p64 and found that kidney cortex showed the highest abundance. We emphasize, however, that even in the kidney p64 is a low-abundance protein. We constructed an expression cDNA library using standard methods and screened it with anti-p64 antisera. Several positive clones were identified and the one which had the longest insert, termed H2B, was analyzed in detail. H2B was used for Northern analysis, where it showed that the kidney expresses a 6.3-kb message, as well as other

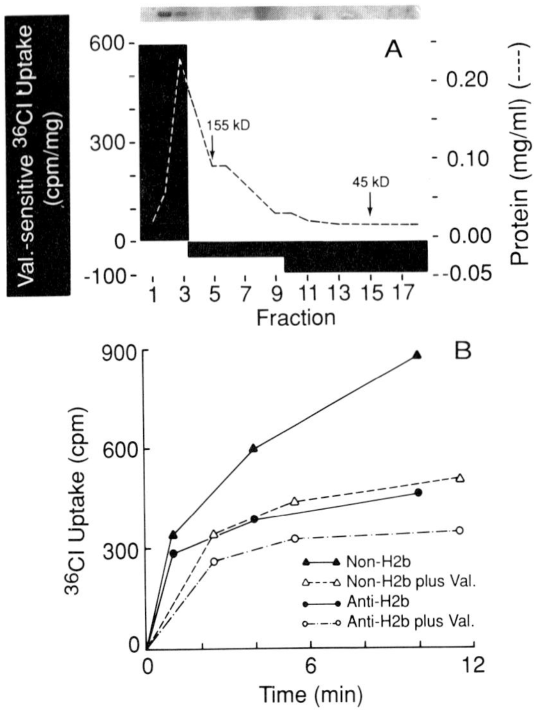

FIGURE 3 (A) Bovine kidney cortex vesicles were solubilized with octyl glucoside and passed on a gel filtration column. The effluent was divided into four fractions and the protein content was measured in each fraction. Equal amounts of protein were revconstituted into asolectin phospholipids and ^{36}Cl uptake was measured in the presence and absence of valinomycin as described previously (Redhead *et al.*, 1992). Each fraction was analyzed by SDS–PAGE and immunoblots using an antibody that was affinity purified against the fusion protein H2B. (B) Anti-p64 antibodies were affinity purified against the fusion protein H2B and the resultant antibodies were incubated with solubilized kidney cortex vesicles. The control antibodies were the same anti-p64 antibodies "affinity purified" against the same region of *E. coli* lysates using bacteria transfected with the vector lacking the H2B insert. Immunodepletion experiments are described in the text.

shorter messages. We identified overlapping clones and using the polymerase chain reaction we were eventually able to obtain a full-length sequence of the cDNA. There was an open reading frame that codes for a protein of 437 amino acids with an M_r of 49,008 (Fig. 4) (Landry *et al.*, 1993).

Neither the nucleotide nor the predicted amino acid sequence of p64 bear significant homology to any known gene or protein. The deduced amino acid sequence is markedly rich in acidic residues and has a predicted pI of 4.14. A hydrophobicity analysis of the predicted amino acid sequence by the method of Kyte and Doolittle showed that there are at least two potential transmembrane domains at amino acids 201–236 and 367–385. No putative signal sequence was discerned, suggesting that the N-terminus of the protein would be predicted to be in the cytoplasm. This large section of the protein contains consensus sequences for phosphorylation by protein kinase C, tyrosine kinase, and casein kinase II. An octapeptide sequence with a high density of negatively charged amino acids [in single letter code, Q(E,A, or G)SD(P or S)EEP] is repeated four times, and partially for a fifth, in this first intracellular domain. There is no significant homology to this specific sequence motif in the data base except for heavily negatively charged proteins. A more focused search regarding calcium binding proteins did not result in any similarity to this motif, and in particular the motif is not homologous to an E–F hand sequence. One possibility is that it represents a site of protein–protein interaction. The sequence between the two putative transmembrane domains is predicted to be extracellular and it contains one potential N-glycosylation site. The C-terminus of the protein is predicted to contain the second intracellular domain and contains a single consensus sequence for protein kinase A phosphorylation at S435.

We generated a β-galactosidase–H2B fusion protein of about 70 kDa in *Escherichia coli* and affinity-purified anti-p64 antibodies over the fusion protein. These antibodies recognized a 64-kDa protein in kidney cortex and other cells as well as recognized partially purified (by gel filtration) Cl channel proteins. To demonstrate that H2B protein is also a component of the IAA-sensitive Cl channel we performed another immunodepletion experiment as described above. We found that anti-H2B antibodies depleted valinomycin-sensitive ^{36}Cl uptake. These results suggest that we have cloned a Cl channel protein (Fig. 4).

RNA from bovine kidney cortex, skeletal muscle, and heart had a relatively abundant transcript, at approximately 6.5 kb, which hybridized with p64 cDNA. A transcript of the same size was also seen in lower amounts in kidney medulla and adrenal. All tissues examined had two distinct smaller transcripts at approximately 5.5 and 4.5 kb. In addition, kidney medulla, adrenal, and brain have a 7-kb transcript and the heart has a smaller transcript of about 2 kb. This diversity of transcript size in various tissues suggests that this gene is subject to alternate splicing or that a family of closely related genes exist. RNA from the human chloride transporting epithelial cell lines T84 and Panc 1 had a single major transcript of approximately 5.5 kb which hybridizes to the p64 probe at high stringency. RNA from the shark rectal gland, a chloride-secreting epithelium rich in chloride channels, contains transcripts of approximately 4 and 6 kb that hybridize with the p64 probe at moderate stringency.

A

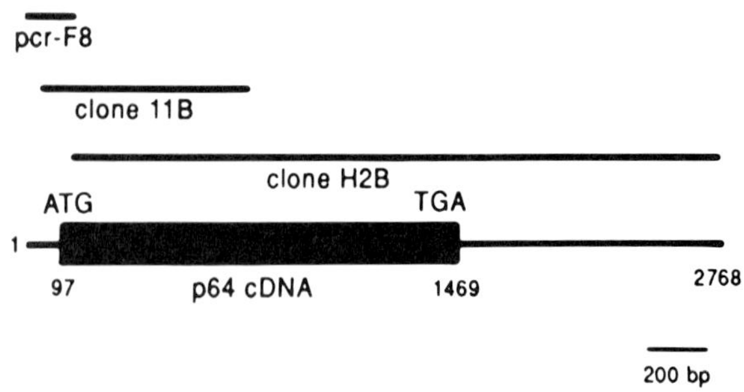

B

```
  1 agattgttagagacctgtcactgatagctgtggctggttgctgtgtttcaaaagcctgtt
 61 gattcctctgcaaaggcggtgcagcatcctctagccctttcctcatctcagacctactcg
    ---------+---------+---------+---------+---------+---------+
                                  M  N  D  E  N  Y  S  T     8
121 ggtctgcttgcgaaaggatctatcagatcaataaacatgaatgacgaaaactacagcacc
    ---------+---------+---------+---------+---------+---------+
    T  I  Y  N  R  V  Q  T  E  R  V  Y  E  D  S  D  P  A  E  N    28
181 accatctacaacagagtccaaacagagagggtgtatgaggattcagacccagcagaaaac
    ---------+---------+---------+---------+---------+---------+
    G  G  P  L  Y  D  E  V  H  E  D  V  R  R  E  D  N  L  Y  V    48
241 ggaggtcccctttacgatgaagtccatgaagatgtgaggagggaagacaacttatatgtc
    ---------+---------+---------+---------+---------+---------+
    N  E  L  E  N  Q  E  Y  D  S  V  A  V  Y  P  V  G  R  Q  G    68
301 aatgagctggaaaaccaagaatatgactctgtggccgtgtaccctgttgggcgtcagggg
    ---------+---------+---------+---------+---------+---------+
    R  T  S  A  S  L  Q  P  E  T  G  E  Y  V  L  P  D  E  P  Y    88
361 agaacctcggcctctttacagccagaaacaggagaatacgtcctgcctgatgaaccatat
    ---------+---------+---------+---------+---------+---------+
    S  K  A  Q  D  P  H  P  G  E  P  T  E  D  E  D  I  S  L  E   108
421 taaggctcaggacccccacccaggggaacccacagaggacgaggacatctccttggaaga
    ---------+---------+---------+---------+---------+---------+
    E  L  L  S  P  T  K  D  Q  E  S  D  S  E  E  P  Q  A  S  D   128
481 attgctctcaccaaccaaggaccaggagtctgactcagaggagccccaggcgtctgaccc
    ---------+---------+---------+---------+---------+---------+
    P  E  E  P  Q  A  S  D  P  E  E  P  Q  G  P  D  P  E  E  P   148
541 ccagaggagccccaggcgtctgacccagaggagccccaggggcctgacccagaggagccc
    ---------+---------+---------+---------+---------+---------+
    Q  E  N  G  N  E  M  E  A  D  L  P  S  P  S  S  F  T  I  Q   168
601 caggagaatggaaacgagatggaggcagacctgccttctccttcgagcttcaccatccag
    ---------+---------+---------+---------+---------+---------+
    N  S  R  A  F  S  T  R  E  I  S  P  T  S  Y  S  A  D  D  V   188
661 aatagcagagccttctccacccgcgagatctcccctaccagctattctgctgacgatgtt
    ---------+---------+---------+---------+---------+---------+
    S  E  G  N  E  S  A  S  A  S  P  E  I  N  L  F  V  K  A  G   208
721 tcagaaggaaacgaatcggcttccgcaagccccgagattaacctctttgtgaaggctgga
    ---------+---------+---------+---------+---------+---------+
    I  D  G  E  S  I  G  N  C  P  F  S  Q  R  L  F  M  I  L  W   228
781 atcgatggtgaaagcattggcaactgtcccttctctcagcgtctctttatgatcctctgg
    ---------+---------+---------+---------+---------+---------+
    L  K  G  V  V  F  N  V  T  T  V  D  L  K  R  K  P  A  D  L   248
841 ctgaaaggagttgtgttcaatgtcaccacggtggatctgaaaagaaagccagctgacctg
    ---------+---------+---------+---------+---------+---------+
    H  N  L  A  P  G  T  H  P  P  F  L  T  F  N  G  D  V  K  T   268
901 cacaaccttgcccctggcactcatccacccttcctgacgttcaacggggatgtgaagaca
    ---------+---------+---------+---------+---------+---------+
```

FIGURE 4 (A) Partial restriction map of H2B. (B) Complete nucleotide and predicted amino acid sequence of H2B. (C) Kyte–Doolittle hydropathy plot of H2B.

C

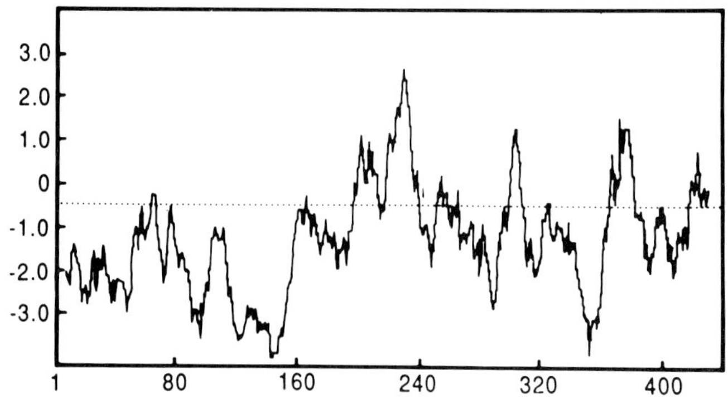


```
     D  V  N  K  I  E  E  F  L  E  E  T  L  T  P  E  K  Y  P  R   288
 961 gacgtcaataagatcgaggagttcctggaggagactttgacccctgaaaagtaccccaga
     ---------+---------+---------+---------+---------+---------+
     L  A  A  K  H  R  E  S  N  T  A  G  I  D  I  F  V  K  F  S   308
1021 ctggctgcgaaacatcgggaatccaacacggcaggcatcgacatcttcgttaagttctcc
     ---------+---------+---------+---------+---------+---------+
     A  Y  I  K  N  T  K  Q  Q  S  N  A  A  L  E  R  G  L  T  K   328
1081 gcctacatcaaaaacacgaagcagcagagcaacgctgcccttgagcgaggcctgaccaag
     ---------+---------+---------+---------+---------+---------+
     A  L  K  K  L  D  D  Y  L  N  T  P  L  P  E  E  I  D  A  D   348
1140 gctctgaagaagctggatgactacctgaacacccctctgccggaggagatcgatgccgac
     ---------+---------+---------+---------+---------+---------+
     T  R  G  D  D  E  K  G  S  R  R  K  F  L  D  G  D  E  L  T   368
1201 actcgcggggatgatgaaaagggctcccggcgcaagttcctggatggggacgagttgacc
     ---------+---------+---------+---------+---------+---------+
     L  A  D  C  N  L  L  P  K  L  H  V  V  K  I  V  A  K  K  Y   388
1261 ctggccgactgcaacctgttgcccaagctccatgtggtcaagattgtggccaagaaatac
     ---------+---------+---------+---------+---------+---------+
     R  N  Y  D  F  P  A  E  M  T  G  L  W  R  Y  L  K  N  A  Y   408
1321 cgcaactacgatttcccggctgagatgacgggtctgtggcggtacctcaagaacgcctac
     ---------+---------+---------+---------+---------+---------+
     A  R  D  E  F  T  N  T  C  A  A  D  S  E  I  E  L  A  Y  A   428
1381 gcccgggacgagttcaccaacacctgcgcggccgacagcgagatcgagctggcctatgcg
     ---------+---------+---------+---------+---------+---------+
     D  V  A  K  R  L  S  R  S                                    437
1441 gacgtggccaagcgcctcagccgatcctgaacttggccgtctcgccccatcgctgctgca
     ---------+---------+---------+---------+---------+---------+
1501 gaaggactccaccgtctccccaaggactctggcttcacacacttctcttccacgccacct
1561 ctggaaacttctgtctctatcaggccacatctctctggcatccttgtaacccccccgcct
1621 gctggctttgaccaaagtcagcgccatctccctctggtcaggcaggaatcccgcaagcag
1681 aaaagagagaaacggtcaagctgtcagcctttcggacctggactgggttccgtgtgtttg
1741 ggggtcggtgccgggggaaccctgtctgtgggattgtcactggccccttttgccaatat
1801 tgaaatatctcctgagatgcattagaaacctacaacatgctgtccttatccacccctcct
1861 ctttgtcagtaactccaaggccaaacgccctcttcgttcttattagaaagaggtgcatgg
1921 ctgtgttagtggagacagagagtggatggggcagacgtgggtctgggcacgtggcagaca
1981 cacctccagtcactcagcactttctagcattgtcacttcgggcactgaaggccagattgg
2041 aaaacaacttcagacagtctcagttttctgcagtgctgggcacagtctgatgtgatcaca
2101 tacctggcttgatgctgatctctattatctgtccagaccttgccaggaagggaaattggc
2161 atgccattccaaactggggtacctggggaggggaggggaggggagaagtgttgatgccaa
2221 aaggcccgtggggtctaccagccatggggtttgcttgcttatgggagtggtttcattgga
2281 gattacatgcctgagtttgactatgtttatccacggttgaattttggtctctgagaaagc
2341 agatgggagtgtggggggtggggggatggatgcaggaggctaagaagacctttgtatgag
2401 gctcagtgttccctgggaatcttctagatagatcttttctcttatgatctaagtctcgaa
2461 ccagtgatgtcatgtagctgccatcttcatagcaaaataccaacaattgaccaaaaatga
2468 aaaaaaa
```

FIGURE 4 *Continued*

D. p64 Is Not Targeted to the Plasma Membrane of Nonepithelial Cells but in Epithelial Cells It Is Located in the Apical Plasma Membrane

p64 was present at the RNA and protein level in all cells tested. To examine its distribution we biotinylated the surface of confluent monolayers of epithelial cells such as MDCK and two nasal polyp cell lines, one obtained from a normal subject and the other obtained from a patient with cystic fibrosis. The cells were solubilized and the biotinylated proteins were precipitated with avidin beads and washed extensively. The proteins were then analyzed by immunoblots using an anti-p64 fusion protein antibody. All these epithelial cell lines had p64 present on their apical membranes. However, when similar studies were performed in three nonepithelial cell lines, 3T3 fibroblasts, COS cells (monkey kidney fibroblatoid cell line), and Sf 9, insect ovary cells, there was no p64 on the surface. Membranes isolated from these cells had p64 in them.

We also injected RNA transcribed from the T7 promoter construct of the full-length clone into *Xenopus* oocytes. No new plasma membrane chloride current appeared on electrophysiological studies. However, a new protein with the appropriate molecular mass appeared in microsomes and was unable to be targetted to the plasma membrane. Biotinylation of the surface membranes of oocytes failed to show the presence of the endogenous or exogenous p64. In addition p64 was overexpressed in Sf 9 cells using a baculovirus system. The results shows that the protein is expressed at very high levels. However, it does not reach the surface. Heterologous expression in COS cells also did not lead to targeting of the protein to the surface.

These results demonstrate that p64 either contains a "retention" signal in nonepithelial cells or it requires an accessory protein to target it to the plasma membrane. Because our studies have identified the presence of several transcripts that cross-hybridize with p64 probes, it is possible that a family of transcripts exists some of which are targeted to the plasma membrane while others are retained in intracellular organelles.

If all p64 transcripts are similar then the epithelial targeting of p64 must require another protein whose function is similar to "molecular chaperones." Studies have identified a group of proteins which direct the targeting of other membrane proteins to their final destination. A mutation in *Drosophila, ninaA,* results in the retention of one of the rhodopsin proteins in the endoplasmic reticulum (Colley *et al.*, 1991). This protein is a resident protein of the ER and is a membrane protein with a large cytoplasmic domain. The cytoplasmic domain is homologous to proteins that bind FK506, the cyclosporin homolog. Since this class of proteins, the cyclophilins, also catalyze *cis trans* isomerization of prolines, it has been suggested that they are involved in proper folding of their substrates. The surprise is that none of the other rhodopsins in *Drosophila* is affected, suggesting that each rhodopsin molecule might have its own folding facilitator. A yeast protein, SH2, has been identified which promotes correct targeting of many amino acid transporters to the plasma membrane (Ljungdahl *et al.*, 1992). When this protein is inactivated, these polytopic membrane proteins are retained in the ER in an improperly folded manner.

Acknowledgments

The work reported here was supported by grants from NIDDK and the Cystic Fibrosis Foundation.

References

Al-Awqati, Q. (1986). Proton translocating ATPases. *Annu. Rev. Cell Biol.* **2,** 179–199.

Barasch, J., Gershon, M. D., Nuñez, E. A., Tamir, H., and Al-Awqati, Q. (1988). Thyrotropin induces the acidification of the secretory graules of parafollicular cells by increasing the chloride conductance of the granular membrane. *J. Cell Biol.* **107,** 2137–2147.

Barasch, J., Kiss, B., Prince, A., Saiman, L., Gruenert, D., and Al-Awqati, Q. (1991). Defective acidification of intracellular organelles in cystic fibrosis. *Nature (London)* **352,** 70–73.

Chen, S. H., Gregory, R. J., Marshall, J., Paul, S., Souza, D. W., White, G. A., O'Riordan, C. R., and Smith, A. E. (1990). Defective intracellular transport and processing of CFTR is the molecular basis of most cystic fibrosis. *Cell (Cambridge, Mass.)* **63,** 827–834.

Colley, N. J., Baker, E. K., Stamnes, M. A., and Zuker, C. S. (1991). The cyclophilin homologue, ninaA is required in the secretory pathway. *Cell (Cambridge, Mass.)* **67,** 255–263.

Denning, G. M., Anderson, M. P., Amara, J. F., Marshall, J., Smith, A. E., and Welsh, M. J. (1992). Processing of mutant CFTR is temperature sensitive. *Nature (London)* **358,** 761–764.

Egan, F., Flotte, T., Afione, S., Solow, B., Zeitlin, P. L., Carter, B. J., and Guggino, W. B. (1992). Defective regulation of an outwardly rectifying Cl channels by protein kinase A corrected by insertion of CFTR. *Nature (London)* **335/338,** 581–584.

Glickman, J., Croen, K., Kelly, S., and Al-Awqati, Q. (1983). Golgi membranes contain an electrogenic H^+ pump in parallel to a chloride conductance. *J. Cell Biol.* **97,** 1303–1308.

Hurtley, S. M., and Helenius, A. (1989). Protein oligomerization in the endoplasmic reticulum. *Annu. Rev. Cell Biol.* **5,** 277–307.

Landry, D. W., Reitman, M., Cragoe, E. J., Jr., and Al-Awqati, Q. (1987). Epithelial chloride channel. Development of inhibitory ligands. *J. Gen. Physiol.* **90,** 779–798.

Landry, D. W., Akabas, M. H., Redhead, C., Edelman, A., Cragoe, E. J., and Al-Awqati, Q. (1989). Purification and reconstitution of chloride channels from kidney and trachea. *Science* **244,** 1469–1472.

Landry, D. W., Sullivan, S., Nicolaides, M., Redhead, C., Edelman, A., Field, M., Al-Awqati, Q., and Edwards, J. (1993). Molecular cloning of p64, a chloride channel protein. *J. Biol. Chem.* **268,** 14948–14955.

Ljungdahl, P. O., Gimeno, C. J., Styles, C. A., and Fink, G. R. (1992). SH3: A novel component of the secretory pathway specifically required for localization of amino acid permeases in yeast. *Cell (Cambridge, Mass.)* **71,** 463–478.

Olsen, R. W., and Tobin, A. J. (1990). Molecular biology of $GABA_A$ receptors. *FASEB J.* **4,** 469–1480.

Paulmichl, M., Li, Y., Wickman, K., Ackerman, M., Peralta, E., and Clapham, D. (1992). New mammalian chloride channel identified by expression cloning. *Nature (London)* **356,** 238–241.

Ran, S., and Benos, D. J. (1991). Purification and reconstitution of a chloride channel from tracheal membranes. *J. Biol. Chem.* **266,** 4782–4788.

Redhead, C. R., Edelman, A., Brown, D., Landry, D. W., and Al-Awqati, Q. (1992). A ubiquitous 64 kDa protein is a component of a chloride channel of plasma and intracellular membranes. *Proc. Natl. Acad. Sci. U.S.A.* **89,** 3716–3720.

Riordan, J. R. (1992). The molecular biology of chloride channels. *Curr. Opin. Nephrol. Hypertens.* **1,** 35–42.

Schoumacher, R. A., Shoemaker, R. L., Halm, D. R., Tallant, E. A., Wallace, R. W., and Frizzell, R. A. (1987). Phosphorylation fails to activate chloride channels from cystic fibrosis airway cells. *Nature (London)* **330,** 752–754.

Steinmeyer, K., Ortland, C., and Jentsch, T. J. (1991). Primary structure and functional expression of a developmentally regulated skeletal muscle chloride channel. *Nature (London)* **354,** 301–304.

Toyoshima, Y., and Thompson, T. E. (1975). Chloride flux in bilayer membranes: The electrically silent chloride flux in semispherical bilayers. *Biochemistry* **14,** 1525.

Weiss, M. S., and Schulz, G. E. (1992). Structure of porin refined at 1.8 A resolution. *J. Mol. Biol.* **227,** 493–509.

Welsh, M. J., and Liedtke, C. M. (1986). Chloride and potassium channels in cystic fibrosis airway epithelia. *Nature (London)* **322,** 467–470.

Welsh, M. J., Anderson, M. P., Rich, D. P., Berger, H. A., Denning, G. M., Ostergaard, L. S., Sheppard, D. N., Cheng, S. H., Gregory, R. J., and Smith, A. E. (1992). CFTR: A chloride channel with novel regulation. *Neuron* **8,** 821–829.

Willumsen, N. J., and Boucher, R. C. (1991). Na transport and intracellular Na activity in cultured human nasal epithelium. *Am. J. Physiol.* **261,** C319–C331.

CHAPTER 4

Anion Channels in the Mitochondrial Outer Membrane

Marco Colombini

Laboratories of Cell Biology, Department of Zoology, University of Maryland, College Park, Maryland 20742

Current Topics in Membranes, Volume 42

I. INTRODUCTION

The notion that the outer mitochondrial membrane contains channels does not raise eyebrows these days when patch clamping of most cell membranes reveals some sort of channel(s). However, when first observed in 1975, the reconstitution of an anion-selective channel from mitochondria was a surprise (Schein *et al.*, 1976). By hindsight, the existence of large channels in the outer membrane favoring anions fits nicely with the major function of mitochondria, energy transduction. Substrates and products are mostly negatively charged molecules: pyruvate, ADP, ATP, phosphate, etc. These must cross the outer membrane at rather high rates in order to supply the energy needs of an active, rapidly respiring cell.

From the early experiments that demonstrated that mitochondria were composed of two layers of membranes (Werkheiser and Bartley, 1957), the evidence indicated that the outer membrane was not merely "leaky" but permselective. Standards of membrane intactness used for other systems did not apply to the outer membrane. Its demonstrated permeability to sucrose did not indicate membrane damage because dextrans and cytochrome C were impermeant if mitochondria were isolated with care (Werkheiser and Bartley, 1957; Wojtczak and Zaluska, 1969). This and electron-microscopic evidence indicated the presence of large channels in the outer mitochondrial membrane (Parsons *et al.*, 1965).

The reconstitution into planar phospholipid membranes of large voltage-gated channels, called VDAC (voltage-dependent anion-selective channel), indicated that these channels were not static structures but dynamic and under regulation (Schein *et al.*, 1976; Colombini, 1979). The voltage dependence of the channels is steep and not unlike that responsible for the electrical excitability of nerves, muscles, and other cells. Although still unmeasured, an electrical potential across the outer membrane capable of altering VDAC's conformation is not only possible but highly likely. Unless the colloidal charge in the intermembrane space and the cytoplasm are identical, a nonzero Donnan potential must exist. Its value could be easily altered by protein phosphorylation or local changes in pH or free $[Ca^{2+}]$.

The importance of VDAC and its properties to mitochondrial function is strongly indicated by the remarkable conservation of its structure and functional properties (Colombini, 1989). The discovery of new regulatory mechanisms (Holden and Colombini, 1988) and their remarkable conservation (Liu and Colombini, 1991) provides further evidence for an important and elaborate regulatory system.

II. BASIC PROPERTIES

VDAC channels form highly conductive pathways (4 to 4.5 nS in 1 *M* KCl) in phospholipid membranes (Colombini, 1979,1989). An example of VDAC channels from rat liver inserting into planar phospholipid membranes (Fig. 1) shows their uniformity. These reflect a large aqueous pore whose radius in the high-conducting (open) state is estimated at 1.2 to 1.5 nm (see below). This pore allows inulin, polyethylene glycol (PEG) 3400, and dextran 2000 to cross the membrane and excludes PEG 6800 and dextran 8000 (Colombini, 1980a,b; Zalman *et al.*, 1980). The channels are quite stable and long-lived in the open state at low membrane potentials. Raising the potential to either positive or negative values causes VDAC to enter low-conducting states termed "closed" because there is evidence that they are essentially impermeable to important metabolites (see below). Each channel can therefore undergo two separate voltage-

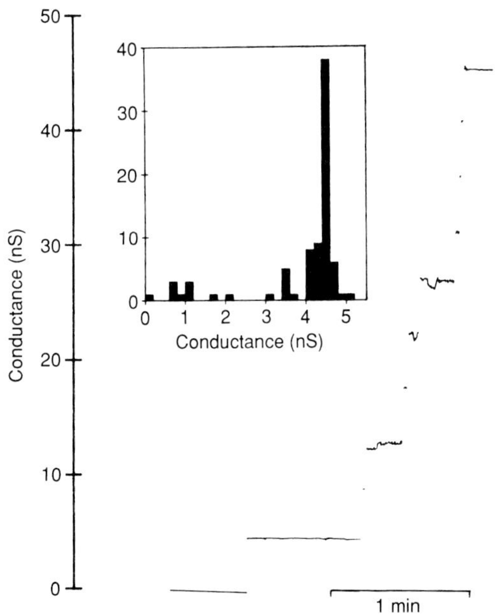

FIGURE 1 Single VDAC channels, isolated from rat liver mitochondria, inserting into planar phospholipid membranes. The increments in membrane conductance (see inset) are clustered at about 4.5 nS (medium was 1 *M* KCl, 5 m*M* $CaCl_2$; the applied voltage was 10 mV). Smaller increments are due to substates.

dependent structural changes, one at positive potentials and the other at negative potentials. The situation is more complex, however, because there exists a family of closed states at both signs of the potential, the significance of which is unclear (Colombini, 1980b,1986). With higher applied potentials and following special treatments (Colombini *et al.*, 1987; Holden and Colombini, 1988; Liu and Colombini, 1992a), VDAC channels can be induced to enter states with extremely low conductances.

Some look at VDAC closure as an inactivation process because the channels ''adapt'' with time to the closed state. The rate of channel opening depends on the length of time they have been held in a closed conformation by means of an electric field. This molecular memory may be important in long-term changes in mitochondrial function that may be induced by VDAC closure. With respect to the study of VDAC's electrophysiological properties, this memory is simply a nuisance.

VDAC channels are not absolutely selective for anions over cations but a clear preference exists (Colombini, 1980b). This preference is seen even with small ions capable of traveling through the channel at some distance from the protein walls. K^+ and Cl^- have almost the same mobility in water but Cl^- can be selected by over 4 : 1 depending on the ionic conditions (Fig. 2). As the size and charge of the ion increases, the selectivity for the anion should increase for ions of comparable size and absolute charge. For example, VDAC is more permeable to the doubly negatively charged

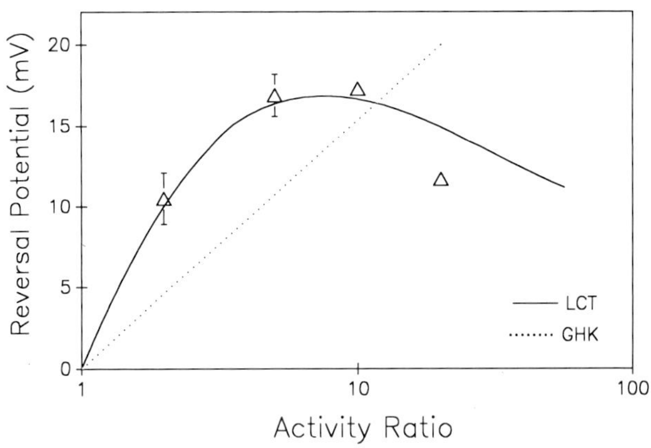

FIGURE 2 The reversal potential of VDAC channels from *N. crassa* reconstituted into planar membranes. The activity of KCl was 0.06 in the low-salt side. The data are indicated by the triangles (mean ± SD) and the theoretical fits are the solid line (large-channel theory; Zambrowicz and Colombini, 1993) and dotted line (Goldman–Hodgkin–Katz equation).

calcium–EDTA complex than it is to the much smaller doubly positively charged calcium ion (M. Colombini, unpublished observation).

III. MOLECULAR STRUCTURE

A. Composition

VDAC channels whose genes have been cloned from mammalian and fungal sources (or the protein sequenced) range in molecular weight from 30k to 32k. So far they fall into two categories: those composed of 282 amino acids (after removal of the N-terminal methionine; Mihara and Sato, 1985; Kleene *et al.*, 1987; Kayser *et al.*, 1989) and those with an 11 amino acid extention beyond the N-terminal methionine (as deduced from the nucleotide sequence; Bureau *et al.*, 1992; Blachly-Dyson *et al.*, 1993). These may be referred to as VDAC1 and VDAC2 following the terminology introduced by Blachly-Dyson *et al.* (1993). However, estimates of polypeptide molecular weight based on SDS–PAGE vary between 29k and 37k (De Pinto and Palmieri, 1992; Bureau *et al.*, 1992; Blumenthal *et al.*, 1993) depending on the source of the material. The rat version of VDAC2 forms a band at 36k on SDS–PAGE indicating possible glycosylation (Bureau *et al.*, 1992) or anomalous mobility. Perhaps these effects account for the variability in molecular weights (measured by SDS–PAGE) of VDAC from the different sources although other VDAC gene families may exist (Mike Forte, personal communications). However, the rat protein just mentioned is reported to interact tightly with the GABA receptor and its role in this interaction is the subject of interesting speculation. The proposed plasma-membrane association of this gene product supports previous reports of VDAC channels located on the surface membrane of animal cells (Blatz and Magleby, 1983; Nelson *et al.*, 1984; Thinnes, 1992).

Despite VDAC's symmetrical behavior (e.g., voltage-dependent channel closure at both positive and negative membrane potentials) strong evidence now favors the conclusion that each VDAC channel is formed by one 30-kDa polypeptide. Estimates of protein present in two-dimensional crystals of VDAC from *Neurospora crassa* viewed as frozen-hydrated specimens indicated that the mass per channel-forming unit might be less than needed for a dimer (Mannella, 1986,1987). Mass measurements (Thomas *et al.*, 1991) based on scanning-transmission electron microscopy of the same kind of two-dimensional crystals of VDAC yielded a mass per unit area of crystal of 1.9 kDa/nm^2. Since, on the average, one channel occupies an area of 23 nm^3, the mass of matter per channel is 44 kDa. Thus there is insufficient mass for a dimer and the results are consistent with a 32% lipid content in the crystal. Further support for the monomer hypothesis came from work with VDAC mutants expressed in yeast (Peng *et al.*, 1992b). The mutants chosen were some of those (see below) that had been shown to alter VDAC's selectivity by changing the charge on the protein walls of the aqueous pore. If two polypeptides contributed to the walls of the pore then the selectivity of the resulting channel would depend on charged residues from both polypeptides. Thus the measured selectivity of an individual channel should be an average of the contributions of each polypeptide. A dimer therefore raises the possibility of hybrid channels. By introducing a plasmid containing a yeast VDAC sequence containing two point mutations that change the reversal potential of VDAC by 10 mV into a yeast cell containing the wild-type VDAC gene, Peng and co-workers tried to obtain hybrid channels, i.e., channels whose selectivity was half way between that of the wild-type VDAC and that of the purely mutant form. The selectivity of individual VDAC channels isolated from this yeast strain revealed only channels with either pure wild-type selectivity or pure mutant selectivity (Fig. 3). A sufficient number of channels was examined such that the probability of missing a hybrid (if such existed) was less than 1 in 10^7. The authors repeated the whole experiment with a different mutant VDAC gene with the same results. Thus, these experiments on functional VDAC channels from yeast agree with the electron-microscopic experiments with *N. crassa* VDAC. Each VDAC channel is most likely composed of one and only one 30-kDa polypeptide.

Early work indicated that detergent-solubilized VDAC was a multimer of 30-kDa polypeptides (Linden and Gellerfors, 1983). This, taken together with the observation that the number of channels inserting into a planar phospholipid membrane was linearly related to the amount of detergent-solubilized VDAC added to the aqueous phase (Colombini, 1980b; Roos *et al.*, 1982), indicated that one VDAC channel was composed of two 30-kDa polypeptides. A reexamination of the data shows that in deducing the molecular weight of VDAC from its sedimentation rate, Linden and Gellerfors did not consider the contribution to the overall mass of the water in the pore. This and tightly bound sterol may account for some of the mass attributed to protein.

Further evidence that VDAC is a monomer comes from indications that the protein is inherently asymmetric in intact mitochondria (De Pinto and Palmieri, 1992).

B. Pore Size

VDAC's pore size has been estimated in a variety of ways each with its own drawbacks. A trivial approach is to simply use the single-channel

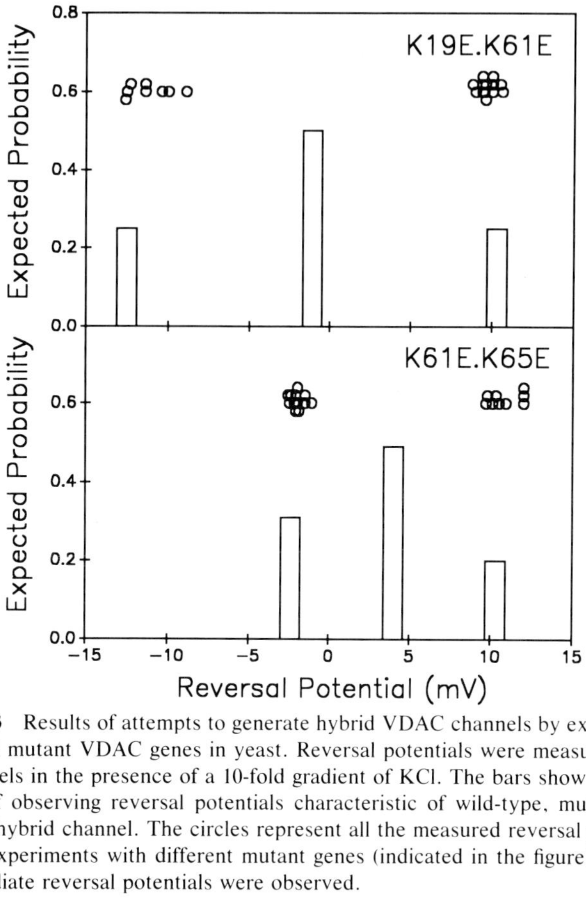

FIGURE 3 Results of attempts to generate hybrid VDAC channels by expressing both wild-type and mutant VDAC genes in yeast. Reversal potentials were measured on single VDAC channels in the presence of a 10-fold gradient of KCl. The bars show the expected probability of observing reversal potentials characteristic of wild-type, mutant, and the intermediate hybrid channel. The circles represent all the measured reversal potentials. In two sets of experiments with different mutant genes (indicated in the figure) no channels with intermediate reversal potentials were observed.

conductance and calculate the radius of a cylinder of solution (whose length is the thickness of the membrane) that would yield the same conductance. This approach has been shown to yield poor estimates for channels of known pore radius. For example the conductance of gramicidin is two times greater than that of nystatin (12 pS vs 6 pS in 2 *M* KCl) even though its pore radius is two times smaller (0.2 nm vs 0.4 nm) (Holz and Finkelstein, 1970; Kasumov *et al.*, 1979; Finkelstein and Anderson, 1981). However, for large channels this approach should yield better estimates since the solution in the channel should behave more like bulk solution. One must, however, at least perform the obvious correction for access resistance. This type of calculation leads to a pore radius of 1.0 nm.

Another approach to estimating the pore radius is to determine the size of the largest nonelectrolyte that can permeate through the pore

(Colombini, 1980a,b). VDAC was shown to be permeable to inulin and PEG 3400 but not permeable to PEG 6800. The Stokes–Einstein radii of the permeants are 1.4 and 1.9 nm, respectively (Scherrer and Gerhardt, 1971). Thus the pore radius was estimated to be 2 nm. The flexibility of these nonelectrolytes may allow them to deform resulting in permeation through narrower openings. Thus the 2-nm estimate is likely an overestimate.

Electron micrographs of negatively stained two-dimensional crystals of VDAC reveal (after computer filtration and averaging) an array of stain-filled pores. An estimated pore radius is 1.2 to 1.5 nm (Mannella *et al.*, 1992). This technique has been used with various negative stains including phosphotungstic acid, uranyl nitrate, and aurothioglucose yielding comparable results. Drawbacks to this approach are positive staining, which would overestimate the size of the pore; grain size, which could underestimate the size of the pore keeping the stain away from the wall; and orientation of the stain, in the case of aurothioglucose, which could underestimate the radius (the glucose could preferentially bind to the channel by hydrogen bonding at sites with no gold attached).

A novel method was developed (Vodyanoy *et al.*, 1992) to estimate the pore size based on the properties of the access resistance. When corrected for the size of the permeating ion, the estimate is 1.2 nm. This approach relies on a macroscopic understanding of the nature of the relationship between pore radius and access resistance that may not apply exactly to structures as small as channels.

Perhaps a good estimate is in the range of 1.2 to 1.5 nm. However, considering the strong evidence that amino acid side chains extend into the channel from a β-sheet wall (Blachly-Dyson *et al.*, 1990), the exact physical size of the pore is, in fact, ill-defined. Most important is the conclusion that VDAC is large enough to allow metabolites the size of ATP to cross the outer membrane.

C. Secondary Structure

VDAC's fundamental properties are highly conserved in mitochondria isolated from all eukaryotic kingdoms (Colombini, 1989). This requires a fundamental conservation of protein structure. However, measuring structural conservation based on primary sequence and antibody cross-reactivity reveals marked changes in the structure over time (De Pinto and Palmieri, 1992). There is a 24% conservation in the primary sequence between yeast and human sequence (Kayser *et al.*, 1989) and only 43% conservation between two fungi, *S. cerevisiae* and *N. crassa* (Kleene *et al.*, 1987). Remarkably, however, the number of amino acids in the protein is identical for VDAC1 from fungi and the human (282 amino acids after removal of the N-terminal methionine). Most importantly, the pattern of VDAC's secondary structure is highly conserved (Fig. 4; Colombini *et al.*, 1992). It is this pattern of VDAC's secondary structure that is believed to be the key to VDAC's structure and function.

VDAC's unusual structure required a different method of analysis than that normally used for membrane proteins. The large size of VDAC's pore and small mass of protein per channel require a thin protein wall, probably one layer of protein. This protein wall must have a polar surface facing the aqueous pore and an apolar surface facing the hydrocarbon tails of the phospholipids. Examinations of the primary sequence reveal amino acid stretches that would have the appropriate character to form such a

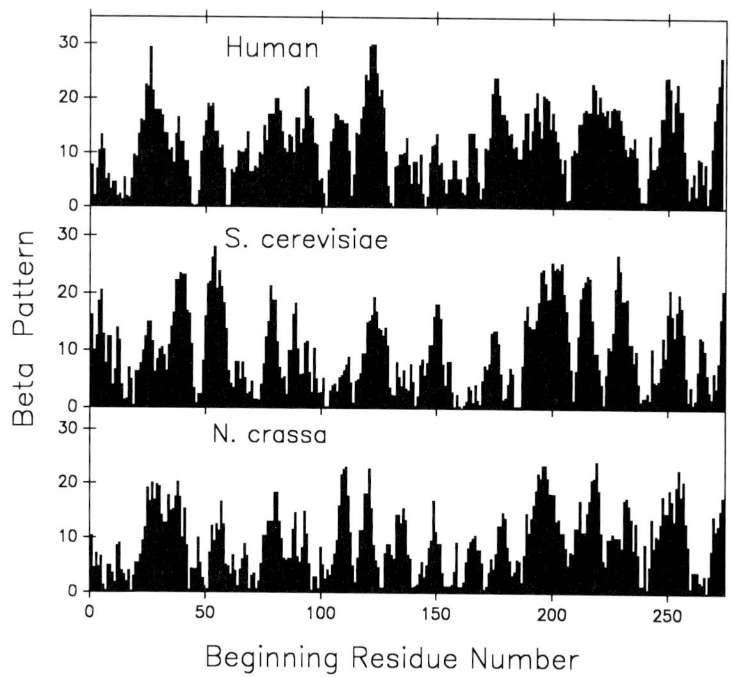

FIGURE 4 Results of the analysis of the primary sequence of VDAC1 from three species. The hydropathy value of each group of 10 adjacent amino acids was summed after every second value was multiplied by −1. These values are plotted against the number of the first amino acid in the summation (starting at the N-terminus). The higher the peak the better the alternating polar–nonpolar pattern, and the more likely the sequence will form part of the wall of the VDAC channel.

wall if the wall had a β structure. One such analysis (Blachly-Dyson *et al.*, 1989) that looks for 10-amino acid stretches that contain alternating polar and nonpolar residues shows that the location of these segments in the primary sequence is remarkably conserved in the known VDAC sequences (Fig. 4). In the representation in the figure, the higher the peak the better the alternating pattern and thus the more likely the 10 amino acids following this point in the sequence are to form part of the wall of the channel. For yeast VDAC, the transmembrane location of all but one of the major peaks has been verified (see below).

It goes without saying that natural selection tends to preserve the essential features of a protein to preserve the desired function. The high degree of interaction between the protein structure of VDAC and its environments (the two membrane surfaces, the aqueous pore, and the lipid bilayer) should provide constraints on the location of polar, charged, and nonpolar residues, but perhaps a weaker constraint on the packing of residues. This would explain the relatively low percentage of amino acid identity in the primary sequence and low antibody cross-reactivity.

D. Structure of the Open State

The amino acid stretches predicted to form the walls of the pore by analysis of the primary sequence were tested to determine if they, in fact, form this wall. The proposed transmembrane strands should form a β structure that would wrap around forming a barrel. Charged residues on the suspected transmembrane strands should all face the lumen side of the barrel and these residues should influence the channel's selectivity. It was found (Blachly-Dyson *et al.*, 1990) that substituting one of these charged residues with a residue of opposite charge altered the channel's selectivity in a discrete and predictable manner (boxed residues in Fig. 5). All but one of the predicted transmembrane strands showed the expected selectivity change. Charge substitutions on the remaining strand and on regions predicted to be outside the channel (facing the environment on one or the other membrane surface) had no effect on selectivity (circled residues in Fig. 5). From these studies, a membrane-folding pattern for the open state of the channel can be put forward that is based on solid functional studies (Fig. 5). It is significant that at the ends of the transmembrane strands one often finds residues that would be expected to break the pattern thus locking the strand in the membrane. These include proline residues and adjacent charged residues. There is also a high probability of finding aromatic residues close to the membrane surface. While strands linked together by short loop regions are likely to be adjacent, the side-by-side location of other strands has not been determined.

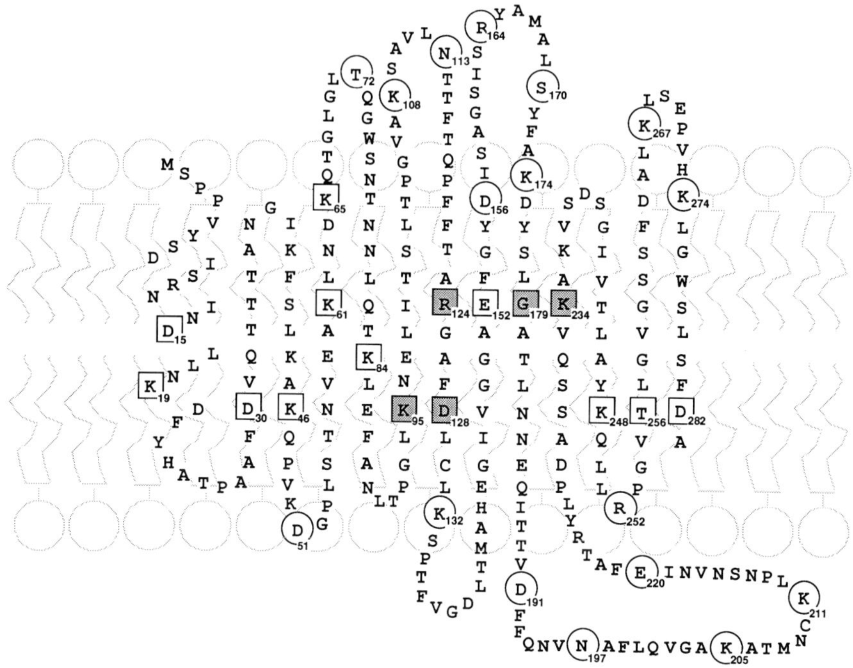

FIGURE 5 Proposed folding pattern of yeast VDAC in a phospholipid membrane. The N-terminal α-helix is followed by 10 β strands. If this sheet were rolled into a cylinder it would form the channel. The boxed residues influence channel activity in the open state and the circled residues do not. The shaded residues have a similar effect on the selectivity of the closed state.

The N-terminus of the VDAC sequence contains a stretch of 20 or so amino acids that form an amphipathic α-helix. This segment has been reported (Mihara and Sato, 1985) to contain the information necessary to target the protein to the mitochondrion. However, unlike many other leader sequences, the segment is not cleaved. Since one side of this putative α-helix is highly charged and the other is highly nonpolar, this could form a transmembrane strand forming part of the wall of the pore. While originally proposed to be located outside the membrane (Forte *et al.*, 1987), site-directed mutation at two sites (Asp 15 and Lys 19) influenced the channel's selectivity in a way consistent with a transmembrane localization (Blachly-Dyson *et al.*, 1990). Thus, in Fig. 5 the α-helix is included as part of the wall of the pore. However, antibodies directed to the α-helical portion of VDAC can bind to the channel in mitochondria. This has been interpreted as evidence that the α-helix is located on the surface (De Pinto and Palmieri, 1992). This interpretation is complicated by evidence that the α-helix and four or five β strands move out of the channel upon voltage-dependent closure (see below). Thus it is possible that antibody binding is occurring on closed or partly closed channels.

The size of the pore formed when these transmembrane strands are allowed to form a cylinder depends on the tilt of the strands. The original analysis proposed a 33° tilt (Forte *et al.*, 1987). This seemed reasonable if a large number of β strands contributed to the wall of the channel (the original proposal was 19 strands). However, if a monomer forms the channel with 13 transmembrane strands, a 33° tilt produces an aqueous pore with a very small diameter. Thus, either more protein segments form the walls of the pore or the protein strands must be tilted more severely. Further testing on regions of yeast VDAC that might form transmembrane strands revealed no effect of charge changes on the channel's ion selectivity (Peng *et al.*, 1992a). Thus other strand tilts were explored. Due to the restriction imposed by interstrand hydrogen bonding, β strands should tilt in a quantized manner: 0°, ≈30°, ≈55°, etc. A 55° tilt would result in a pore of the right size. The only endocytic protein channel whose structure has been solved by X-ray diffraction is porin from *Rhodobacter capsulatus* (Weiss *et al.*, 1991). It forms a β barrel with strands that tilt over a variety of angles. Thus while VDAC shows no homology with this channel, the tilt in VDAC may be more complex than the first-order theoretical expectations.

IV. STRUCTURE FROM ELECTRON-MICROSCOPIC IMAGING

The large size of VDAC's pore has facilitated imaging of the channel with the aid of the electron microscope. Large ordered arrays of VDAC channels in the outer membranes of fungal and plant cells have allowed detailed images to be produced through the use of computer filtration and averaging methods. Techniques have been developed (Mannella, 1982) to increase the size and number of these arrays in the isolated mitochondrial outer membranes.

A. Negative Staining

Negatively stained arrays of channels yield high-contrast images of the channel's aqueous pore filled with electron-dense stain. The packing of channels in the array shown in Fig. 6 is typical but can vary somewhat with conditions of preparation (Mannella *et al.*, 1983). The packing of the channels is not what one might expect from simple cylinders. The

FIGURE 6 An image of an array of VDAC channels from *N. crassa* after negative staining. The channels were in a large 2D crystal. The image was subjected to computer filtration and averaging. (Unpublished work of Lorie Thomas.)

crystallographic repeating unit contains six channels but that does not define the functional repeating unit. The first investigator to observe this chose the six channels adjacent to each other to be the repeating unit (Mannella, 1982). However, more recent observations indicate that the repeating unit may be centered in the region of low electron density (light regions in Fig. 6). Thus the functional repeating unit may be the six stain-filled pores around the island of low density. These structures could then pack in a hexagonal fashion to yield the observed pattern.

B. Freeze-Drying and Shadowing

Detailed images obtained with specimens that were rapidly frozen, dried, and shadowed with platinum (Fig. 7) show the six channel openings surrounding an elevated central area that corresponds to the low-density region in the negatively stained images. The elevated region may correspond to protein domains coming from each of the six channels (see also

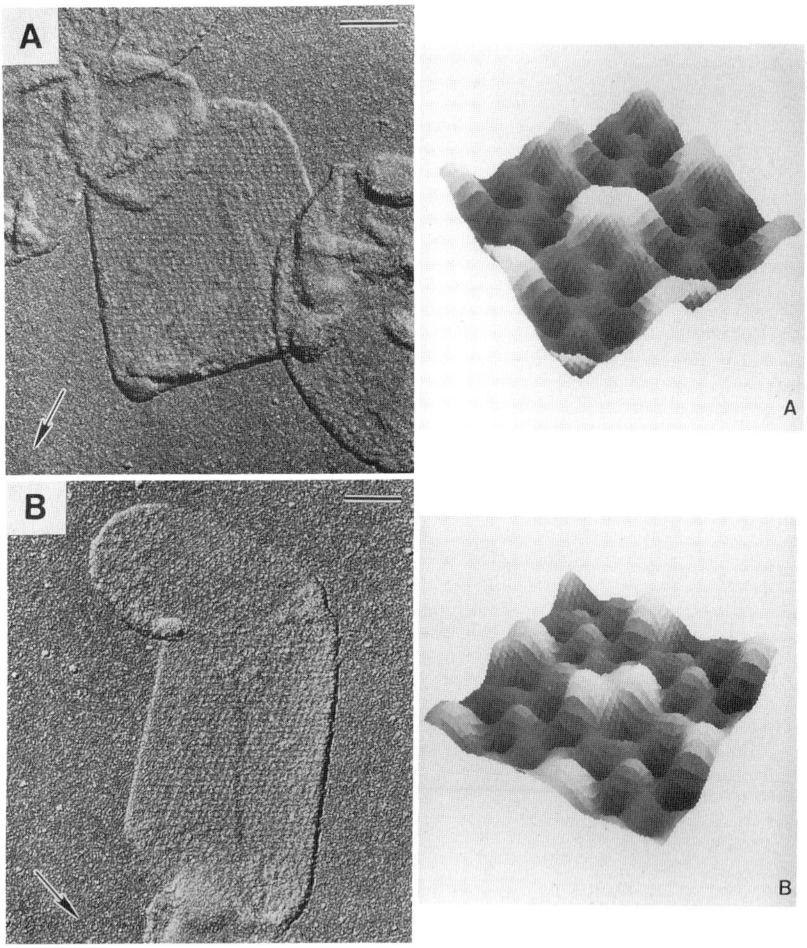

FIGURE 7 Images of the surface structure of VDAC arrays after freeze-drying. Electron micrographs, such as those shown on the left, were used to generate the averaged and filtered images on the right. (A and B) The two surfaces of the VDAC 2D crystals formed in the outer membranes of *N. crassa*. From Thomas *et al.* (1991). Scale bars are 100 nm.

Fig. 9). These may be resting on small lipid regions in the channel array. The protein–protein interactions in these domains may hold the six-channel unit together.

C. Freeze-Fracturing and Shadowing

Freeze-fracture of outer membranes works well except when large arrays are generated by phospholipase treatment. Thus, only small VDAC arrays have been visualized by freeze-fracture (Lorie Thomas, unpublished observations). However, these are very revealing because they show an almost perfect hexagonal array of particles (the work of Lorie Thomas summarized in Colombini, 1994). Each particle (Fig. 8) is about the size of one six-channel unit. The deep grooves between the particles are likely to correspond to the openings of the channels as visualized in the freeze-dry images. Thus, despite their lower resolution, the freeze-fracture images indicate the same elevated region surrounded by the channel openings as being the functional unit. Indeed, the elevated region may be the site that binds soluble proteins such as the VDAC modulator and hexokinase.

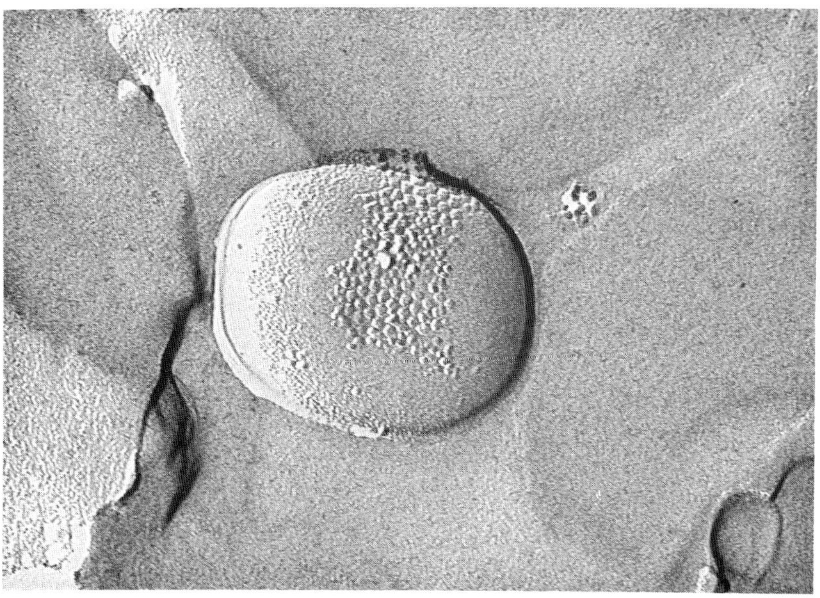

FIGURE 8 Electron micrograph of an outer membrane from *N. crassa* after freeze-fracture. (Courtesy of Lorie Thomas.)

D. Freezing in Vitreous Ice

The protein wall of VDAC's pore was visualized by rapidly freezing the two-dimensional arrays in a thin layer of vitreous ice (Mannella *et al.*, 1989). These unstained specimens should be closest to the native structure. Like the results from negative staining, images from these frozen-hydrated specimens (Fig. 9) show that VDAC channels are essentially right cylinders perpendicular to the plane of the membrane. They indicate a cylinder diameter from the center of one protein wall to another of 3.8 nm. If the average length of amino acid side chains is 0.5 nm, this is consistent with a pore radius of 1.4 nm.

The image in Fig. 9 also reveals the presence of "protein arms" (circled area) extending into the region surrounded by six VDAC channels. This region corresponds to the elevated regions seen after freeze-fracture and freeze-drying.

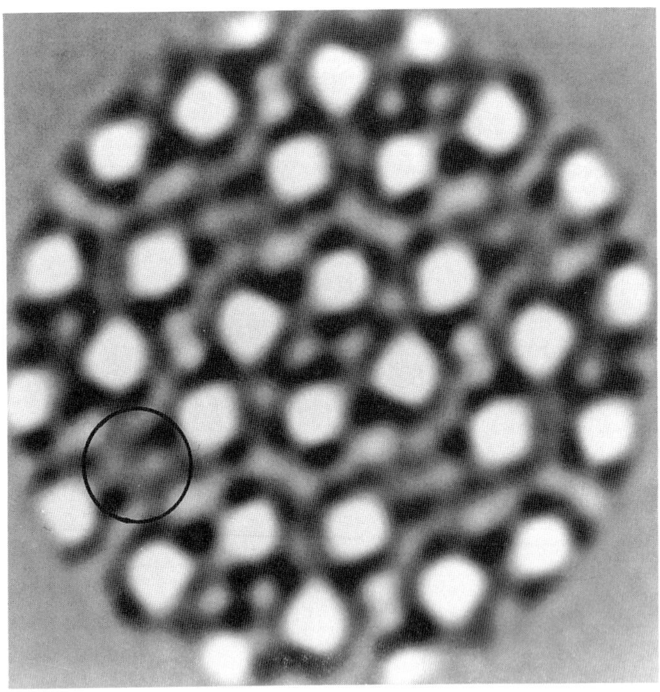

FIGURE 9 Correlation average of an electron micrograph of a VDAC array from *N. crassa* after rapid freezing in vitreous ice. From Mannella (1990), Fig. 3, with permission from Birkhäuser Verlag.

E. Two Surfaces of the Crystal

The VDAC arrays analyzed by electron microscopy are usually located in closed vesicles derived from the mitochondrial outer membrane. After phospholipase treatment, the vesicles shrink in size somewhat as some phospholipids are cleaved and removed by dialysis and the channel arrays grow in size. The vesicles flatten out on the electron microscope grid (Fig. 7) and even under the best conditions only half of them have channel arrays while the rest have only a random distribution of membrane proteins. Could it be that small preexisting seed crystals grow in two dimensions upon phospholipase treatment and these are only present in a portion of the mitochondrial population?

When these collapsed vesicles are visualized by transmission electron microscopy after negative staining, most often the vesicles that contain two-dimensional crystals have these crystals on both membranes (Mannella, 1982; Lorie Thomas, unpublished). The crystal lattices of these structures are mirror images of each other indicating that one crystal is upside down compared to the other (Thomas *et al.*, 1991). This is expected from a structure lacking a plane of symmetry. When these structures are viewed by the freeze-dry/shadow technique, where only the top crystal is visible, both types of crystals are seen (Fig. 7) indicating inside-out and right side-out vesicles (note that when negative stain reveals two superimposed crystals, one on the upper membrane and one on the lower membrane, the two crystal lattices are mirror images). The structure of both surfaces of the crystal looks very similar (Fig. 7). Both have a raised central area surrounded by six channels.

F. Structure of the Open and Closed States

It is generally assumed that the structures visualized by electron microscopy represent the open state of VDAC. Successful attempts to close VDAC without electrical potentials by using chemical agents (Colombini *et al.*, 1987) led to attempts to visualize the closed VDAC channels by electron microscopy. By using König's polyanion, images of negatively stained arrays with a smaller pore opening were obtained (Mannella and Guo, 1990). The pore size was estimated to be 0.85 nm in radius. This is very similar to pore-size estimates based on the permeability of nonelectrolytes [0.9 nm, based on slight permeability to γ-cyclodextrin (Colombini *et al.*, 1987)].

V. THE VOLTAGE-GATING PROCESS

Voltage gating of membrane channels has two aspects: (1) the gating process—the structural change by which the protein goes from a state of high permeability to one of low permeability and (2) the voltage dependence—the process by which gating is coupled to changes in the membrane electrical potential. In the case of VDAC, the structural change is large. The pore radius has been estimated to decrease from about 1.2–1.5 nm to 0.8–0.9 nm. This is unlikely to be due simply to a local obstruction because the volume of the aqueous pore has been estimated to decrease by 20 to 40 nm^3 based on water volume inaccessible to dextrans but accessible to sucrose (Zimmerberg and Parsegian, 1986). The channel's selectivity is dramatically changed not only in magnitude but also in sign. The channel goes from weakly anion selective in the open state to weakly cation selective in the closed state (Colombini, 1980b; Zhang and Colombini, 1990; Benz and Brdiczka, 1992). This combination of steric and electrostatic changes indicates drastic changes in VDAC's permeability to organic anions such as the metabolites that cross the outer membrane during mitochondrial function ($MgATP^{2-}$, $MgADP^{-}$, succinate, phosphate, citrate, etc.).

Portions of the VDAC channel that move upon channel closure were identified by the use of site-directed mutagenesis. Two independent lines of evidence point to the same region of the protein as being the part that moves upon voltage-dependent channel closure. Amino acid substitutions resulting in charge changes at particular locations on the VDAC channel could affect the selectivity of the open state and not the closed state (Peng *et al.*, 1992a). Those particular sites must be close to the ionic flow in the open state but not in the closed state. Other sites had weaker than expected effects on the closed state selectivity indicating some displacement. Still others has comparable effects on both the open and closed state selectivities thus showing no evidence of displacement (shaded and boxed residues in Fig. 5). The sites that had no or weaker than expected effects on the closed state selectivity (but normal effect on the open state) were located primarily on the N-terminal and some on the C-terminal portions of the channel. These areas were proposed to move out of the channel upon channel closure. Since the net charge on these areas is positive, this motion would account not only for the reduction in pore size of the channel but also for the voltage dependence of the conformational change (due to the movement of charge through the electric field) and the change in selectivity because the remaining transmembrane strands would result in a net negative charge in the pore.

Remarkably, most of the sites that seemed to move out of the channel based on selectivity measurements also influenced the voltage dependence of the channel (Thomas *et al.*, 1993). An increase in positive charge at the sites increased the voltage dependence while a decrease in positive charge had the opposite effect. The changes in voltage dependence were discrete increases or decreases in the steepness of the voltage dependence as expected from an increase or decrease in the amount of charge moving through the membrane potential upon channel closure. Little change was observed in the conformational energy difference between the open and closed state indicating that the mutations were not simply favoring a particular structural state. One site (K248E) is indicated as moving out of the membrane by the selectivity measurements but had essentially no effect on the voltage dependence. However, since this site was close to the end of the transmembrane strand, it may move out of the pore without moving through a significant part of the membrane potential and thus charge change at this site would not affect overall voltage dependence.

The lack of effect of mutations at positions 30 and 51 on the voltage dependence make it difficult to see what regions of the protein should be moving upon channel closure at both positive and negative potentials. While a clear overall picture is still elusive, the α-helix and the adjacent four β strands and the β strand at the C-terminus seem to be the parts that move in response to the electric field, although some of these may only move in response to a potential of one particular sign.

A working model for voltage gating in VDAC (Fig. 10) involves the motion of a large portion of the protein through the membrane. This proposed model accounts for: (1) voltage dependence by proposing the movement of a charged domain through the membrane potential; (2) voltage gating at both positive and negative potentials; (3) the measured changes in channel diameter and volume; and (4) the change in ion selectiv-

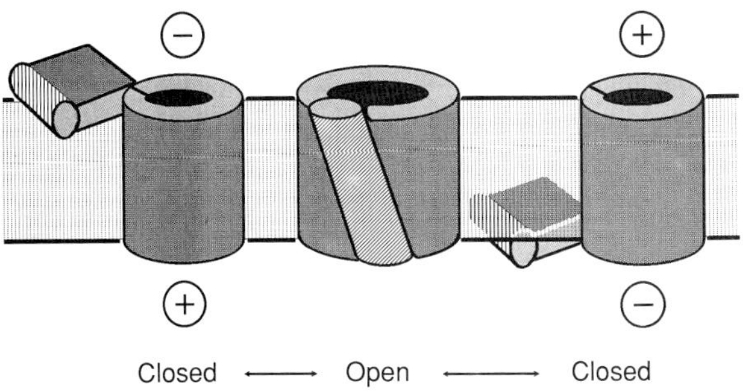

FIGURE 10 Model of the voltage gating of VDAC. The open channel is a barrel composed of one tilted α-helix and a cylindrical β-sheet wall. A membrane potential induces closure by driving part of the wall of the channel out toward the negative side of the membrane. Reopening involves the reincorporation of this domain into the channel.

ity upon channel closure by moving a charged domain out of the membrane. An alternative proposal (Mannella, 1990) suggests that the α-helix might close the channel by entering the pore thus reducing its diameter and volume. While aesthetically pleasing, this notion does not account for: (1) voltage-dependent closure (the α-helix in yeast VDAC has no net charge and a dipole moment largely canceled out by the location of charged side chains); (2) selectivity changes associated with closure; and (3) the fact that charges at positions 15 and 19 on the helix influence channel selectivity in the open state and not at all or less than expected in the closed state.

VI. THE PROPERTIES OF VDAC CAN BE TUNED AND MODULATED

Since VDAC's properties have been studied largely in reconstituted systems, it is natural to wonder how different the properties are from the "physiological" state. This would include both changes in environment (lipids, ionic strength, etc.) and level of regulation or modulation. True *in situ* experiments examining VDAC's electrophysiological properties have not been performed for obvious reasons. However, one can begin to gain a sense of the complexity of such changes in properties by considering the many ways VDAC has been found to be influenced and its properties markedly altered.

The fact that VDAC reconstituted from widely different sources behaves almost the same way after reconstitution in planar membranes may yield the wrong impression that these properties are those found in the intact cell. Rather, the conserved behavior upon reconstitution may reflect the basic properties of the conserved structure of the channel. Yet, unlike many proteins, VDAC's fundamental property of voltage dependence is affected very little by ionic strength and pH. Dramatic changes in voltage dependence occur only above pH 10 and below pH 4 (Bowen *et al.*, 1985; Ermishkin and Mirzabekov, 1990). Indeed, steep voltage dependence can be easily observed in solutions as unphysiological as 80 mM $CaCl_2$ (Schein *et al.*, 1976). However, this insensitivity to these environmental factors is simply the result of using the wrong factors.

A. Osmotic Pressure

The voltage range at which VDAC switches between open and closed states can be shifted along the voltage axis by increasing the colloidal osmotic pressure of the medium (Zimmerberg and Parsegian, 1986). Thus the protein content of the cytoplasm and intermembrane space of the mitochondrion will tend to cause VDAC to close at lower voltages. These authors speculated that the cell could use this property to regulate the total protein content of the cytoplasm by restricting mitochondrial ATP production through closure of VDAC in the outer membrane. Be that as it may, VDAC's ability to close at low potentials is enhanced by the presence of proteins in the environment.

B. The VDAC Modulator

In addition to the nonspecific osmotic effect of proteins, mitochondria contain at least one protein that specifically changes VDAC's properties (Holden and Colombini, 1988). This protein, dubbed the VDAC modulator, increases VDAC's voltage dependence (Liu and Colombini, 1992b) and changes VDAC's kinetic properties (Holden and Colombini, 1988). Closing rates are enhanced and opening rates are reduced (Fig. 11). The channels are only sensitive to the modulator when the modulator-containing side is made negative. If modulator is added to both sides then both gating processes are affected. At higher doses of modulator, VDAC channels tend to remain closed even in the absence of a membrane potential (Fig. 11). Positive potentials on the modulator-containing side are often capable of reopening such channels.

The importance of this protein is indicated by its remarkable conservation. Modulator preparations from plants and fungi act on mammalian preparations of VDAC and vice versa (Liu and Colombini, 1991). This conserved recognition coupled with sensitivity estimated to be in the nanomolar range indicates the presence of a regulatory mechanism which tunes the protein to the state of the cell.

C. Ultrasteep Voltage Dependence

VDAC's voltage dependence is dramatically increased by the presence of soluble polyanions. Under "normal" reconstitution conditions, VDAC is steeply voltage dependent like that seen in the more widely studied channels in nerve and muscle. However, dextran sulfate (8 kDa) can increase VDAC's voltage dependence 20-fold (Mangan and Colombini, 1987). The population of channels can switch from mostly open to mostly closed over a voltage range of less than 0.5 mV. This high voltage dependence is unique in all of biology. There is strong evidence that the polyanion partitions into the access resistance region of the channel in a

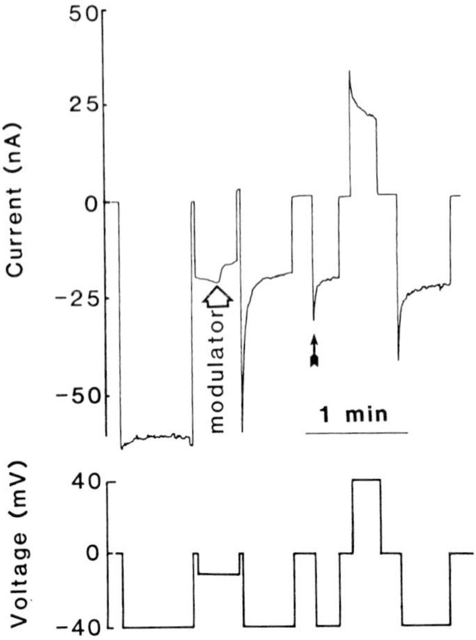

FIGURE 11 The VDAC modulator induces an increase in the rate of voltage-dependent VDAC closure. *Neurospora crassa* VDAC channels were inserted into a membrane made from diphytanoyl phosphatidyl choline in which the channel closes at higher potentials. The addition of a VDAC modulator-containing protein extract is indicated. The upper trace shows the current recorded in response to the voltages applied (lower trace). Reprinted from Liu and Colombini (1991), Fig. 1, with permission of the American Physiological Society.

voltage-dependent manner effectively amplifying the effect of the electric field. It probably interacts electrostatically with the positively charged voltage sensor resulting in channel closure.

The production of this "ultrasteep" voltage dependence is not limited to dextran sulfate but has been observed with polyaspartic acid and even with highly negatively charged proteins and nucleic acids but at higher doses (Colombini *et al.*, 1989). Perhaps the organized structure of the latter interferes with optimal interaction. A highly negatively charged polymer, referred to as König's polyanion, which contains a mixture of carboxyl groups and nonpolar groups acts at much lower concentrations perhaps as a result of binding through the nonpolar residues (Colombini *et al.*, 1987). The effect of the VDAC modulator may be partially through this process although it is not a highly negatively charged protein (pI≈5).

D. Donnan Potential

The presence of charged macromolecules in the cytoplasm and the mitochondrial intermembrane space must result in a Donnan potential across the outer mitochondrial membrane unless the colloidal charge in the two aqueous compartments happens to be exactly the same. The well-known conversion between the condensed and orthodox versions of the mitochondria (Hackenbrock, 1968) changes the volume of the intermembrane space and therefore perhaps the protein concentration in this space. Such a change would be expected to change the Donnan potential across the outer membrane. Perhaps when mitochondria go from the rapidly respiring condensed state to the orthodox state they change the Donnan potential across the outer membrane and thus change the fraction of VDAC channels that are open. Naturally, the net charge on proteins can also be changed by such things as protein phosphorylation and pH changes. Indeed, Donnan potential changes must result in changes in the pH of one or both compartments and this can in turn change the net charge on protein. Thus possibilities exist for both positive and negative feedback.

E. Aluminum Inhibition

Micromolar amounts of aluminum hydroxide strongly inhibit VDAC's voltage dependence (Dill *et al.*, 1987). While aluminum's role in biology is unclear and is perhaps largely introduced as a result of industrialization, aluminum is a normal constituent found in human blood at micromolar levels. Thus, its presence may influence VDAC's voltage dependence *in vivo*. Unlike many other reported effects of aluminum, its action on VDAC is through a form found at physiological pH ($Al(OH)_3$) (Zhang and Colombini, 1989) and not through species more common at acid pH such as Al^{3+}. $Al(OH)_3$ and other neutral metal trihydroxides act in a way that appears to be consistent with the neutralization of VDAC's voltage sensor but seems in fact to be some kind of indirect effect (Zhang and Colombini, 1990). The aluminum binding site does seem to translocate through the membrane in a way that is coupled to the channel gating process.

F. NADH-Induced Increase in Voltage Dependence

Micromolar quantities of NADH can double VDAC's voltage dependence (Zizi *et al.*, 1994). The oxidized form (NAD^+) has no significant effects. These results indicate that the permeability of the outer membrane is sensitive to the glycolytc rate by sensing one of its most important products, NADH. This may be one of the molecular mechanisms behind the Crabtree effect.

VII. FUNCTION

The mitochondrion can be looked at as a debilitated organism now totally dependent on the host for survival. Yet the host is also dependent on the mitochondrion for many critical metabolic functions. Indeed there are no known eucaryotic cells that lack mitochondria or some endosymbiote. It seems reasonable that this symbiotic relationship must involve strict regulation, probably at various levels. VDAC channels are the primary pathway for the flow of metabolites across the mitochondrial outer membrane. Their highly conserved properties include a variety of regulatory mechanisms that could restrict the flow of metabolites between the cytoplasm and the mitochondrion. Such a bottle-neck could limit such things as energy production and mitochondrial growth and reproduction. Thus VDAC may play an important role in these and other processes.

In addition to the host–symbiont relationship, the whole question of the regulation of energy metabolism in a cell may be influenced by the metabolic state of VDAC and VDAC's interaction with other proteins. The Pasteur and Crabtree effects can be looked at as phenomena resulting from the interaction of two very different energy-producing systems in a cell (naturally the situation should be even more complex in chloroplast-containing cells as well as in cells with more exotic energy-transducing mechanisms). The fact that the two different genes so far identified in humans that contribute VDAC differ in their ability to bind hexokinase indicates the importance of regulation (Blachly-Dyson *et al.*, 1993). While the function of hexokinase binding to mitochondria via VDAC is controversial (Nelson and Kabir, 1985), this binding has been linked to adaptation to malignancy (Arora *et al.*, 1992) and metabolic regulation (BeltrandelRio and Wilson, 1992).

There is evidence that VDAC channels tend to be associated with sites of tight contact between the outer and inner membranes (Brdiczka *et al.*, 1989). The role(s) of these contact sites is unclear but may well be related to metabolic regulation or protein import. These contact areas may well subdivide the intermembrane space into different regions with different composition and properties.

The role of VDAC channels reportedly located on the plasma membrane of cells (Blatz and Magleby, 1983; Nelson *et al.*, 1984; Thinnes, 1992) is much more speculative. To dismiss the possibility that these large channels have any role because they are so conductive that they could kill the cell is to forget that sodium channels if left open for extended periods could kill cells. The key, naturally, is regulation, as it is for most cellular processes. These large channels seem to be kept closed until needed to perform some function. Situations that might require large channels are those in which high permeability is needed for short periods of time and those that require large molecules to cross the membrane.

VIII. PROSPECTS

The view of mitochondrial structure, function, and regulation that evolved in the 1960s and 1970s is too simple and does not account for more recent observations. The discovery of channels in the inner membrane and regulated channels in the outer membrane requires the development of theories and models of mitochondrial function that are more complex, imaginative, and daring. The ability of channels to respond to changes in environmental conditions and the levels of metabolic intermediates points to intricate regulation that may not be fully appreciated without the performance of difficult whole-cell experiments. Yet, much can still be learned by studying the properties and structures of the isolated components.

Acknowledgments

I thank Dr. Martin Zizi for many stimulating discussions. This work was supported by ONR Grant N00014-90-J-1024.

References

Arora, K. K., Parry, D. M., and Pedersen, P. L. (1992). Hexokinase receptors: Preferential enzyme binding in normal cells to non-mitochondrial sites and in transformed cells to mitochondrial sites. *J. Bioenerg. Biomembr.* **24,** 47–53.

BeltrandelRio, H., and Wilson, J. E. (1992). Coordinated regulation of cerebral glycolytic and oxidative metabolism, mediated by mitochondrially bound hexokinase dependent on intramitochondrially generated ATP. *Arch. Biochem. Biophys.* **296,** 667–677.

Benz, R., and Brdiczka, D. (1992). The cation-selective substate of the mitochondrial outer membrane pore: Single-channel conductance and influence on intermembrane and peripheral kinases. *J. Bioenerg. Biomembr.* **24,** 33–39.

Blachly-Dyson, E., Peng, S. Z., Colombini, M., and Forte, M. (1989). Probing the structure of the mitochondrial channel, VDAC, by site-directed mutagenesis: A progress report. *J. Bioenerg. Biomembr.* **21,** 471–483.

Blachly-Dyson, E., Peng, S. Z., Colombini, M., and Forte, M. (1990). Alteration of the selectivity of the VDAC ion channel by site-directed mutagenesis: Implications for the structure of a membrane ion channel. *Science* **247,** 1233–1236.

Blachly-Dyson, E., Zambrowicz, E. B., Yu, W. H., Adams, V., McCabe, E. R. B., Adelman,

J., Colombini, M., and Forte, M. (1993). Cloning and functional expression in yeast of two human isoforms of the outer mitochondrial membrane channel, VDAC. *J. Biol. Chem.* **268,** 1835–1841.

Blatz, A. L., and Magleby, K. L. (1983). Single voltage-dependent chloride-selective channels of large conductance in cultured rat muscle. *Biophys. J.* **43,** 237–241.

Blumenthal, A., Kahn, K., Beja, O., Galun, E., Colombini, M., and Breiman, A. (1993). Purification and characterization of the voltage-dependent anion-selective channel (VDAC) protein from wheat mitochondrial membranes. *J. Plant Physiol.* **101,** 579–587.

Bowen, K. A., Tam, K., and Colombini, M. (1985). Evidence for titratable charges controlling the voltage dependence of the outer mitochondrial membrane channel, VDAC. *J. Membr. Biol.* **86,** 51–60.

Brdiczka, D., Adams, V., Kottke, M., and Benz, R. (1989). Topology of peripheral kinases: Its importance in transmission of mitochondrial energy. *In* "Anion Carriers of Mitochondrial Membranes" (A. Azzi, K. A. Nałęcz, M. J. Nałęcz, and L. Wojtczak, eds.), pp. 215–224. Springer-Verlag, New York.

Bureau, M. H., Khrestchatisky, M., Heeren, M. A., Zambrowicz, E. B., Kim, H., Grisar, T. M., Colombini, M., Tobin, A. J., and Olsen, R. W. (1992). Isolation and cloning of a voltage-dependent anion channel-like M_r 36,000 polypeptide from mammalian brain. *J. Biol. Chem.* **267,** 8679–8684.

Colombini, M. (1979). A candidate for the permeability pathway of the outer mitochondrial membrane. *Nature (London)* **279,** 643–645.

Colombini, M. (1980a). The pore size and properties of channels from mitochondria isolated from *Neurospora crassa*. *J. Membr. Biol.* **53,** 79–84.

Colombini, M. (1980b). Structure and mode of action of a voltage-dependent anion-selective channel (VDAC) located in the outer mitochondrial membrane. *Ann. N.Y. Acad. Sci.* **341,** 552–563.

Colombini, M. (1986). Voltage gating in VDAC: Toward a molecular mechanism. *In* "Ion Channel Reconstitution" (C. Miller, ed.), pp. 533–552. Plenum, New York.

Colombini, M. (1989). Voltage gating in the mitochondrial channel, VDAC. *J. Membr. Biol.* **111,** 103–111.

Colombini, M. (1994). The mitochondrial channel, VDAC. *In* "Membrane Electrochemistry" (M. Blank and I. Vodyanoy, eds.), Vol. 235, Ch. 12. ACS Books, Washington, D.C.

Colombini, M., Yeung, C. L., Tung, J., and König, T. (1987). The mitochondrial outer membrane channel, VDAC, is regulated by a synthetic polyanion. *Biochim. Biophys. Acta* **905,** 279–286.

Colombini, M., Mangan, P. S., and Holden, M. J. (1989). Modulation of the mitochondrial channel, VDAC, by a variety of agents. *In* "Anion Carriers of Mitochondrial Membranes" (A. Azzi, K. A. Nałęcz, M. J. Nałęcz, and L. Wojtczak, eds.), pp. 215–224. Springer-Verlag, New York.

Colombini, M., Peng, S., Blachly-Dyson, E., and Forte, M. (1992). Probing the molecular structure and structural changes of a voltage-gated channel. *In* "Methods in Enzymology" (B. Rudy and L. Iversen, eds.), vol. 207, pp. 432–444. Academic Press, London.

De Pinto, V., and Palmieri, F. (1992). Transmembrane arrangement of mitochondrial porin or voltage-dependent anion channel (VDAC). *J. Bioenerg. Biomembr.* **24,** 21–26.

Dill, E. T., Holden, M. J., and Colombini, M. (1987). Voltage gating in VDAC is markedly inhibited by micromolar quantities of aluminum. *J. Membr. Biol.* **99,** 187–196.

Ermishkin, L. N., and Mirzabekov, T. A. (1990). Redistribution of the electric field within the pore contributes to the voltage dependence of mitochondrial porin channel. *Biochim. Biophys. Acta* **1021,** 161–168.

Finkelstein, A., and Anderson, O. S. (1981). The gramicidin A channel: A review of its

permeability characteristics with special reference to the single-file aspect of transport. *J. Membr. Biol.* **59,** 155–171.
Forte, M., Guy, H. R., and Mannella, C. A. (1987). Molecular genetics of the VDAC ion channel: Structural model and sequence analysis. *J. Bioenerg. Biomembr.* **19,** 341–350.
Hackenbrock, C. R. (1968). Ultrastructural basis for metabolically-linked mechanical activity of mitochondria. II. Electron-transport linked ultrastructural transformations in mitochondria. *J. Cell Biol.* **37,** 345–357.
Holden, M. J., and Colombini, M. (1988). The mitochondrial outer membrane channel, VDAC, is modulated by a soluble protein. *FEBS Lett.* **241,** 105–109.
Holz, R., and Finkelstein, A. (1970). The water and non-electrolyte permeability induced in thin lipid membranes by the poloyene antibiotics nystatin and amphotericin B. *J. Gen. Physiol.* **56,** 125–135.
Kasumov, Kh.M., Borisova, M. P., and Vainshtein, V. A. (1979). How do ionic channel properties depend on the structure of polyene antibiotic molecules. *Biochim. Biophys. Acta* **551,** 229–237.
Kayser, H., Dratzin, H. D., Thinnes, F. P., Götz, H., Schmidt, W. E., Eckart, K., and Hilschmann, N. (1989). Characterization and primary structure of a 31-kD porin from human B lymphocytes (Porin 31HL). *Biol. Chem. Hoppe-Seyler* **370,** 1265–1278.
Kleene, R., Pfanner, N., Pfaller, R., Link, T. A., Sebald, W., Neupert, W., and Tropschug, M. (1987). Mitochondrial porin of *Neurospora crassa:* cDNA cloning, *in vitro* expression and import into mitochondria. *EMBO J.* **6,** 2627–2633.
Linden, M., and Gellerfors, P. (1983). Hydrodynamic properties of porin isolated from outer membrane of rat liver mitochondria. *Biochim. Biophys. Acta* **736,** 125–129.
Liu, M., and Colombini, M. (1991). Voltage gating in the mitochondrial outer membrane channel, VDAC, is regulated by a very conserved protein. *Am. J. Physiol.* **260,** C371–C374.
Liu, M., and Colombini, M. (1992a). Regulation of mitochondrial respiration by controlling the permeability of the outer membrane through the mitochondrial channel, VDAC. *Biochim. Biophys. Acta* **1098,** 255–260.
Liu, M. Y., and Colombini, M. (1992b). A soluble protein increases the voltage dependence of the mitochondrial channel, VDAC. *J. Bioenerg. Biomembr.* **24,** 41–46.
Mangan, P. S., and Colombini, M. (1987). Ultra-steep voltage dependence in a membrane channel. *Proc. Natl. Acad. Sci. U.S.A.* **84,** 4896–4900.
Mannella, C. A. (1982). Structure of the outer mitochondrial membrane: Ordered arrays of pore-like subunits in outer-membrane fractions from *Neurospora crassa* mitochondria. *J. Cell Biol.* **94,** 680–687.
Mannella, C. A. (1986). Mitochondrial outer membrane channel (VDAC,porin): Two-dimensional crystals from *Neurospora. In* "Methods in Enzymology" (S. Fleischer and B. Fleischer, eds.), Vol. 125, pp. 595–610. Academic Press, Orlando, FL.
Mannella, C. A. (1987). Electron microscopy and image analysis of the mitochondrial outer membrane channel, VDAC. *J. Bioenerg. Biomembr.* **19,** 329–340.
Mannella, C. A. (1990). Structural analysis of mitochondrial pores. *Experientia* **46,** 137–145.
Mannella, C. A., and Guo, X.-W. (1990). Interaction between the VDAC channel and a polyanionic effector. *Biophys. J.* **57,** 23–31.
Mannella, C. A., Colombini, M., and Frank, J. (1983). Structural and functional evidence for multiple channel complexes in the outer membrane of *Neurospora crassa* mitochondria. *Proc. Natl. Acad. Sci. U.S.A.* **80,** 2243–2247.
Mannella, C. A., Guo, X.-W., and Cognon, C. (1989). Diameter of the mitochondrial outer membrane channel: Evidence from electron microscopy of frozen-hydrated membrane crystals. *FEBS Lett.* **253,** 231–234.

Mannella, C. A., Forte, M., and Colombini, M. (1992). Toward the molecular structure of the mitochondrial channel, VDAC. *J. Bioenerg. Biomembr.* **24,** 7–19.

Mihara, K., and Sato, R. (1985). Molecular cloning and sequencing of cDNA for yeast porin, an outer mitochondrial membrane protein: A search for targeting signal in the primary structure. *EMBO J.* **4,** 769–774.

Nelson, B. D., and Kabir, F. (1985). Adenylate kinase is the source of ATP for tumor mitochondrial hexokinase. *Biochim. Biophys. Acta* **841,** 195–200.

Nelson, D. J., Tang, J. M., and Palmer, L. G. (1984). Single-channel recording of apical membrane chloride conductance in A6 epithelial cells. *J. Membr. Biol.* **80,** 81–89.

Parsons, D. F., Bonner, W. D., Jr., and Verboon, J. G. (1965). Electron microscopy of isolated plant mitochondria and plastids using both thin-section and negative staining techniques. *Can. J. Bot.* **43,** 647–655.

Peng, S., Blachly-Dyson, E., Forte, M., and Colombini, M. (1992a). Large scale rearrangement of protein domains is associated with voltage gating of the VDAC channel. *Biophys. J.* **62,** 123–135.

Peng, S., Blachly-Dyson, E., Forte, M., and Colombini, M. (1992b). Determination of the number of polypeptide subunits in a functional VDAC channel from *Saccharomyces cerevisiae*. *J. Bioenerg. Biomembr.* **24,** 27–31.

Roos, N., Benz, R., and Brdiczka, D. (1982). Identification and characterization of the pore-forming protein in the outer membrane of rat liver mitochondria. *Biochim. Biophys. Acta* **686,** 204–214.

Schein, S. J., Colombini, M., and Finkelstein, A. (1976). Reconstitution in planar lipid bilayers of a voltage-dependent anion-selective channel obtained from *Paramecium* mitochondria. *J. Membr. Biol.* **30,** 99–120.

Scherrer, R., and Gerhardt, P. (1971). Molecular sieving by the *Bacillus megaterium* cell wall and protoplast. *J. Bacteriol.* **107,** 718–735.

Thinnes, F. P. (1992). Evidence for extra-mitochondrial localization of the VDAC/porin channel in eucaryotic cells. *J. Bioenerg. Biomembr.* **24,** 71–75.

Thomas, L., Kocsis, E., Colombini, M., Erbe, E., Trus, B. L., and Steven, A. C. (1991). Surface topography and molecular stoichiometry of the mitochondrial channel, VDAC, in crystalline arrays. *J. Struct. Biol.* **106,** 161–171.

Thomas, L., Blachly-Dyson, E., Colombini, M., and Forte, M. (1994). Mapping of residues forming the voltage sensor of the VDAC ion channel. *Proc. Natl. Acad. Sci. U.S.A.* **90,** 5446–5449.

Vodyanoy, I., Bezrukov, S. M., and Colombini, M. (1992). Measurement of ion channel access resistance. *Biophys. J.* **61,** A114.

Weiss, M. S., Kreusch, A., Schiltz, E., Nestel, U., Welte, W., Weckesser, J., and Schulz, G. E. (1991). The structure of porin from *Rhodobacter capsulatus* at 1.8 Å resolution. *FEBS Lett.* **280,** 379–382.

Werkheiser, W. C., and Bartley, W. (1957). The study of steady-state concentrations of internal solutes of mitochondria by rapid centrifugal transfer to a fixation medium. *Biochem. J.* **66,** 79–91.

Wojtczak, L., and Zaluska, H. (1969). On the permeability of the outer mitochondrial membrane to cytochrome c. I. Studies on whole mitochondria. *Biochim. Biophys. Acta* **193,** 64–72.

Zalman, L. S., Nikaido, H., and Kagawa, Y. (1980). Mitochondrial outer membrane contains a protein producing nonspecific diffusion channels. *J. Biol. Chem.* **255,** 1771–1774.

Zambrowicz, E. B., and Colombini, M. (1993). Zero-current potentials in a large membrane channel: A simple theory accounts for complex behavior. *Biophys. J.* **65,** 1093–1100.

Zhang, D. W., and Colombini, M. (1989). Inhibition by aluminum hydroxide of the voltage-dependent closure of the mitochondrial channel, VDAC. *Biochim. Biophys. Acta* **991,** 68–78.

Zhang, D.-W., and Colombini, M. (1990). Group IIIA-metal hydroxides indirectly neutralize the voltage sensor of the voltage-dependent mitochondrial channel, VDAC, by interacting with a dynamic binding site. *Biochim. Biophys. Acta* **1025,** 127–134.

Zimmerberg, J., and Parsegian, V. A. (1986). Polymer inaccessible volume changes during opening and closing of a voltage-dependent ionic channel. *Nature (London)* **323,** 36–39.

Zizi, M., Forte, M., Blachly-Dyson, E., and Colombini, M. (1994). NADH regulates the gating of VDAC, the mitochondrial outer membrane channel. *J. Biol. Chem.* **269,** 1614–1616.

CHAPTER 5

Regulation of Chloride Channels in Lymphocytes

Michael D. Cahalan and Richard S. Lewis*

Department of Physiology and Biophysics, University of California at Irvine, Irvine, California 92717; and *Department of Molecular and Cellular Physiology, Stanford University, Stanford, California 94305

I. FUNCTIONS OF T AND B LYMPHOCYTES

Lymphocytes provide protection against the onslaught of innumerable pathogens. The molecular and cellular bases of immunity reside within three conceptually separate but interlocking cellular responses. The recognition of antigens by specific receptors sets off an intracellular cascade of signal transduction, which in turn leads to a variety of effector functions. Antigen specificity is provided during primary differentiation from bone marrow stem cells, giving rise through the somatic rearrangement of T- and B-cell antigen receptor genes to specific membrane receptors. The T-cell receptor for a given T cell then is able to recognize a single, specific antigen bound to a major histocompatibility (MHC) protein at the surface of an antigen-presenting cell. The B-cell antigen receptor, or membrane immunoglobulin, recognizes its corresponding antigen without the need for antigen presentation. Ligand binding brings receptors laterally into proximity with other receptors in the membrane and results in the activation of kinases, phosphatases, and phospholipase C. The resulting effector functions depend upon transcriptional activation of additional sets of genes which vary according to the specific lymphocyte subset. In general, these later responses include secretion of lymphokines, cell proliferation, and further differentiation into memory cells or antibody-secreting plasma cells, for example.

For the immune response to work, lymphocytes must be able to disperse throughout the body, borne in part by the flow of blood and lymph and also by adhering to and moving along tissue surfaces. They must also be able to regulate their volume and survive up to decades. Watching the interaction of T and antigen-presenting cells at the microscopic level, we can visualize the immunosurveillance without which we would succumb to infection and tumors.

These remarkable cells, each unique in having undergone differentiation by genetic recombination to yield a specific receptor that can recognize a single, foreign antigen, are also endowed with an array of ion channels that helps them to perform essential cellular functions. The purpose of this review is to focus on chloride channels in T and B lymphocytes. We discuss their distribution, regulation, biophysical properties, and similarities with channels in other cell types, as determined primarily by patch-clamp experiments, and discuss their possible role in lymphocyte functions. Before describing properties of the chloride channel subset, we present a brief overview of the other types of channels found (and not found) in lymphocytes.

II. ION CHANNEL PHENOTYPE OF LYMPHOCYTES

Patch-clamp techniques have revealed a diverse set of ion channels in lymphoid cells, including several distinct types of potassium, calcium, and chloride channels (reviewed by Lewis and Cahalan, 1990; Premack and Gardner, 1991; Cahalan *et al.*, 1992). Biochemical assays of lymphocyte function following mitogen stimulation have illuminated some of the essential roles that these channels play in the activation and effector functions of T and B lymphocytes. It has become clear that ion channels mediate a variety of cellular behaviors, including the Ca^{2+} signal during antigen activation of lymphocytes, the secretion of lymphokines, killing of target cells, progression through the cell cycle, and volume regulation in response to osmotic stress. Ion channels in cells of the immune system offer promising targets for therapeutic agents in the treatment of a variety of disorders, including immunodeficiency, autoimmunity, leukemias and lymphomas, graft rejection, and inflammation.

A. Voltage-Gated K^+ {K(V)} Channels

Except for occasional sightings in lymphoid cell lines, voltage-gated Na^+ and Ca^{2+} channels have not been observed; thus lymphocytes and related cell lines are generally not electrically excitable. Nevertheless, the most commonly seen and dominant channel under resting conditions is a voltage-activated K^+ channel. Originally described in quiescent human T lymphocytes (DeCoursey *et al.*, 1984; Matteson and Deutsch, 1984), the "type *n*" channel has been the subject of intensive investigation, in part because agents which block the type *n* channel also inhibit mitogen-regulated cell proliferation in both T and B cells (DeCoursey *et al.*, 1984; Chandy *et al.*, 1984; Deutsch *et al.*, 1986; Amigorena *et al.*, 1990). A *Shaker*-related gene (Kv1.3) encoding the type *n* channel has been cloned, sequenced, and expressed in *Xenopus* oocytes, as well as in mammalian cell lines (Grissmer *et al.*, 1990a). Two additional voltage-gated K^+ channels have been described in murine thymocytes and T cells: the *n′* channel, with properties only slightly different from those of type *n*, and the type *l* channel, with expression restricted normally to $CD8^+$ cells (DeCoursey *et al.*, 1987a; Lewis and Cahalan, 1988). The type *l* channel is abnormally expressed in $CD4^-$ $CD8^-$ T cells with autoimmune pathologies, including murine models of lupus erythematosus, arthritis, diabetes, and multiple sclerosis (Chandy *et al.*, 1986,1990; Grissmer *et al.*, 1988,1990b). The gene (Kv3.1) encoding the type *l* channel has also been identified (Grissmer *et al.*, 1992a). The type *n* channel is normally the predominant conductance in quiescent human T lymphocytes and is responsible for establishing the resting membrane potential of -50 to -70 mV, near the foot of the conductance–voltage relation, where only a small fraction of the available 500 or so channels per cell are open (Cahalan *et al.*, 1985).

B. Calcium-Activated K^+ {K(Ca)} Channels

In addition to the three voltage-gated K^+ {K(V)} channels, at least two types of Ca^{2+}-activated K^+ {K(Ca)} channels have been described in T and B cells (Mahaut-Smith and Schlichter, 1989; Mahaut-Smith and Mason, 1991). A low-conductance channel, especially prevalent in Jurkat T cells, is selectively blocked by apamin, a toxin from bee venom, at subnanomolar concentrations (Grissmer *et al.*, 1992b). A second, higher conductance channel is blocked by charybdotoxin (CTX), a scorpion venom, at nanomolar levels (Grissmer *et al.*, 1993). The latter channel is found in normal human T and B cells and becomes abundant in proliferating cells following mitogenic activation (Grissmer *et al.*, 1993; Partiseti *et al.*, 1992,1993). Both of these channels are opened by a rise in cytosolic Ca^{2+}, with steep Ca^{2+} dependence (Hill coefficient of 4, midpoint ~400 n*M*) at levels seen following mitogen stimulation. The Ca^{2+}-activated K^+ channels serve to hyperpolarize the membrane potential during the mitogen-stimulated rise in cytosolic $[Ca^{2+}]$.

C. Mitogen-Regulated Ca^{2+} Channels

Shortly after mitogen stimulation, cytosolic $[Ca^{2+}]$ rises from its resting level of about 100 n*M* and often begins to oscillate (Lewis and Cahalan, 1989), reaching levels >1 μM at the peak of each oscillation. This "Ca^{2+} signal" is believed to be one of several early triggering signals in the activation of T lymphocytes by mitogens. The measured influx of Ca^{2+} ions during this time translates to an estimated whole-cell current carried by Ca^{2+} ions on the order of 1 pA. Two candidates have been proposed to underlie the Ca^{2+} influx: a 7-pS, Ca^{2+} permeable channel activated directly by IP_3 (Kuno and Gardner, 1987) and a very low-conductance (10 fS), Ca^{2+}-selective channel activated indirectly by depletion of intracellular Ca^{2+} stores (Lewis and Cahalan, 1989). The Ca^{2+}-selective channel is likely identical to a conductance found in mast cells, termed Icrac for "calcium-release-activated conductance" (Hoth and Penner, 1992). Investigations in Jurkat T cells have demonstrated that this low-conductance Ca^{2+}-selective pathway is the major influx mechanism activated (McDonald *et al.*, 1993; Zweifach and Lewis, 1993).

We have analyzed the contributions of Ca^{2+} and K^+ channels to mitogen-stimulated Ca^{2+} signaling in Jurkat T cells (Lewis and Cahalan, 1989; Grissmer *et al.*, 1992b). Activation of voltage-dependent Ca^{2+}-selective channels provides the Ca^{2+} influx which generates the rising phase of the $[Ca^{2+}]_i$ signal. The amount of Ca^{2+} released from intracellular stores is limited; the Ca^{2+} signal depends primarily upon influx driven by a negative membrane potential. At elevated $[Ca^{2+}]_i$, the Ca^{2+} channels are inhibited, providing negative feedback to help terminate an oscillation. Potassium channels provide the electrical driving force for Ca^{2+} entry by maintaining the membrane potential. In Jurkat cells, $[Ca^{2+}]_i$ oscillations are abolished by simultaneous addition of apamin and CTX to block both calcium-activated K(Ca) and voltage-gated K(V) channels (Grissmer *et al.*, 1992b). The relative contributions of K(V) and K(Ca) channels may vary in different T-cell subsets, since their levels of expression can change dramatically depending upon the state of activation. In principle, activity of other channels that influence the membrane potential could play an indirect role in modulating the Ca^{2+} signal. The possible involvement of Cl^- channels in lymphocyte activation and volume regulation is reviewed at the conclusion of this chapter. Table I summarizes the major types and properties of ion channels described to date.

TABLE I

Lymphocyte Ion Channels

Channel	Gating (midpoint)	Size(pS)	Pharmacology (K_i value)	
K(V)				
Type *n* (Kv1.3)	−35 mV	18	TEA (10 m*M*)	CTX (3 n*M*)
Type *n'*	−35 mV	18	TEA (100 m*M*)	CTX (3 n*M*)
Type *l* (Kv3.1)	−5 mV	27	TEA (0.1 m*M*)	
K(Ca)				
Mini	400 n*M*	5[a]	TEA (2 m*M*)	Apamin (<1 n*M*)
Midi	300 n*M*	35[a]	TEA (40 m*M*)	CTX (3 n*M*)
Ca^{2+}				
IP_3-gated	IP_3	7		
"Icrac"	Ca^{2+} stores depletion	0.01	Ni^{2+} (1 m*M*)	
Cl^-				
Mini	Volume	2–3[a]	DIDS (17 μ*M*)[b]	NPPB (50 μ*M*)
Midi	Kinase Depolarization	40–50[a]		
Maxi	Patch excision Voltage	400 pS[a] Many sublevels	SITS[b] (5 m*M*)	Zn^{2+}

[a] Determined under conditions of symmetrical permeant ion concentration.

[b] Block is voltage dependent. K_i was evaluated at +40 mV for external DIDS and SITS block of the mini-Cl^- channel (Lewis *et al.*, 1993), and at 0 mV for internal SITS block of the maxi-Cl^- channel (Bosma, 1989).

III. PATCH-CLAMP STUDIES OF LYMPHOCYTE Cl^- CHANNELS

A. Volume-Regulated Mini-Cl^- Channel

Cells of the lymphocytic lineage (T and B lymphocytes and related cell lines) express a class of volume-activated, low-conductance Cl^- channels that we refer to as mini-Cl^- channels. Such channels have been characterized in murine thymocytes and splenic T cells, human peripheral blood T cells, and the human leukemic Jurkat T-cell line (Cahalan and Lewis, 1988; Lewis *et al.*, 1993). Mini-Cl^- channels outnumber other types of channels in lymphocytes. Based on an estimate of unitary conductance, lymphocytes express $\sim 10^4$ mini-Cl^- channels per cell. Despite their abundance, the channels are inactive in resting cells and only contribute significantly to membrane conductance if the cells are swollen, e.g., during exposure to hyposmotic solutions. These and other properties suggest a role as the swelling sensor for the regulatory volume decrease (RVD) response. The properties and activation mechanism of the mini-Cl^- channel are summarized below.

1. Regulation

Mini-Cl^- channels in lymphocytes are controlled by the transmembrane osmotic gradient. Conditions leading to cell swelling, such as exposure to hyposmotic extracellular or hyperosmotic intracellular solutions, promote channel activation (Cahalan and Lewis, 1988; Lewis *et al.*, 1993). The chloride conductance, g_{Cl}, can reach values of several nanosiemens, making a major contribution to the overall membrane conductance at voltages near the resting potential. The relationship between cell volume and Cl^- channel activation has been studied quantitatively using whole-cell recording and videomicroscopy (Ross *et al.*, 1994). In Jurkat T cells exposed to a 100-mOsm stimulus, Cl^- channel activation began ~60 sec following the onset of swelling. During this period, the cell volume increased by an average of 19%. This corresponds well with the apparent threshold for activation of Cl^- flux by hyposmotic solutions (Sarkadi *et al.*, 1984b).

The activation mechanism for the mini-Cl^- channel is not yet completely understood. Channel gating is not detectably voltage dependent in the range −100 to +100 mV, as steps to these potentials elicit a constant current with no sign of time-dependent activation or inactivation (Lewis *et al.*, 1993). Suction applied to the pipette's interior reversibly opposes channel opening by an osmotic stimulus, suggesting that membrane stretch is part of the activation mechanism. However, the amount of suction required is much smaller than the equivalent osmotic pressure provided by the 100-mOsm gradient (Lewis *et al.*, 1993). Intracellular second messengers such as Ca^{2+} or cAMP are not sufficient to activate the channels, as neither ionomycin nor inclusion of cAMP in the recording pipette has an effect. Intracellular EGTA (10 m*M*) does not alter g_{Cl} induction, indicating that a Ca^{2+} rise is not necessary for channel activation. However, these results do not rule out a possible modulatory role for intracellular Ca^{2+}. One important clue to the activation mechanism is that channel opening requires a source of intracellular ATP. During whole-cell recordings with ATP-free pipette solutions, the inducibility of g_{Cl} declines over several minutes. In contrast, responsiveness to an osmotic gradient is preserved for >10 min by including 4 m*M* ATP in the pipette, or by using the less-invasive perforated-patch recording technique (Lewis *et al.*, 1993). The mechanism by which ATP exerts its effect is not known, but the possibilities include roles as an energy or phosphate donor or an allosteric effector. A poorly hydrolyzable analogue, ATPγS, can also support the activation of Cl^- channels during osmotic swelling (Cahalan and Lewis, 1988).

Insertion of "new" Cl^- channels from an intracellular vesicular pool is one mechanism for the slow modulation of Cl^- conductance during the cell cycle in developing ascidian embryos (Block and Moody, 1990). Should such a mechanism underlie g_{Cl} activation in T cells, an increase in membrane area and hence membrane capacitance (C_m) would accompany the increase in Cl^- current. However, C_m actually decreases slightly during the induction of g_{Cl}, indicating that Cl^- channels preexist in the plasma membrane (Ross *et al.*, 1994). Based on a comparison between measured values of C_m and theoretical values predicted from cell diameter and specific membrane capacitance, 40–70% of the surface membrane area of Jurkat T cells may be from ruffles and/or microvilli. Swelling-induced activation of g_{Cl} may therefore involve the resorption of these structures, leading to the opening of Cl^- channels already present in the plasma membrane. A decrease in the number of microvilli in hypotonically treated T cells has been observed using scanning electron microscopy (Cheung *et al.*, 1982; Grinstein *et al.*, 1984).

2. Biophysical Properties and Pharmacology

Mini-Cl^- channels are ~20-fold more permeable to Cl^- than Na^+ or K^+. The channel selects anions, with the permeability sequence *I* (1.35) > SCN (1.23) > NO_3, Br (1.17) > Cl^- (1.0) > $MeSO_3$ (0.43) > acetate (0.32), propionate (0.27) > ascorbate (0.19) > aspartate (0.11) and gluconate (0.10), with the permeability of each ion relative to Cl^- in parentheses

(Lewis *et al.*, 1993). In symmetrical Cl^--containing solutions, the current–voltage relation shows modest outward rectification ($I_{Cl, +50mV}/I_{Cl, -50 mV} \sim 1.4$).

The size and kinetic behavior of mini-Cl^- channels in lymphocytes has been estimated using fluctuation analysis of whole-cell Cl^- current (Lewis *et al.*, 1993). The estimated ratio of variance to mean current during the slow induction of g_{Cl} is -0.2 pA at a membrane potential of -80 mV, corresponding to a chord conductance of ~2–3 pS. The Cl^- noise spectrum under conditions of low open probability is approximated by a Lorentzian function with a cutoff frequency of 386 Hz. These results predict a mean open time of ~0.4 msec. Osmotically activated Cl^- current has not yet been detected at the single-channel level in membrane patches.

Mini-Cl^- channels are blocked in a voltage- and time-dependent manner by the stilbene disulfonates DIDS and SITS (Lewis *et al.*, 1993). Blockade increases with depolarization, showing an *e*-fold increase every 19–22 mV and K_i values of 17 μM (DIDS) or 89 μM (SITS) at +40 mV, the voltage at which blockade is maximal. The voltage dependence makes the blocking action of moderate doses (100–500 μM) of DIDS or SITS undetectable at voltages near the resting potential of these cells, a factor to take into account when using these agents to probe channel function in intact cells. The time dependence of channel blockade is biexponential, suggesting that the drugs may bind to multiple sites within the membrane electric field (Lewis *et al.*, 1993). Mini-Cl^- channels are also blocked by NPPB, CPZ, and niflumic acid, with K_i values near 50 μM (G. Ehring and M. D. Cahalan, unpublished observations). The mini-Cl^- channel is insensitive to verapamil or 1,9-dideoxyforskolin (Lewis *et al.*, 1993), agents reported to block the P-glycoprotein-associated Cl^- current (Valverde *et al.*, 1992).

3. Comparison with Cl^- Channels in Lymphocytes and Other Cells

The osmotically sensitive mini-Cl^- channel of lymphocytes is distinct from other Cl^- channels present in lymphocytes, but shares important biophysical characteristics with selected Cl^- channels of several other cell types, including cells of both neuroectodermal and hematopoietic lineages. Mini-Cl^- channels in chromaffin cells, lymphocytes, mast cells, and neutrophils are regulated differently, as described below, raising the possibility that the mini-Cl^- channel of lymphocytes is a member of a larger group of low-conductance Cl^- channels whose regulation is linked to cell-type-specific functions.

Mini-Cl^- channels are distinct from cAMP-activated Cl^- channels in T cells. Like the mini-Cl^- conductance, the whole-cell Cl^- conductance activated by cAMP or PGE_1 in Jurkat T cells lacks voltage dependence and displays a mild degree of outward rectification (Maldonado *et al.*,

1991). However, the two differ in that the cAMP/PGE_1-activated current is not consistently blocked by 100 μM DIDS, and cAMP has no detectable effect on mini-Cl^- channels. Furthermore, single cAMP-activated Cl^- channels in T cells have a much larger unitary conductance (~40 pS at 0 mV) than the mini-Cl^- channels.

Mini-Cl^- channels do not appear to underlie other volume-activated Cl^- currents reported in transformed normal and cystic fibrosis (CF) B cells (McDonald *et al.*, 1992); possibly lymphocytes possess multiple volume-activated Cl^- pathways. McDonald *et al.* (1992) reported a hypotonically activated Cl^- current that inactivates at potentials greater than +50 mV. The current appears similar in its regulation and kinetic properties to volume-sensitive Cl^- conductances described in greater detail in the human colonic cell line T84 (Worrell *et al.*, 1989; Solc and Wine, 1991), in human airway epithelial cells (J. D. McCann *et al.*, 1989), and in a human intestinal epithelial cell line (Kubo and Okada, 1992). The epithelial Cl^- currents are similar to mini-Cl^- channels in their dependence on osmolarity and selectivity for anions over cations (P_{Cl}/P_{Na}~20; Worrell *et al.*, 1989). They also show a similar outward rectification ($I_{+50\ mV}/I_{-50\ mV}$ = 1.5 to 2) and similar, but not identical, selectivity among anions (Kubo and Okada, 1992). However, these epithelial Cl^- currents display pronounced inactivation at potentials > +40 mV and, at least in T84 cells, a significantly larger single-channel conductance (50–75 pS; Worrell *et al.*, 1989; Solc and Wine, 1991). Thus, this class of volume-activated Cl^- channels differs in both gating and conductance properties from the mini-Cl^- channel.

Mini-Cl^- channels have been found in neutrophils, another cell of the hematopoietic lineage (Stoddard *et al.*, 1993). As are their counterparts in T cells, the neutrophil channels are activated by hyperosmotic internal or hyposmotic external solutions, a process which is opposed by suction through the pipette; are not voltage- or Ca^{2+}-dependent; have a unitary conductance of 1.5 pS by variance analysis; show a similar, though not identical, anion selectivity; and are blocked in a voltage-dependent manner by SITS. These channels may be involved in the RVD response of neutrophils (Stoddard *et al.*, 1993), in a manner analogous to the proposed function of mini-Cl^- channels in T cells (Cahalan and Lewis, 1988; Lewis *et al.*, 1993).

Nonhematopoietic cells also express mini-Cl^- channels, although with additional modes of regulation. In adrenal chromaffin cells, whole-cell Cl^- current can be activated by a positive transmembrane osmotic gradient (Doroshenko and Neher, 1992) or by intracellular GTPγS (Doroshenko *et al.*, 1991). Like the T cell's mini-Cl^- channel, these currents lack voltage-dependent gating, have a low permeability to aspartate (P_{asp}/P_{Cl}~0.1), show modest outward rectification, are blocked reversibly by DIDS in a voltage-dependent manner, and have a unitary conductance of 1–2 pS as estimated from noise analysis. With ATP-free internal solutions, a sustained osmotic gradient induced a transient I_{Cl} in chromaffin cells (Doroshenko and Neher, 1992), much like the transient behavior of the osmotically activated Cl^- current in lymphocytes under ATP_i-free conditions (Lewis *et al.*, 1993). However, the Cl^- current in chromaffin cells does not show the same dependence on pipette pressure: a brief pressure pulse elicits the same increase in g_{Cl} as a sustained osmotic stimulus, and negative pressure of up to −55 cm H_2O (~5.4 kPa) does not inhibit g_{Cl} (Doroshenko and Neher, 1992). These results suggest that, in chromaffin cells, additional steps may intervene between cell membrane distention and channel activation, such that termination of swelling does not immediately curtail channel activation. Interestingly, inhibitors of phospholipase A2 and lipoxygenase prevent the induction of g_{Cl} by GTPγS in chromaffin cells. Thus, activation of Cl^- channels in these cells may involve a second messenger generated by arachidonic acid metabolism (Doroshenko, 1991; Doroshenko and Neher, 1992). The possible involvement of this pathway in activation of mini-Cl^- channels in lymphocytes has not been explored.

A small-conductance Cl^- channel described in mast cells is also similar in many respects to the lymphocyte mini-Cl^- channel, excluding its mode of regulation. The mast cell Cl^- channel is activated by cell-surface receptors (substance P, compound 48/80) as well as by intracellular application of cAMP, high Ca^{2+}, GTPγS, and possibly other mediators invoked by cell-surface receptors. The induction of g_{Cl} by intracellular second messengers is slow, particularly in the case of Ca^{2+}, indicating that they are unlikely to activate the channel directly (Penner *et al.*, 1988; Matthews *et al.*, 1989). In these cells, swelling is not tightly correlated with the induction of the current, and direct manipulation of cell volume by positive or negative pressure on the pipette causes only small and slow changes in I_{Cl} (Matthews *et al.*, 1989). Despite these obvious differences in its regulation, the mast cell Cl^- channel shares permeation and pharmacological properties with mini-Cl^- channels in lymphocytes, including a unitary conductance of 1–2 pS and a sensitivity to blockade by DIDS.

Osmotically activated Cl^- currents recorded in fibroblasts and cardiac cells also resemble the mini-Cl^- channel currents of T cells, although they have not been characterized in great detail. In human dermal fibroblasts, hyposmotic extracellular or hyperosmotic intracellular solutions induce a Cl^- conductance with moderate outward rectification and voltage-independent gating (Estacion, 1991). Intracellular ATP enhances the induction rate, duration, and amplitude of g_{Cl}, as in lymphocytes. In cardiac myocytes, a transmembrane hyposmotic gradient activates a Cl^- conductance that is voltage independent, shows moderate outward rectification, and has a permeability sequence of $NO_3 > Cl^- > MeSO_3 >$ aspartate (Tseng, 1992). Neither the cardiac nor the fibroblast Cl^- current has been analyzed at the microscopic level, but these results provide additional evidence that the prevalence of volume-activated mini-Cl^- extends beyond the borders of the hematopoietic lineage.

What gene or genes encode volume-regulated mini-Cl^- channels? Jentsch and colleagues have shown that the ClC-2 gene induces a volume-regulated Cl^- conductance in *Xenopus* oocytes (Thiemann *et al.*, 1992; Grunder *et al.*, 1992). Like the lymphocyte mini-Cl− channel, the ClC-2 gene product is opened by cell swelling and lacks voltage-dependent gating, but the open channels rectify inwardly and have a different sequence of permeant ions than the outwardly rectifying mini-Cl^- channel. Furthermore, strong hyperpolarization activates ClC-2 channels. It is not known whether lymphoid cells express mRNA for ClC-2, although the expression of this gene is widespread. In other cell types, another candidate gene is *MDR1* which codes for P-glycoprotein, an active transporter responsible for the resistance of cancer cells to chemotherapeutic agents. Valverde and colleagues (1992) reported that fibroblasts transfected with the *MDR1* gene expressed a volume-regulated Cl− channel, with many properties similar to the lymphocyte mini-Cl^- channel, including the lack of voltage dependence and requirement for ATP. Unlike the mini-Cl^- channel, the volume-regulated channel in *MDR* transfectants tends to inactivate at depolarizing potentials and is sensitive to block by verapamil and dideoxyforskolin. Furthermore, the induction of the volume-regulated channel is inhibited by including the anti-mitotic chemotherapeutic agents doxorubicin or vinblastine in the pipette (Gill *et al.*, 1992). Attempts to reproduce several of these findings have been unsuccessful in our lab. For example, control fibroblasts lacking P-glycoprotein express volume-sensitive Cl^- channels at the same levels as *MDR1*-transfected fibroblasts (Ehring *et al.*, 1994). The identity of the gene(s) encoding the volume-regulated mini-Cl^- channel remains unknown, although its dependence on ATP suggests that it may belong to the ABC family of proteins.

B. Kinase-Regulated Midi-Cl^- Channel

Not long after the discovery of volume-regulated Cl^- channels in lymphocytes, a second type of Cl^- channel was described, initially at the single-channel level. Using protocols that mimicked a series of experiments performed on epithelial cells (reviewed by Frizzell and Halm, 1990), Chen *et al.* (1989) characterized an outwardly rectifying Cl^- channel of moderate conductance (40–50 pS) in leukemic Jurkat T cells and in transformed B lymphoblasts. This channel, which we have termed the "midi-Cl^-" channel for its intermediate conductance, is closely similar to a channel which has been referred to in other cell types as ORDIC (for "outwardly rectifying, depolarization-induced conductance"). The regulation of this channel in both epithelia and lymphocyte cell lines has been investigated extensively, in part because abnormal regulation of this channel in epithelia might contribute to defective Cl^- secretion in cystic fibrosis patients. A biophysically related channel has also been described in fibroblasts (Bear, 1988).

1. Regulation

The midi-Cl^- channel exhibits a very complex regulation of channel opening (Chen *et al.*, 1989). As with the ORDIC channel in epithelia, different stimuli have been shown to induce channel activity. Patch excision followed by strong depolarization for a prolonged period (> +40 mV for 10 sec to 10 min) is the most reliable way to activate the channel (excision/depolarization protocol). This unphysiological stimulus has been used as a method for determining whether a channel is in fact present in a given patch. Once opened, the lymphocyte midi-Cl^- channel continues to chatter between open and closed states. In "normal" (non-CF) cells—either Jurkat or control B lymphoblasts—the channel can be induced to open during four different protocols that point toward phosphorylation of a membrane substrate by cAMP-dependent protein kinase: First, in cell-attached patch recordings, channel activity is often observed if cells have been exposed to a permeable cAMP analogue added to the bath (Chen *et al.*, 1989). Second, in whole-cell recordings, prostaglandin E_1, which is known to increase cellular cAMP levels, induces a Cl^- conductance (Maldonado *et al.*, 1991). A cAMP antagonist and a peptide inhibitor of the enzyme block the ability of PGE_1 to induce this conductance. Third, photorelease of cAMP from a caged cAMP analogue induced a whole-cell Cl^- conductance following flash stimulation (McDonald *et al.*, 1992). Finally, and perhaps most directly, in excised inside-out patch recordings and in whole-cell dialysis experiments, application of purified catalytic subunit of cAMP-dependent protein kinase, along with ATP, promoted channel activity (Chen *et al.*, 1989). Thus, phosphorylation of the channel protein or a closely associated substrate can open the channel.

The regulation of channel opening by cAMP-dependent protein kinase appears to be very different in CF cells (Chen *et al.*, 1989). In B lymphoblasts derived from a CF patient, the channel could not be activated by cAMP-dependent protein kinase applied to excised inside-out patches, although it could still be opened by excision/depolarization. Bath exposure of the CF lymphocytes to a permeable cAMP analogue also failed to elicit channel activity in cell-attached patches. These initial results raised the possibility of using lymphocytes as a more tractable and accessible cell system in which to investigate the CF defect.

Cytosolic [Ca^{2+}] may be a physiologically important regulator of the lymphocyte midi-Cl^- channel. Channel activation of cAMP requires a basal level of Ca^{2+} (10^{-7} or greater), but, once activated, the channel is not directly dependent upon [Ca^{2+}]. Nevertheless, raising [Ca^{2+}] to higher levels appears to activate the channel without involving the cAMP pathway. In cell-attached patches and in whole-cell recordings, adding Ca^{2+} ionophore activated a conductance closely similar or identical to that induced by cAMP (Nishimoto *et al.*, 1991). A similar conductance was also activated by flash photolysis of intracellular DM-nitrophen (caged Ca^{2+}) to elevate cytosolic Ca^{2+} (McDonald *et al.*, 1993). As with other activation methods (cAMP, patch excision/depolarization), once the channel is activated it is not sensitive to Ca^{2+}, suggesting that activation by Ca^{2+} is not direct, but regulated by a Ca^{2+}-dependent enzyme. This alternative regulatory pathway appears to be mediated through calcium-calmodulin-dependent (CaM) kinase. A peptide inhibitor containing the autoinhibitory region of CaM kinase partially inhibits the activation of Cl^- conductance by Ca^{2+} ionophore in whole-cell measurements (Nishimoto *et al.*, 1991). Moreover, in excised inside-out patches, direct application of purified CaM kinase plus ATP elicits Cl^- channel activity in 100% of the patches with channels, as defined by the excision/depolarization test.

Several aspects of midi-Cl^- channel regulation have proven to be difficult to reproduce, including regulation by cAMP-dependent protein kinase, regulation by Ca^{2+} in CF cells, and several detailed biophysical parameters of channel gating and conduction. Differences in cell lines, culture conditions, or experimental protocols may contribute to the variability. Whether changes in Cl^- conductance during the cell cycle account for some of the variability is controversial. Bubien and colleagues (1990) have reported that Cl^- conductance, measured with video imaging of SPQ fluorescence in populations of cells, is highest during the G_1 phase of the cell cycle. For normal resting B cells isolated from tonsils or peripheral blood, the cell cycle was initiated by addition of monoclonal antibody to the F_c portion of surface IgM, along with B-cell growth factor. B-cell lines or activated B cells were synchronized by treating the cells with hydroxyurea to arrest the cells. Cells in G_0 or in S phase had low Cl^- conductances, while those in the G_1 phase had uniformly higher Cl^- conductances. In contrast, G_1 phase cells from CF patients had very low Cl^- conductances. Gardner and colleagues, using patch-clamp methods, have been unable to confirm the cell-cycle dependence of g_{Cl} (McDonald *et al.*, 1992). There may be a different Cl^- transport system regulated during the cell cycle seen in SPQ imaging that does not show up in patch recordings.

Alternatively, experimental differences in the method of cell-cycle synchronization may account for the discrepancy. Bubien and colleagues have also observed an increase in cAMP-regulated Cl^- conductance in CF cells transfected with wild-type cystic fibrosis transmembrane receptor (CFTR) (Krauss *et al.*, 1992a). In these transfectants, cells in G_1 showed the largest resting and cAMP-dependent Cl^- conductance, compared with cells in S phase. Comparing imaging and patch-clamp experiments as methods to evaluate Cl^- conductance, the cell-cycle dependence seems more pronounced with the imaging assay. SPQ imaging experiments indicate that expression of wild-type CFTR is required for cAMP regulation of the Cl^- conductance and that the ability of cAMP to increase Cl^- conductance is cell-cycle dependent. In addition, antisense CFTR oligonucleotides inhibit the Cl^- conductance in G_1 cells from normal B lymphoblasts (Krauss *et al.*, 1992b). Patch-clamp experiments on freshly isolated, "real" lymphocytes from CF and non-CF individuals will be informative.

The role of Ca^{2+} in regulating g_{Cl} in CF lymphocytes is disputed. Gardner and colleagues have observed activation of g_{Cl} by elevating Ca^{2+} in both "normal" and CF cell lines (McDonald *et al.*, 1992), whereas Bubien and colleagues (1990) report that neither cAMP nor Ca^{2+} ionophore caused increased Cl^- conductance in CF cells.

2. Biophysical Properties and Pharmacology

Several biophysical aspects of the midi-Cl^- channel await further characterization. The cAMP-activated Cl^- conductance appears to rectify outwardly to a greater degree in patches than in whole-cell measurements. This may reflect activation of multiple populations of channels by cAMP in whole-cell measurements, in addition to the midi-Cl^- channel that is easily recognized in single-channel patch measurements. The midi-Cl^- channel displays 10:1 Cl^- selectivity over Na^+ (Chen *et al.*, 1989), but so far a thorough study of ion permeation has not been reported. The channel is sensitive to several typical Cl^- channel blockers, including IAA, NPPB, and SITS (McDonald *et al.*, 1992; Garber, 1992).

3. Cl^- Channels in Other Cell Types

The lymphocyte midi-Cl^- channel, in most respects, behaves similarly to the outwardly rectifying Cl^- conductance in epithelia. Both channels are activated by excision/depolarization or by cAMP-dependent protein kinase in normal, but not CF cells (Chen *et al.*, 1989; reviewed by Frizzell and Halm, 1990). In epithelial and lymphoid cells, the ORDIC channel is regulated by more than one kinase: by PKA, PKC, or CaM kinase in epithelia (see Chapter 8, this volume); and by PKA or CaM kinase in Jurkat T cells (Chen *et al.*, 1989; Nishimoto *et al.*, 1991). According to Garber (1992), channel activation by excision/depolarization is highly temperature sensitive in Jurkat T cells: more than twice as many patches show channel activity, with much more rapid induction, at 30°C than at 20°C. The epithelial ORDIC displays a similar temperature dependence (Greger *et al.*, 1989), as does the maxi-Cl^- chanel in T lymphocytes (see below; Pahapill and Schlichter, 1992). However, careful investigation of single-channel conductance and open times led Garber to conclude that the biophysical properties of the excision/depolarization-induced Cl^- conductance are different in lymphocytes (Garber, 1992). The clearest level of single-channel conductance is smaller in lymphocytes than in epithelial cells studied under identical conditions in the same laboratory. Furthermore, the lymphocyte channel's open probability is not significantly voltage dependent. In contrast, the epithelial ORDIC has an open probability that is significantly voltage dependent (reviewed by Frizzell and Halm, 1990). Possibly the lymphocyte midi-Cl^- channel is related, but not structurally identical, to the epithelial ORDIC.

The cloning of the CFTR gene and the subsequent reconstitution of the CFTR gene product into planar lipid bilayers has led to reevaluation of the properties and regulation of the midi-Cl^- channel. Bear *et al.* (1992) concluded that the CFTR itself functions as a kinase-regulated, 8–10 pS, nonrectifying Cl^- channel with properties distinct from the midi-Cl^- channel (Bear *et al.*, 1992). To date, a channel matching the description of the CFTR-Cl^- channel has not been described in lymphocytes, although the CFTR is reportedly expressed in lymphocytes. PCR techniques are required to detect the low levels of expression of CFTR transcript in Jurkat cells (McDonald *et al.*, 1992) and in other lymphoid lines (Krauss *et al.*, 1992a). Antisense oligonucleotides to the CFTR inhibited expression and abolished the Cl^- conductance normally present in G_1 B lymphoblasts, determined by the SPQ fluorescence quench assay (Krauss *et al.*, 1992b). Transfection of CFTR into CF cells restored the cell-cycle and cAMP-regulated Cl^- conductance (Krauss *et al.*, 1992a). As suggested for other cell types, the CFTR appears to regulate the midi-Cl^- channel. It is unclear whether expression levels are high enough for the Cl^- channel function of the CFTR to contribute significantly to the lymphocyte Cl^- conductance. It may be that the CFTR regulates Cl^- channels whose structures have not yet been identified.

C. Maxi-Cl^- Channels

Extremely large-conductance "maxi"-Cl^- channels have been described in T and B lymphocytes, thymocytes, and antibody-secreting hybridoma cells. These channels can be induced to conduct by patch excision and, once opened, the channel activity is complex, displaying voltage-dependent gating kinetics with multiple open and closed states, as well as several subconductance levels. The channel, although normally dormant, is quite abundant and distributed nonuniformly, judging from the probability of observing the channel in excised patches. The regulation and biophysical properties of maxi-Cl^- channels present several unusual features, as described below.

1. Regulation

The highly complex regulation of maxi-Cl^- induction is poorly understood mechanistically. Two aspects can be distinguished: an initial induction phase that leads to channel activity and a more rapid, voltage-dependent channel gating process. Maxi-Cl^- channel activity can be induced with high frequency by excising an inside-out patch: 65% of patches from hybridoma cells (Bosma, 1989), 50% of patches from human T cells (Schlichter *et al.*, 1990), and 40% of patches from mouse splenic B cells (F. V. McCann *et al.*, 1989) exhibited channel activity after excision. In most cases, at least two channels per patch were observed, suggesting clustering of channel proteins in the membrane. The channel may be inhibited by cytoplasmic factors; in intact cells channel activity appeared after some delay. Supporting this view, Bosma (1989) found that patches excised into solution containing Mg-ATP had maxi-Cl^- channels with steeper voltage dependence and a greater tendency to remain in a low subconductance level. In silent patches, prolonged depolarization tended to encourage channel activity.

The maxi-Cl^- channel usually is not observed under normal cell-attached or whole-cell recording conditions. However, a low level of channel activity can be induced in these recording configurations by periods of prolonged depolarization, in a manner similar to the ORDIC described above (Bosma, 1989; Schlichter *et al.*, 1990). In macrophages and myotubes, adding the Ca^{2+} ionophore A23187 increased channel activity in cell-attached patches (Schwarze and Kolb, 1984). However, this method of induction has not been reported in lymphocytes. Once the patch is excised and channel activity initiated, the channel no longer depends upon $[Ca^{2+}]$ at the exposed cytoplasmic membrane surface. Channel activity occurs spontaneously in cell-attached recordings at temperatures above 32°C; the channel is more likely to contribute to the membrane's conductance under physiological conditions (Pahapill and Schlichter, 1992). However, in whole-cell recordings, Cl^- conductances commensurate with the degree of channel activity in excised patches or in cell-attached patches at body temperature have not been observed. Further investigation is needed to clarify the mechanisms of channel induction.

Once induced, the maxi-Cl^- channel's open probability is strongly dependent upon membrane potential. Most investigators agree that the channel is open near 0 mV and can be closed by either negative or positive potentials, but voltage gating may depend upon the cell type, the patch configuration, the time after excision, and conditioning pulses to varying potentials. Bosma (1989) reports that once the channel is induced, its voltage dependence depends upon whether it is in a cell-attached or an excised patch. In cell-attached patches, the channel's voltage dependence is nearly bell-shaped with an *e*-fold change in conductance per 10.4 or 15.8 mV, at negative or positive potentials, respectively, with a midpoint near +10 mV, relative to the resting potential. In excised patches, the channel no longer closes with depolarizing voltage steps and exhibits weaker voltage dependence in the negative range of potentials (*e*-fold per 25 mV). In human T cells, Schlichter and colleagues (1990) described a steeper voltage dependence in excised patches, with *e*-fold changes in the open probability per 4.4 or 2.6 mV at negative and positive potentials, respectively. Time constants for channel closure were in the range of 100–150 msec. Pahapill and Schlichter (1992) reported that the channel's voltage dependence is influenced by both the patch configuration and the temperature. In cell-attached recordings at room temperature, very few (< 2%) patches displayed maxi-Cl^- activity, but 69% of previously quiescent patches showed channel activity upon warming to >32°C; upon return to room temperature, channel activity slowly diminished. In warm cell-attached patches, maxi-Cl^- channels lost the ability to close with membrane hyperpolarization. In addition to patch configuration and temperature, conditioning depolarizing pulses allow the channel to be more active at negative potentials (F. V. McCann *et al.*, 1989).

2. Biophysical Properties and Pharmacology

For maxi-Cl^- channel conductance levels, an enormous single-channel conductance of 350–400 pS is the most frequent. The number of subconductance levels observed varies according to cell type and lab. F. V. McCann and colleagues (1989) described four evenly spaced levels of conductance that can interconvert in murine B cells, while Schlichter and colleagues (1990) saw uniform sublevels of ~45 pS in human T cells. In hybridoma cells, Bosma (1989) noted four nonuniformly distributed subconductance levels and a supraconductance level, in addition to the 400-pS conductance. This most frequent conductance sublevel saturates at 580 pS as [Cl^-] is raised, with a K_m of 120 mM, suggesting weak binding interactions with sites in the pore (Schlichter *et al.*, 1990).

Channel selectivity favors anions over cations by at least 10:1 (F. V. McCann *et al.*, 1989; Schlichter *et al.*, 1990), but the channel is promiscuous in choosing anions: even large anions such as aspartate permeate rather freely. F. V. McCann *et al.* (1989) report a $P_{Cl}/P_{aspartate}$ of 1.25, while Schlichter's group find a value of about 12. The selectivity sequence determined from measured reversal potentials in human T-cell patches, according to Schlichter and colleagues (1990) is I (1.38) > NO_3 (1.14) > Br (1.04) > Cl^- (1) > H_2PO_4/HPO_4 > F (0.57) > isethionate (0.56) > HCO_3 (0.56) > SO_4 (0.49) > propionate (0.30) > gluconate (0.29) > aspartate (0.08). Bosma's (1989) selectivity sequence in hybridoma cells differed somewhat: F (1.25) > I (1.18) > SCN (1.1) > Br (1.07) > Cl^- (1.0) > glucuronate (0.78) > NO_3 (0.68) > aspartate (0.62).

The pharmacological sensitivity of maxi-Cl^- channels in lymphocytes has not been extensively investigated. Bosma (1989) reports that SITS, applied internally, blocks the channel with a K_i of 5 m*M* at 0 mV; the voltage dependence indicates a binding site at an electrical distance 37% across the membrane field. DIDS was fivefold less potent; 1 m*M* 9-AC had no effect. Zn^{2+} or Ni^{2+} (1 m*M*) added internally produced a reversible, voltage-dependent block (Schlichter *et al.*, 1990).

3. Comparison with Cl^- Channels in Other Cell Types

Similar maxi-Cl^- channels are present in macrophages, skeletal muscle, Schwann cells, hippocampal neurones, and a variety of epithelial cells (reviewed in Chapter 8, this volume). The molecular configuration of the maxi-Cl^- channel may be related to bacterial porin, the structure of which has been resolved to 1.8 Å (Schulz, 1993). The large conductance and voltage-dependent gating are properties shared by a voltage-dependent anion channel (VDAC) of the outer mitochondrial membrane; VDAC's tertiary structure resembles that of porin from gram-negative bacteria (Mannella *et al.*, 1992). Interestingly, a porin-like molecule is present in the plasma membrane of lymphocytes (Thinnes, 1992). Following reconstitution into planar lipid bilayers, the lymphocyte porin exhibits properties similar to those of the maxi-Cl^- channel (Benz *et al.*, 1992).

IV. FUNCTIONAL ROLES OF Cl^- CHANNELS

A. Contribution of P_{Cl} to Resting Potential

Ion channels can initiate or modulate functional responses of T and B cells by altering the membrane potential and by gating the flow of specific ions across the plasma membrane. Normally, in a quiescent lymphocyte with low cytosolic [Ca^{2+}], the membrane potential is primarily established by voltage-gated K^+ channels, in particular by the type *n* channel encoded by Kv1.3 (Cahalan *et al.*, 1985; Grinstein and Smith, 1990; Freedman *et al.*, 1992; Leonard *et al.*, 1992). Blocking type *n* channels usually results in membrane depolarization; murine T cells are an exception: their membrane potential is governed by an electrogenic Na^+/K^+ pump (Ishida and Chused, 1993). K^+ channel expression is extraordinarily low in mature, resting $CD4^+$ and $CD8^+$ T cells from mice—only 5–10 K(V) channels per cell (DeCoursey *et al.*, 1987b; Lewis and Cahalan, 1988). Human T cells typically express about 500 K(V) channels per cell, and, of these, a few open channels suffice to maintain the resting potential. If K^+ channels are blocked, the resulting membrane depolarization limits Ca^{2+} entry through mitogen-regulated Ca^{2+} channels (Grissmer *et al.*, 1992b; Lin *et al.*, 1993).

Do Cl^- ions contribute to the membrane potential and thereby modulate functional responses? Here, the answer is not as straightforward. Available evidence suggests that there is a degree of Cl^- permeability which keeps the cells depolarized away from the K^+ equilibrium potential (Wilson and Chused, 1985). The Cl^- concentration inside T lymphocytes is 30–40 m*M*, implying an equilibrium potential for Cl^- of -30 to -40 mV (Felber and Brand, 1982). Even rather bulky organic anions such as aspartate can permeate the Cl^- channels described above. Thus, the reversal potential for the Cl^- channels under physiological conditions may be closer to 0 mV than the equilibrium potential for Cl^-. In any event, if Cl^- channels open, the membrane potential depolarizes.

B. Mitogenesis

Cl^- channels may act during mitogenic activation and in the killing of target cells by cytotoxic T lymphocytes and natural killer cells. The Cl^- channel blockers SITS and DIDS inhibit target cell lysis by cytotoxic T cells or NK cells (Gray and Russell, 1986; Prochazka *et al.*, 1988). Furthermore, treatment with DIDS (but not SITS) or removal of external Cl^- inhibits antibody-stimulated Ca^{2+} influx (Rosoff *et al.*, 1988). Possibly Cl^- channels influence cell proliferation by upregulating, for example, as seen in B-lymphoblastoid cells during the cell cycle (Bubien *et al.*, 1990). Despite these tantalizing findings, the role of Cl^- channels in lymphocyte activation remains unclear, primarily because highly potent and selective blockers for these channels are not available.

In mast cells, a cAMP-dependent Cl^- conductance opens during stimuli that result in degranulation (Penner *et al.*, 1988). The opening of Cl^- channels may facilitate Ca^{2+} influx by maintaining a negative membrane potential. Mast cells lack K^+ channels and are highly susceptible to depolarization. In contrast, lymphocytes generally express both K(V) and K(Ca) channels; these channels maintain a negative membrane potential in both the resting and activated state. If Cl^- channels were to open in lymphocytes, the membrane potential would depolarize rather than hyperpolarize. As discussed above, depolarization inhibits both Ca^{2+} influx and subsequent gene expression leading to cell proliferation (Grinstein and Smith, 1990; Leonard *et al.*, 1992; Freedman *et al.*, 1992).

Might opening Cl^- channels be one mechanism for depressing immune responses? Agents such as PGE_1 or PGE_2 that elevate cAMP levels can suppress mitogenic signaling in lymphocytes (Coffey and Hadden, 1985; Kammer, 1988; Bastin *et al.*, 1990; Maldonado *et al.*, 1991). Perhaps opening kinase-regulated Cl^- channels depolarizes the membrane potential, thereby suppressing mitogen-stimulated Ca^{2+} infux. Using a *lac-Z* reporter gene construct driven by the interleukin 2 response element which binds the nuclear factor of activated T cells (NF-AT), Negulescu and Cahalan (1994) followed Ca^{2+} signals (induced by T-cell receptor stimulation) that lead to gene expression in single cells. Adding a permeable cAMP analogue suppressed both $[Ca^{2+}]$ signals and gene expression in response to T-cell receptor engagement. Thapsigargin, a microsomal Ca-ATPase inhibitor, leads to Ca^{2+} influx. The resultant cytosolic $[Ca^{2+}]$ level can be "clamped" to varying, stable levels, by altering the ionic gradients of K^+ and Ca^{2+}, thus bypassing receptors and pre-Ca^{2+} elements of signal transduction. Thapsigargin protocols have made it possible to determine quantitatively the Ca^{2+} dependence of gene expression in single cells. Adding a permeable cAMP analogue inhibited NF-AT-driven interleukin-2 gene expression at both pre- and post-Ca^{2+} steps. Ca^{2+} oscillations induced by T-cell receptor engagement were blocked by cAMP, *and* gene expression was suppressed, even when $[Ca^{2+}]$ was artificially elevated using thapsigargin (Negulescu and Cahalan, 1994). The action of kinase-regulated Cl^- channels could partly explain inhibition of Ca^{2+} signaling by cAMP analogues.

C. Volume Regulation

Cell swelling as a consequence of osmotic challenge induces RVD, as Cl^-, K^+, and water leave cells via separate, conductive pathways (Grinstein *et al.*, 1982; Sarkadi *et al.*, 1984a,1985). Cahalan and Lewis (1987,1988) and Deutsch and Lee (1988) have proposed that cell swelling activates mini-Cl^- channels, resulting in membrane depolarization and the consequent activation of K(V) channels. According to this hypothesis, the opening of Cl^- channels provides the initial trigger for RVD; the resulting depolarization opens K(V) channels. Cl^- and K^+ conductances become simultaneously active, allowing Cl^- and K^+ to leave the cell. Water follows the osmolytes, bringing cellular volume back toward normal. The mini-Cl^- channel is an attractive candidate as the RVD trigger because it is opened by cell swelling. Furthermore, the channel's sequence of ion selectivity parallels the cell's ability to volume regulate in solutions containing different anions. In addition, pharmacological agents that block the mini-Cl^- channels also inhibit RVD (reviewed in Cahalan and Lewis, 1988; Lewis *et al.*, 1993).

The type *n* K(V) channel, encoded by Kv1.3, could mediate K^+ efflux during RVD for three reasons: pharmacology, voltage dependence, and cellular distribution. First, agents that are known to block type *n* K^+ channels inhibit ^{86}Rb or ^{42}K fluxes induced by hypotonicity (Chandy *et al.*, 1984; Sarkadi *et al.*, 1985; DeCoursey *et al.*, 1985; Sands *et al.*, 1989; Grinstein and Smith, 1990). Second, according to the Cl^- trigger hypothesis for RVD (Cahalan and Lewis, 1988; Deutsch and Lee, 1988), the voltage dependence of type *n* channels allows them to open during the depolarization induced by Cl^- channel activation. Third, the expression of type *n* channels parallels the ability of particular cells to volume regulate. For example, human B lymphocytes have fewer type *n* K^+ channels than do T cells and volume regulate less effectively (Deutsch *et al.*, 1986). An interleukin-2-dependent T-cell line undergoes RVD in parallel with the expression levels of type *n* channels (Lee *et al.*, 1988). By transfecting the Kv1.3 gene into a lymphoid cell (CTLL) lacking both K(V) channels and the ability to volume regulate, Deutsch and Chen (1993) have further tested Kv1.3 as a mediator of the K^+ efflux limb of RVD. Transfected cells expressing Kv1.3 (but not Kv3.1) acquired the ability to regulate volume in response to hypotonic challenge, nicely confirming the hypothesis that type *n* channels encoded by Kv1.3 carry K^+ efflux during RVD.

V. SUMMARY AND PROSPECTS

Patch-clamp techniques have been instrumental in identifying at least three types of Cl^- channel in lymphocytes, categorized according to their single-channel conductances as mini-, midi-, and maxi-Cl-channels. These channels are normally closed in the plasma membrane of quiescent and proliferating cells, but are among the most abundant channels found in lymphocytes. Channel regulation is complex. The opening of each channel subtype depends upon at least two factors: ATP and cell swelling for the mini-Cl^- channel, phosphorylation or excision/depolarization for the midi-Cl^- channel, and excision and voltage for the maxi-Cl^- channel. Fascinating questions remain. For dedicated patch clampers: What factors determine the ion permeation properties which enable relatively bulky anions to pass through both low- and high-conductance channels? Mechanistically, how do channels open? For persistent molecular biologists: Do any of the Cl^- channels seen in patch-clamp experiments coincide with any known Cl^- channel gene products? Having reviewed the data, we must conclude that the molecular identity of all three lymphocyte Cl^- channels is still unknown. For tenacious cell physiologists and immunologists: What functional roles are played by the lymphocyte Cl^- channels? Rapid answers to these questions may depend upon the development of highly potent and selective Cl^- channel blockers. Perhaps the world of toxins may yet provide such ligands, as the recent characterization of chlorotoxin from scorpion venom promises (DeBin *et al.*, 1993).

References

Amigorena, S., Choquet, D., Teillaud, J. L., Korn, H., and Fridman, W. H. (1990). Ion channel blockers inhibit B cell activation at a precise stage of the G_1 phase of the cell cycle. *J. Immunol.* **144,** 2038–2045.

Bastin, B., Payet, M. C., and Dupuis, G. (1990). Effects of modulators of adenylyl cyclase on interleukin-2 production, cytosolic Ca^{2+} elevation, and K^+ channel activity in Jurkat T cells. *Cell. Immunol.* **128,** 385–399.

Bear, C. E. (1988). Phosphorylation-activated chloride channels in human skin fibroblasts. *FEBS Lett.* **237,** 145–149.

Bear, C. E., Li, C., Kartner, N., Bridges, R. J., Jensen, T. J., Ramjeesingh, M., and Riordan, J. R. (1992). Purification and functional reconstitution of the cystic fibrosis transmembrane conductance regulator (CFTR). *Cell (Cambridge, Mass.)* **68,** 809–818.

Benz, R., Maier, E., Thinnes, F. P., Goetz, H., and Hilschmann, N. (1992). The channel properties of the human B-lymphocyte membrane-derived "porin 31HL" are similar to those of mitochondrial porins. *Biol. Chem. Hoppe-Seyler* **373,** 295–303.

Block, M. L., and Moody, W. J. (1990). A voltage-dependent chloride current linked to the cell cycle in ascidian embryos. *Science* **247,** 1090–1092.

Bosma, M. M. (1989). Anion channels with multiple conductance levels in a mouse B lymphocyte cell line. *J. Physiol. (London)* **410,** 67–90.

Bubien, J. K., Kirk, K. L., Rado, T. A., and Frizzell, R. A. (1990). Cell cycle dependence of chloride permeability in normal and cystic fibrosis lymphocytes. *Science* **248,** 1416–1419.

Cahalan, M. D., and Lewis, R. S. (1987). Ion channels in T lymphocytes: Role in volume regulation. *J. Gen. Physiol.* **90,** 7a.

Cahalan, M. D., and Lewis, R. S. (1988). Role of potassium and chloride channels in volume regulation by T lymphocytes. *In* "Cell Physiology of Blood" (R. Gunn and J. Parker, eds.), pp. 281–301. Rockefeller Univ. Press, New York.

Cahalan, M. D., Chandy, K. G., DeCoursey, T. E., and Gupta, S. (1985). A voltage-gated K^+ channel in human T lymphocytes. *J. Physiol. (London)* **358,** 197–237.

Cahalan, M. D., Chandy, K. G., and Grissmer, S. (1992). Potassium channels in development, activation, and disease in T lymphocytes. *Curr. Top. Membr.* **39,** 357–394.

Chandy, K. G., DeCoursey, T. E., Cahalan, M. D., McLaughlin, C., and Gupta, S. (1984). Voltage-gated potassium channels are required for human T lymphocyte activation. *J. Exp. Med.* **160,** 369–385.

Chandy, K. G., DeCoursey, T. E., Fischbach, M., Talal, N., Cahalan, M. D., and Gupta, S. (1986). Altered K^+ channel expression in abnormal T lymphocytes from mice with the *lpr* gene mutation. *Science* **233,** 1197–1200.

Chandy, K. G., Cahalan, M. D., and Grissmer, S. (1990). Autoimmune diseases linked to abnormal K^+ channel expression in double-negative $CD4^-$ $CD8^-$ T cells. *Eur. J. Immunol.* **20,** 747–751.

Chen, J. H., Schulman, H., and Gardner, P. (1989). A cAMP-regulated chloride channel in lymphocytes that is affected in cystic fibrosis. *Science* **243,** 657–660.

Cheung, R. K., Grinstein, S., Dosch, H.-M., and Gelfand, E. W. (1982). Volume regulation by human lymphocytes: Characterization of the ionic basis for regulatory volume decrease. *J. Cell. Physiol.* **112,** 189–196.

Coffey, R. G., and Hadden, J. W. (1985). Neurotransmitters, hormones, and cyclic nucleotides in lymphocyte activation. *Fed. Proc., Fed. Am. Soc. Exp. Biol.* **44,** 112–117.

DeBin, J. A., Maggio, J. E., and Strichartz, G. R. (1993). Purification and characterization of chlorotoxin, a chloride channel ligand from the venom of the scorpion. *Am. J. Physiol.* **264,** C361–C369.

DeCoursey, T. E., Chandy, K. G., Gupta, S., and Cahalan, M. D. (1984). Voltage-gated K^+ channels in human T lymphocytes: a role in mitogenesis? *Nature (London)* **307,** 465–468.

DeCoursey, T. E., Chandy, K. G., Gupta, S., and Cahalan, M. D. (1985). Voltage-dependent ion channels in T lymphocytes. *J. Neuroimmunol.* **10,** 71–85.

DeCoursey, T. E., Chandy, K. G., Gupta, S., and Cahalan, M. D. (1987a). Two types of potassium channels in murine T lymphocytes. *J. Gen. Physiol.* **89,** 379–404.

DeCoursey, T. E., Chandy, K. G., Gupta, S., and Cahalan, M. D. (1987b). Mitogen induction of ion channels in murine T lymphocytes. *J. Gen. Physiol.* **89,** 405–420.

Deutsch, C., and Chen, L.-Q. (1993). Heterologous expression of specific K^+ channels in T lymphocytes: Functional consequences for volume regulation. *Proc. Natl. Acad. Sci. U.S.A.* **90,** 10036–10040.

Deutsch, C., and Lee, S. C. (1988). Cell volume regulation in lymphocytes. *Renal Physiol. Biochem.* **11** (3–5), 260–276.

Deutsch, C., Krause, D., and Lee, S. C. (1986). Voltage-gated potassium conductance in human T lymphocytes stimulated with phorbol ester. *J. Physiol. (London)* **372,** 405–423.

Doroshenko, P. (1991). Second messengers mediating activation of chloride current by intracellular GTPγS in bovine chromaffin cells. *J. Physiol. (London)* **436,** 725–738.

Doroshenko, P., and Neher, E. (1992). Volume-sensitive chloride conductance in bovine chromaffin cell membrane. *J. Physiol. (London)* **449,** 197–218.

Doroshenko, P., Penner, R., and Neher, E. (1991). Novel chloride conductance in the membrane of bovine chromaffin cells activated by intracellular GTPγS. *J. Physiol. (London)* **436,** 711–724.

Ehring, G., Osipchuk, Y., and Cahalan, M. D. (1994). Volume-activated chloride channels in drug-sensitive and -resistant cell lines. *Biophys. J.* **66,** A412.

Estacion, M. (1991). Characterization of ion channels seen in subconfluent human dermal fibroblasts. *J. Physiol. (London)* **436,** 579–601.

Felber, S. M., and Brand, M. D. (1982). Factors determining the plasma membrane potential of lymphocytes. *Biochem. J.* **204,** 577–585.

Freedman, B. D., Price, M. A., and Deutsch, C. J. (1992). Evidence for voltage modulation of IL-2 production in mitogen-stimulated human peripheral blood lymphocytes. *J. Immunol.* **149,** 3784–3794.

Frizzell, R. A., and Halm, D. R. (1990). Chloride channels in epithelial cells. *Curr. Top. Membr. Transp.* **37,** 247–282.

Garber, S. S. (1992). Outwardly rectifying chloride channels in lymphocytes. *J. Membr. Biol.* **127,** 49–56.

Gill, D. R., Hyde, S. C., Higgins, C. F., Valverde, M. A., Mintenig, G. M., and Sepulveda, F. V. (1992). Separation of drug transport and chloride channel functions of the human multidrug resistance P-glycoprotein. *Cell (Cambridge, Mass.)* **71,** 23–32.

Gray, L. S., and Russell, J. H. (1986). Cytolytic T lymphocyte effector function requires plasma membrane chloride flux. *J. Immunol.* **136,** 3032–3037.

Greger, R., Kunzelmann, K., and Gerlach, L. (1989). Mechanisms of chloride transport in secretory epithelia. *Ann. N.Y. Acad. Sci.* **574,** 403–415.

Grinstein, S., and Smith, J. D. (1990). Calcium-independent volume regulation in human lymphocytes: Inhibition by charybdotoxin. *J. Gen. Physiol.* **95,** 97–120.

Grinstein, S., Clarke, C. A., Dupré, A., and Rothstein, A. (1982). Volume-induced increase of anion permeability in human lymphocytes. *J. Gen. Physiol.* **80,** 801–823.

Grinstein, S., Rothstein, A., Sarkadi, B., and Gelfand, E. W. (1984). Responses of lymphocytes to anisotonic media: Volume-regulating behavior. *Am. J. Physiol.* **246,** C204–C215.

Grissmer, S., Cahalan, M. D., and Chandy, K. G. (1988). Abundant expression of type l K^+ channels, a marker for lymphoproliferative diseases? *J. Immunol.* **141,** 1137–1142.

Grissmer, S., Dethlefs, B., Wasmuth, J., Goldin, A. L., Gutman, G. A., Cahalan, M. D., and Chandy, K. G. (1990a). Expression and chromosomal localization of a lymphocyte K^+ channel gene. *Proc. Natl. Acad. Sci. U.S.A.* **87,** 9411–9415.

Grissmer, S., Hanson, D., Natale, P., Cahalan, M. D., and Chandy, K. G. (1990b). CD4-CD8-T cells from mice with collagen arthritis display aberrant expression of type l K^+ channels. *J. Immunol.* **145,** 2105–2109.

Grissmer, S., Ghanshani, S., Dethlefs, B., McPherson, J. D., Wasmuth, J. J., Gutman, G. A., Cahalan, M. D., and Chandy, K. G. (1992a). The *Shaw*-related potassium channel gene, Kv3.1, on human chromosome 11, encodes the type *l* K^+ channel in T cells. *J. Biol. Chem.* **267,** 20971–20979.

Grissmer, S., Lewis, R. S., and Cahalan, M. D. (1992b). Ca^{2+}-activated K^+ channels in human leukemic T cells. *J. Gen. Physiol.* **99,** 1–23.

Grissmer, S., Nguyen, A. N., and Cahalan, M. D. (1993). Ca^{2+}-activated K^+ channels in resting and activated human T cells. *J. Gen. Physiol.* **102,** 601–630.

Grunder, S., Thiemann, A., Pusch, M., and Jentsch, T. J. (1992). Regions involved in the opening of ClC-2 chloride channel by voltage and cell volume. *Nature (London)* **360,** 759–762.

Hoth, M., and Penner, R. (1992). Depletion of intracellular calcium stores activates a calcium current in mast cells. *Nature (London)* **355,** 353–356.

Ishida, Y., and Chused, T. M. (1993). Lack of voltage-sensitive potassium channels and generation of membrane potential by sodium potassium ATPase in murine T lymphocytes. *J. Immunol.* **151,** 610–620.

Kammer, G. M. (1988). The adenylate cyclase-cAMP-protein kinase A pathway and regulation of the immune response. *Immunol. Today* **9,** 222–229.

Krauss, R. D., Bubien, J. K., Drumm, M. L., Zheng, T., Peiper, S. C., Collins, F. S., Kirk, K. L., Frizzell, R. A., and Rado, T. A. (1992a). Transfection of wild-type CFTR into cystic fibrosis lymphocytes restores chloride conductance at G_1 of the cell cycle. *EMBO J.* **11,** 875–883.

Krauss, R. D., Berta, G., Rado, T. A., and Bubien, J. K. (1992b). Antisense oligonucleotides to CFTR confer a cystic fibrosis phenotype on B lymphocytes. *Am. J. Physiol.* **263,** C1147–C1151.

Kubo, M., and Okada, Y. (1992). Volume-regulatory Cl^--channel currents in cultured human epithelial cells. *J. Physiol. (London)* **456,** 351–371.

Kuno, M., and Gardner, P. (1987). Ion channels activated by inositol 1,4,5-trisphosphate in plasma membrane of human T lymphocytes. *Nature (London)* **326**, 301–304.
Lee, S. C., Price, M., Prystowsky, M. B., and Deutsch, C. (1988). Volume response of quiescent and interleukin 2-stimulated T lymphocytes to hypotonicity. *Am. J. Physiol.* **254**, C286–C296.
Leonard, R. J., Garcia, M. L., Slaughter, R. S., and Reuben, J. P. (1992). Selective blockers of voltage-gated K^+ channels depolarize human T lymphocytes: Mechanism of the antiproliferative effect of charybdotoxin. *Proc. Natl. Acad. Sci. U.S.A.* **89**, 10094–10098.
Lewis, R. S., and Cahalan, M. D. (1988). Subset-specific expression of potassium channels in developing murine T lymphocytes. *Science* **239**, 771–775.
Lewis, R. S., and Cahalan, M. D. (1989). Mitogen-induced oscillations of cytosolic Ca^{2+} and transmembrane Ca^{2+} current in human leukemic T cells. *Cell Regul.* **1**, 99–112.
Lewis, R. S., and Cahalan, M. D. (1990). Ion channels and signal transduction in lymphocytes. *Annu. Rev. Physiol.* **52**, 415–430.
Lewis, R. S., Ross, P. E., and Cahalan, M. D. (1993). Chloride channels activated by osmotic stress in T lymphocytes. *J. Gen. Physiol.* **101**, 801–826.
Lin, C. S., Boltz, R. C., Blake, J. T., Nguyen, M., Talento, A., Fischer, P. A., Springer, M. S., Sigal, N. H., Slaughter, R. S., and Garcia, M. L. (1993). Voltage-gated potassium channels regulate calcium-dependent pathways involved in human T lymphocyte activation. *J. Exp. Med.* **177**, 637–645.
Mahaut-Smith, M. P., and Mason, M. J. (1991). Ca^{2+}-activated K^+ channels in rat thymic lymphocytes: Activation by concanavalin A. *J. Physiol. (London)* **439**, 513–528.
Mahaut-Smith, M. P., and Schlichter, L. C. (1989). Ca^{2+}-activated K^+ channels in human B lymphocytes and rat thymocytes. *J. Physiol. (London)* **415**, 69–83.
Maldonado, D., Schumann, M., Nghiem, P., Dong, Y., and Gardner, P. (1991). Prostaglandin E_1 activates a chloride current in Jurkat T lymphocytes via cAMP-dependent protein kinase. *FASEB J.* **5**, 2965–2970.
Mannella, C. A., Forte, M., and Colombini, M. (1992). Toward the molecular structure of the mitochondrial channel, VDAC. *J. Bioenerg. Biomembr.* **24**, 7–19.
Matteson, D. R., and Deutsch, C. (1984). K^+ channels in T lymphocytes: A patch-clamp study using monoclonal antibody adhesion. *Nature (London)* **307**, 468–471.
Matthews, G., Neher, E., and Penner, R. (1989). Chloride conductance activated by external agonists and internal messengers in rat peritoneal mast cells. *J. Physiol. (London)* **418**, 131–144.
McCann, F. V., McCarthy, D. C., Keller, T. M., and Noelle, R. J. (1989). Characterization of a large conductance non-selective anion channel in B lymphocytes. *Cell. Signal.* **1**, 31–44.
McCann, J. D., Li, M., and Welsh, M. J. (1989). Identification and regulation of whole cell chloride currents in airway epithelium. *J. Gen. Physiol.* **94**, 1015–1036.
McDonald, T. V., Nghiem, P. T., Gardner, P., and Martens, C. L. (1992). Human lymphocytes transcribe the cystic fibrosis transmembrane conductance regulator gene and exhibit CF-defective cAMP-regulated chloride current. *J. Biol. Chem.* **267**, 3242–3248.
McDonald, T. V., Premack, B. A., and Gardner, P. (1993). Flash photolysis of caged inositol 1,4,5-trisphosphate activates plasma membrane calcium current in human T cells. *J. Biol. Chem.* **268**, 3889–3896.
Negulescu, P., and Cahalan, M. D. (1994). Intracellular calcium dependence of gene expression in single T lymphocytes. *Proc. Natl. Acad. Sci. U.S.A.* **91**, 2873–2877.
Nishimoto, I., Wagner, J. A., Schulman, H., and Gardner, P. (1991). Regulation of Cl^- channels by multifunctional CaM kinase. *Neuron* **6**, 1–20.

Pahapill, P. A., and Schlichter, L. C. (1992). Cl^- channels in intact T lymphocytes. *J. Membr. Biol.* **125,** 171–183.

Partiseti, M., Choquet, D., Diu, A., and Korn, H. (1992). Differential regulation of voltage- and calcium-activated potassium channels in human B lymphocytes. *J. Immunol.* **148,** 3361–3368.

Partiseti, M., Korn, H., and Choquet, D. (1993). Pattern of potassium channel expression in proliferating B lymphocytes depends upon the mode of activation. *J. Immunol.* **151,** 2462–2470.

Penner, R., Matthews, G., and Neher, E. (1988). Regulation of calcium influx by second messengers in rat mast cells. *Nature (London)* **334,** 499–504.

Premack, B. A., and Gardner, P. (1991). Role of ion channels in lymphocytes. *J. Clin. Immunol.* **11,** 225–238.

Prochazka, G., Landon, C., and Dennert, G. (1988). Transmembrane chloride flux is required for target cell lysis but not for Golgi reorientation in cloned cytolytic effector cells. *J. Immunol.* **141,** 1288–1294.

Rosoff, P. M., Hall, C., Gramates, L. S., and Terlecky, S. R. (1988). 4,4′-diisothiocyanatostilbene-2,2′-disulfonic acid inhibits CD3-T cell antigen receptor-stimulated Ca^{2+} influx in human T lymphocytes. *J. Biol. Chem.* **263,** 19535–19540.

Ross, P., Garber, S., and Cahalan, M. D. (1994). Membrane chloride conductance and capacitance in Jurkat T lymphocytes during osmotic swelling. *Biophys. J.* **66,** 169–178.

Sands, S. B., Lewis, R. S., and Cahalan, M. D. (1989). Charybdotoxin blocks voltage-gated K^+ channels in human and murine T lymphocytes. *J. Gen. Physiol.* **93,** 1061–1074.

Sarkadi, B., Mack, E., and Rothstein, A. (1984a). Ionic events during the volume response of human peripheral blood lymphocytes to hypotonic media. I. Distinctions between volume-activated Cl^- and K^+ conductance pathways. *J. Gen. Physiol.* **83,** 497–512.

Sarkadi, B., Mack, E., and Rothstein, A. (1984b). Ionic events during the volume response of human peripheral blood lymphocytes to hypotonic media. II. Volume- and time-dependent activation and inactivation of ion transport pathways. *J. Gen. Physiol.* **83,** 513–527.

Sarkadi, B., Cheung, R., Mack, E., Grinstein, S., Gelfand, E. W., and Rothstein, A. (1985). Cation and anion transport pathways in volume regulatory response of human lymphocytes to hyposmotic media. *Am. J. Physiol.* **248,** C480–C487.

Schlichter, L. C., Grygorczyk, R., Pahapill, P. A., and Grygorczyk, C. (1990). A large, multiple-conductance chloride channel in normal human T lymphocytes. *Pfluegers Arch.* **416,** 413–421.

Schulz, G. E. (1993). Bacterial porins: Structure and function. *Curr. Opin. Cell Biol.* **5,** 701–707.

Schwarze, W., and Kolb, H.-A. (1984). Voltage-dependent kinetics of an anionic channel of large unit conductance in macrophages and myotube membranes. *Pfluegers Arch.* **402,** 281–291.

Solc, C. K., and Wine, J. J. (1991). Swelling-induced and depolarization-induced Cl^--channels in normal and cystic fibrosis epithelial cells. *Am. J. Physiol.* **261,** C658–C674.

Stoddard, J. S., Steinbach, J. H., and Simchowitz, L. (1993). Whole cell Cl^--currents in human neutrophils induced by cell swelling. *Am. J. Physiol.* **265,** C156–C165.

Thiemann, A., Grunder, S., Pusch, M., and Jentsch, T. J. (1992). A chloride channel widely expressed in epithelial and non-epithelial cells. *Nature (London)* **356,** 57–60.

Thinnes, F. P. (1992). Evidence for extra-mitochondrial localization of the VDAC/Porin channel in eucaryotic cells. *J. Bioenerg. Biomembr.* **24,** 71–75.

Tseng, G.-Y. (1992). Cell swelling increases membrane conductance of canine cardiac cells: Evidence for a volume-sensitive Cl^- channel. *Am. J. Physiol.* **262,** C1056–C1068.

Valverde, M. A., Diaz, M., and Sepulveda, F. V. (1992). Volume-regulated chloride channels associated with the human multidrug-resistance P-glycoprotein. *Nature (London)* **355,** 830–833.

Wilson, H. A., and Chused, T. M. (1985). Lymphocyte membrane potential and Ca^{2+}-sensitive potassium channels described by oxonal dye fluorescence measurements. *J. Cell. Physiol.* **125,** 72–81.

Worrell, R. T., Butt, A. G., Cliff, W. H., and Frizzell, R. A. (1989). A volume-sensitive chloride conductance in human colonic cell line T84. *Am. J. Physiol.* **256,** C1111–C1119.

Zweifach, A., and Lewis, R. S. (1993). Mitogen-regulated Ca^{2+} current of T lymphocytes is activated by depletion of intracellular Ca^{2+} stores. *Proc. Natl. Acad. Sci. U.S.A.* **90,** 6295–6299.

CHAPTER 6

Chloride Channels in Skeletal Muscle and Cerebral Cortical Neurons

Andrew L. Blatz
Department of Physiology, University of Texas Southwestern Medical Center at Dallas, Dallas, Texas 75235

I. INTRODUCTION

Ion channels selective for chloride ions (Cl) have been found in a variety of mammalian tissues, including muscle, nerve, endocrine and exocrine gland, and epithelium. It is probably not an exaggeration to say that, in fact, all cells contain Cl channels. Because of technical difficulties, the study of Cl channels lagged behind that of the cation-selective ion channels, but with the advent of the patch voltage-clamp technique many of these technical problems have been overcome and the Cl channel field has been advancing steadily. In this chapter, I discuss the properties of Cl channels that have been identified at the single-channel level in cells from two excitable mammalian tissues, skeletal muscle and cerebral cortical neurons. For an excellent, comprehensive review of vertebrate and invertebrate muscle Cl conductances see Bretag (1987).

II. SKELETAL MUSCLE CHLORIDE CHANNELS

Vertebrate skeletal muscle plasma membranes are highly permeable to Cl^- (Hodgkin and Horowicz, 1959; Hutter and Noble, 1960; Palade and Barchi, 1977). If this large anion permeability is pharmacologically or genetically reduced, the muscles exhibit hyperexcitability and myotonic-like trains of action potentials (Adrian and Bryant, 1974). This suggests that muscle Cl channels may serve to reduce excitability to a level that maintains the normal one-to-one relationship between nerve impulses and contraction of most skeletal muscle. Although microelectrode voltage-clamp techniques have provided some information about the properties of the ion channels underlying this large muscle anion conductance, the electrophysiological properties of the channels underlying the large resting membrane Cl conductance have not yet been convincingly demonstrated at the single-channel level. During the search for the channels that give rise to the large resting Cl conductance, three Cl channels were demonstrated to be present in muscle-surface membranes; two channels with single-channel conductances of 40–60 pS (with 140 [KCl] on both membrane surfaces), which we called the "fast" and "slow" Cl channels; and one channel, called the "megachannel," with the very large single-channel conductance of about 400 pS under similar conditions (Blatz and Magleby, 1983,1985).

A. Slow and Fast Cl Channels in Skeletal Muscle.

Two Cl-selective ion channels were discovered in tissue-cultured rat skeletal muscle that potentially could contribute to the resting Cl conductance (Blatz and Magleby, 1985). Both of these channels were selective for Cl and could be active at normal negative membrane potentials. Single-channel current traces, obtained with the patch-clamp technique (Hamill *et al.*, 1981), for both of these channels are shown in Fig. 1. One channel, called the "slow Cl channel" exhibited a single-channel conductance of about 60 pS in the presence of symmetrical (i.e., $[KCl]_o = [KCl]_i$) 100 m*M* KCl, and mean open lifetimes on the order of 10 msec. These slow Cl channels were present in only about 2% of excised, inside-out membrane patches and so their properties were not studied in great detail. The "fast Cl channel," on the other hand, was found to be present in

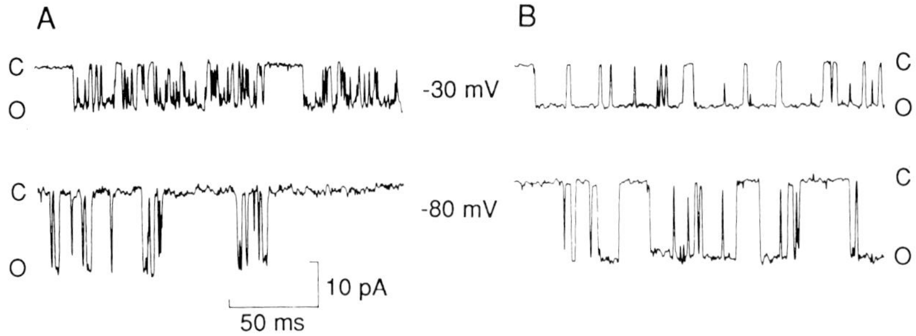

FIGURE 1 Fast and slow Cl channels recorded from tissue-cultured rat skeletal muscle using the patch-clamp technique. Single-channel currents were recorded from excised inside-out patches of muscle-surface membrane. (A) Fast Cl channels recorded at holding potentials of -30 mV (top) and -80 mV (bottom) in the presence of 1.6 *M* $[KCl]_i$ and 140 m*M* $[KCl]_o$. (B) Slow Cl channels recorded at the same membrane potentials in the presence of 1.4 m*M* $[KCl]_i$ and 100 m*M* $[KCl]_o$. All traces were low-pass filtered at 1.6 kHz (-3 dB). (Reproduced by copyright permission of the Biophysical Society, Blatz and Magleby, 1985).

most inside-out patches studied. These channels exhibited unitary conductances of about 45 pS in the presence of symmetrical 100 m*M* KCl and mean open lifetimes on the order of 1 msec.

1. Ion Selectivity

Both the fast and the slow Cl channels were selective for Cl over cations, but the selectivity was much less than perfect, as shown for the slow Cl channel in Fig. 2. For this experiment the shift in reversal potential for a slow Cl channel was measured as the solution bathing the formerly intracellular membrane surface was changed from 100 m*M* KCl (which was present at the formerly extracellular membrane surface throughout) to 1.4 *M* KCl. As expected, under symmetrical conditions, single-channel currents reversed at 0 mV (not shown). As the [KCl] on the intracellular membrane surface was raised, the reversal potential shifted in a depolarizing direction such that, in the presence of 1.4 *M* KCl, the single-channel currents shifted from inward to outward at about +35 mV, as expected for a channel that is selective for anions over cations. The magnitude of the observed reversal potential shift is much less than would be expected for perfect discrimination between anions and cations. The Nernst equation predicts that, under the conditions in Fig. 2, currents through a perfectly anion-selective channel would reverse at a potential of +60 mV. The Goldman–Hodgkin–Katz voltage equation can be used to calculate a permeability ratio of K permeability to Cl permeability for the slow Cl channel of 0.2, indicating that slow Cl channels are actually fairly permeable to K as well as to Cl. That the nonideal shift in reversal potential

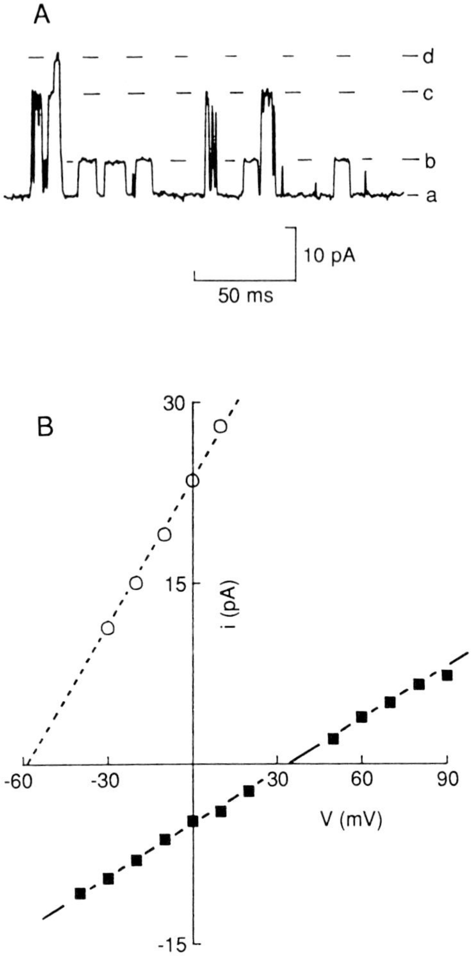

FIGURE 2 Cation permeability of the slow Cl channel from rat skeletal muscle. (A) Patch-clamp currents from an excised inside-out membrane patch containing a slow Cl channel (open channel level b) and a large-conductance Ca-activated K channel (open channel level c). Level a indicates current level when both channels are closed. Level d indicates current magnitude when both the slow Cl channel and the BK channel are open. Bathing solutions: 1.4 M $[KCl]_i$, 100 mM $[KCl]_o$. Holding potential of −10 mV. (B) Current–voltage plot for channels in (A). Open symbols represent current–voltage relationship of BK channels recorded in the same patch. Filled squares represent data from the slow Cl channel. BK channel extrapolated reversal potential of −60 mV, while the slow Cl channels reversed direction at about +30 mV. (Reproduced by copyright permission of the Biophysical Society, Blatz and Magleby, 1985).

was not due to errors in solution composition, voltage measurements, or drift was confirmed by the −60 mV shift in reversal potential for a large-conductance Ca-activated K channel (BK channel) that was also present in the same membrane patch. Skeletal muscle BK channels are known to be perfectly selective for cations over anions (Blatz and Magleby, 1984), and so the observed ideal −60-mV reversal potential shift could be used as a valuable control in this experiment. Similar results were obtained using NaCl, suggesting that slow Cl channels do not discriminate well between monovalent cations. The relative permeability of the fast Cl channels to anions over cations was also found to be about 0.8 (Blatz and Magleby, 1985). Other "Cl channels" seem to be measurably permeant, as measured by reversal potential shifts, to cations (Franciolini and Nonner, 1987). If these Cl channels are 20% permeable to cations then one would expect to be able to record single-channel currents carried exclusively by cations but this turns out not to be the case. No currents through fast or slow Cl channels are observed when Cl is replaced by impermeant anions (Blatz and Magleby, 1985, and unpublished observations). Franciolini and Nonner (1987) proposed a molecular solution for this phenomenon in a neuronal Cl channel which invokes a cation binding site located within the channel pore. When a cation is bound to this site, the channel allows the flux of anions, but when the bound cation leaves the channel it does so as an ion pair coupled to an anion and therefore carries no current. Permeability measurements using reversal potential shifts would still indicate a significant cation permeability, even though conductance measurements would demonstrate no current carried by the cation.

2. Single-Channel Kinetics

Before the advent of single-channel recording techniques like patch clamp and lipid bilayer methods, it could only be speculated that ion channels open and close as rectangular pulses of current. Only when these modern electrophysiological approaches arrived was it actually possible to confirm these speculations. Analysis of the durations and amplitudes of these pulses of currents gives us a valuable window into the underlying conformational rearrangements undergone by ion channel proteins.

One of the Cl channels in skeletal muscle, the fast Cl channel, has proven to be a valuable resource for the study of ion channel kinetic behavior. These channels exhibit many features that are desirable for kinetic studies. The amplitude of the single-channel currents is large and can be increased further by making measurements in the presence of large Cl gradients and high Cl concentrations (Blatz and Magleby, 1985, 1986b). The channels exhibit few excursions into subconducting states, very much unlike another popular Cl-selective channel, the γ-amino butyric acid-activated Cl channel (GABA channel). Fast Cl channels can remain active for extended periods without the washout, rundown, or desensitization problems that plague other channels. These properties make the fast Cl channel an ideal model for the study of ion channel kinetics.

a. Unconditional Dwell Time Distributions. As mentioned above, information about the molecular reaction scheme underlying single fast Cl channel kinetic behavior can be obtained from the durations of the open and closed channel intervals. The first attempt to characterize the kinetic behavior of these channels in terms of underlying reaction schemes used the unconditional distributions of these interval durations (Blatz and Magleby, 1986b). Unconditional distributions are composed of the open and closed interval durations, without regard to their temporal position in the experimental record. In other words, unconditional distributions do not take into account the durations of adjacent open and closed intervals, but assume that the open and closed interval durations are truly not correlated in time. The framework for the analysis of unconditional open and closed interval durations is provided by Colquhoun and Hawkes (1983) and Colquhoun and Sigworth (1983). Using the computational techniques described by these authors as modified by Blatz and Magleby (1985,1986a) to account for the effects of missed events due to filtering, it is possible to determine which kinetic reaction schemes are most likely able to fit the experimental distributions. Using these unconditional distributions, two of which are shown in Fig. 3, Blatz and Magleby (1986b) found that, in fact, a variety of kinetic models were able to fit the experimentally observed distributions and that the unconditional distributions could not identify the best one. Although some models were better fits to the experimental data than were others, an unequivocal determination of the absolute best model was impossible based on analysis of the unconditional distributions. Although the assignment of the best kinetic scheme was not possible, Blatz and Magleby (1986b) were able to determine that, whatever model was correct, it must contain at least two open kinetic states and at least five closed kinetic states. It was the connections between these seven states that was unclear.

b. Conditional Distributions: Adjacent State Analysis. Much more information is contained in the open and closed interval durations then simply the individual durations. The temporal sequence of these open and closed events contains this extra information. When the sequence of interval durations is preserved, the resultant distributions are called conditional distributions. Thus, with conditional distributions one can look at all open intervals that occurred adjacent to short-duration closed

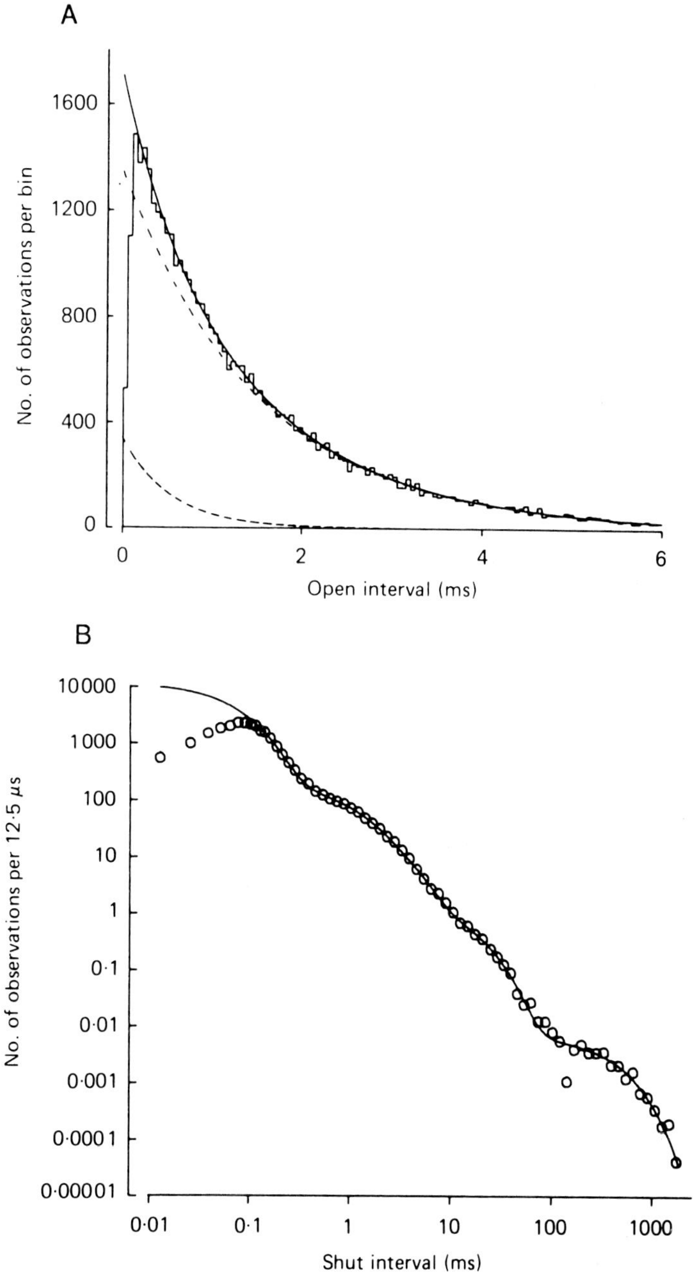

FIGURE 3 Distributions of open (A) and closed (B) interval durations for a single fast Cl channel from tissue-cultured rat skeletal muscle. Holding potential was −40 mV. Bathing solutions were 1*M* $[KCl]_i$, 140 m*M* $[KCl]_o$. Low-pass filtering of 3.2 kHz. Solid line is the best fit sum of two (open) or five (closed) exponential components. (From Blatz and Magleby, 1986b.)

intervals, open intervals next to intermediate-duration closed intervals, the durations of open intervals that occurred next to long closed events, etc. In order to maintain a sufficient number of events in each distribution it is necessary to compare, for example, open intervals that occurred next to *ranges* of closed event durations rather than look at open intervals next to closed events of a specific duration. Using these conditional distributions, the "connectivity" of the underlying kinetic states can be assessed, as well as the absolute numbers of states. Unconditional distributions can be analyzed in a manner such that many fast Cl channel models that are acceptable based on the unconditional distributions can be ruled out based on the conditional distributions. Blatz and Magleby (1989) divided fast Cl channel shut interval durations into several ranges, from those with durations of the order of less than a millisecond, to those with durations of hundreds of milliseconds and analyzed the distributions of open intervals that occurred adjacent to these closed interval ranges. The experimentally observed result is shown in Fig. 4. An "adjacent state" relationship clearly exists between the relative proportion of short and long open interval durations and the durations of adjacent shut intervals. This is manifested as an increase in the relative area of the kinetic open state with the shorter mean lifetime as the mean duration of adjacent closed intervals increases. Thus, the relationship is an inverse one, such that open intervals with shorter mean durations tend to occur next to closed intervals with longer mean durations, and, conversely, open intervals with longer durations tend to occur adjacent to closed intervals with shorter durations. This observation imposes some serious constraints on the interconnections of the seven kinetic states already known to exist for these channels: (1) there must be at least two direct "gateway" transition pathways between the open and closed channel kinetic states, and (2) the open kinetic states with short mean lifetimes must be connected to the closed kinetic states with longer mean lifetimes, and the open kinetic states with longer mean lifetimes must be directly connected with the closed kinetic states with shorter mean lifetimes. Two seven-state kinetic models that fit the unconditional distributions equally well are

$$C5 \rightleftharpoons C4 \rightleftharpoons C3 \rightleftharpoons C2 \rightleftharpoons C1 \rightleftharpoons O1 \rightleftharpoons O2 \qquad \text{(I)}$$

$$\begin{array}{ccccccccc} C5 & \rightleftharpoons & C4 & \rightleftharpoons & C3 & \rightleftharpoons & C2 & \rightleftharpoons & C1 \\ & & & & & & \upharpoonleft\downharpoonright & & \upharpoonleft\downharpoonright \\ & & & & & & O2 & & O1 \end{array} \qquad \text{(II)}$$

Scheme I contains a single gateway transition pathway (between closed state 1 and open state 1) while Scheme II contains two gateway transition pathways (between closed state 1 and open state 1, and between closed state 2 and open state 2). The adjacent state analysis result predicted for

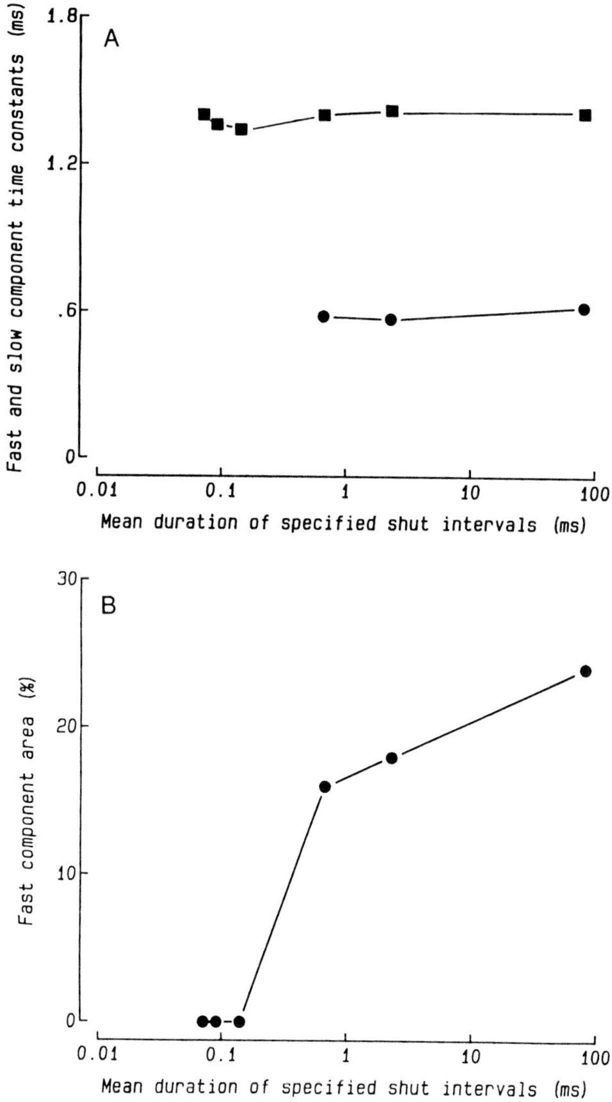

FIGURE 4 Adjacent state analysis of conditional one-dimensional distributions of open intervals occurring adjacent to closed intervals of specified durations. (A) Time constants of the fast (filled circles) and slow (filled squares) components of the distributions of muscle fast Cl channel open intervals occurring next to closed interval durations specified on the *x*-axis. (B) Relative areas of the fast component of the distribution of open intervals occurring next to the closed interval durations specified by the *x*-axis. Since the unconditional distribution of all open intervals can be fit by the sum of two exponentials there is no need to plot the area of the slow component as it is equal to one minus the area of the fast component. (From Blatz and Magleby, 1989.)

these two models is shown in Fig. 5. It is clear that the linear model (Scheme I) cannot account for the observed inverse relationship between the durations of the open and closed channel kinetic states.

c. Two-Dimensional Distributions. In reality, the analysis of the conditional distributions of adjacent open and closed interval durations is a special case of a much more general form of distribution analysis that has more recently been applied to the muscle fast Cl channel. The general form of the adjacent state analysis is called two-dimensional distribution analysis. Two-dimensional distribution analysis of the durations of open and closed interval durations allows the use of all of the kinetic information contained in the open and closed interval durations. A drawback to the use of two-dimensional distribution analysis is the difficulty in compensating for the effects of limited time resolution due to filtering. With one-dimensional unconditional distributions, several methods are available to partially compensate for the corrupting effects of filtering (Wilson and Brown, 1985; Roux and Sauve, 1985; Blatz and Magleby, 1986a). Magleby and Weiss (1990) developed a sophisticated, computationally intense method of analyzing two-dimensional interval duration distributions that utilizes all of the kinetic information contained in the open and closed interval durations and completely accounts for the effects, not only of missed events due to filtering, but of noise as well. The method developed by Magleby and Weiss (1990) uses brute-force simulation of sequences of open and closed channel events from a given kinetic reaction scheme (such as Schemes I and II) followed by digital filtering and the addition of actual patch-clamp noise. The simulated data are then compared, using the method of maximum likelihood, to the experimentally observed two-dimensional distributions of open and closed interval durations. The parameters of the fit, which are the transition rate constants between the kinetic states, are adjusted and the simulation process is repeated until the best fitting set of rate constants for a particular reaction scheme are obtained. Other combinations of open and closed kinetic states are then compared and, after compensating for different numbers of parameters, the reaction scheme with the greatest likelihood is determined. This method would have been somewhat impractical a few years ago, but now, with the advent of faster (and cheaper) microcomputers, the simulation approach is the preferred approach for determining the kinetic reaction scheme underlying single-channel activity.

3. Voltage Dependence of Muscle Fast Cl^- Channel Kinetics

The currents through most, if not all, ion channels are dependent on the magnitude and sign of the membrane electrical field. The magnitude and direction of the ionic current flowing through all ion channels is, of

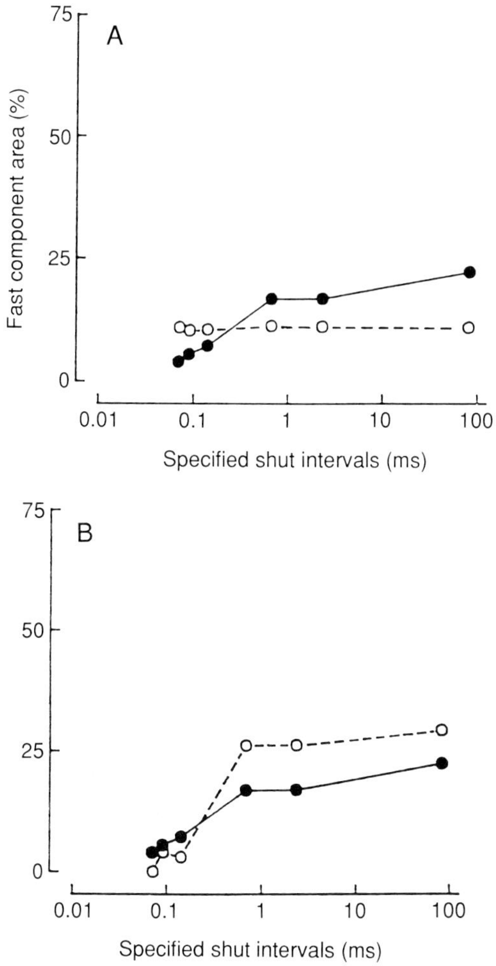

FIGURE 5 Adjacent state predictions of Schemes I (A) and II (B). In both (A) and (B) the experimentally observed relationship between the relative area of the muscle fast Cl channel kinetic state of short mean lifetime is plotted against the mean lifetime of the specified ranges of adjacent closed interval durations (filled circles). Open circles represent the results of an analysis of data simulated using Scheme I (A) or Scheme II (B). From Blatz and Magleby (1989).

course, related to the driving force on the permeant ions with a driving force equal to the membrane potential minus the electrochemical equilibrium potential given by the Nernst equation. The magnitude of ionic current is also dependent on the kinetic activity of the particular ion channel. Given equal driving forces, the net ionic current is proportional to the percentage of time that an ion channel spends in the open conformation. The percentage of time that a channel is open is calculated simply as 100 times the amount of time that the channel is open divided by the total observation time. A more commonly used parameter is called the open probability which is simply the fraction of the total observed time that the channel is in the open state. When a membrane patch contains a single ion channel, the open probability is the percentage of time open divided by 100. When more than one channel is present, the open probability can be estimated from the relative time that zero, one, two, three, or more channels are simultaneously open (Barrett *et al.*, 1982).

Fast Cl channels from tissue-cultured rat skeletal muscle exhibit a voltage dependence such that the open probability is low at hyperpolarized membrane potentials and increases as the membrane potential is made more positive (Fig. 6A). The major effect of depolarization is to decrease the mean duration of shut intervals, while the mean open duration increases to a lesser extent (Fig. 6B). From the steady-state voltage dependence of muscle fast Cl channel kinetics, Weiss and Magleby (1990) calculated that a 17 ($\pm$4)-mV depolarization produced an *e*-fold increase in open probability which is equivalent to an "effective gating charge" (Hodgkin and Huxley, 1952; Hille, 1984) of 1.6 ± 0.32. This voltage dependence falls between the extremely strong voltage dependence of action potential-generating channels such as the rat brain sodium channels (Keller *et al.*, 1986) and the squid axon delayed rectifier K channel (Behrens *et al.*, 1989) and the less voltage-dependent agonist-activated channels such as the GABA-activated Cl channel (Weiss, 1988) and the neuromuscular junction acetylcholine-activated cation channel (Magleby and Stevens, 1972).

B. Large-Conductance Cl Channels in Skeletal Muscle

The third Cl channel that was found to be present in the surface membranes of tissue-cultured skeletal muscle exhibited the extremely large single-channel conductance of over 400-pS in symmetrical 140 m*M* KCl (Blatz and Magleby, 1983). These channels, which we called "megachannels" have some properties which are similar to those of the voltage-dependent anion channels (VDAC) found in mitochondria outer membranes (Colombini, 1979). These channels are not restricted to muscle tissue, but have been found in such tissues as rabbit urinary bladder (Hanrahan *et al.*, 1985), mouse macrophage and chicken myotubes (Schwarze and Kolb, 1984), pulmonary epithelia (Schneider *et al.*, 1985), rat Schwann cells (Gray *et al.*, 1984), and rat mast cells (Lindau and Fernandez, 1986). At the time of their discovery the similarity between megachannels and mitochondrial VDAC prompted the speculation that

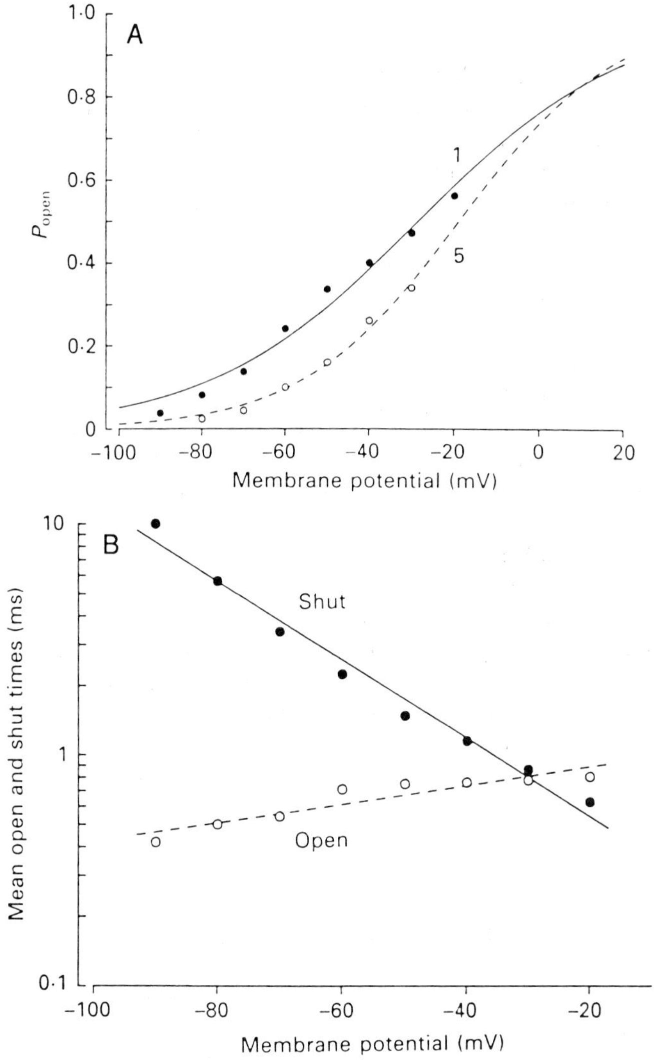

FIGURE 6 Voltage dependence of the fast Cl channel from rat skeletal muscle. (A) The open probability verses holding potential for two fast Cl channels recorded from two separate excised inside-out membrane patches. Different symbols represent the different channels. Solid and dashed lines are the best fits of a Boltzmann distribution to the voltage dependence of the two channels considered separately. (B) Dependence of mean open and closed interval durations on holding potential. The mean open (open circles) and mean closed (filled circles) interval durations were plotted against membrane potential for a single fast Cl channel [channel No. 1 in (A)]. The majority of the voltage dependence of open probability can be explained by the decrease in the mean shut duration. Solid and dashed lines are linear regressions to the logarithm of the mean durations. (From Weiss and Magleby, 1990.)

the megachannels recorded with the patch-clamp technique were actually "lost" channels that had originated from mitochondria and had somehow wandered over to the plasma membrane and had been incorporated into the sarcolemma. Localization to the plasma membrane and the sarcoplasmic reticulum of large-conductance Cl channels has been confirmed using immunohistochemical methods (Babel *et al.*, 1991). The properties of these channels have been described in detail elsewhere (Blatz, 1990) and are only briefly discussed here.

1. Ion Selectivity

Similar to the fast and slow Cl channels in skeletal muscle, muscle megachannels exhibit a Cl to K or Na permeability ratio of about 4:6 (Blatz and Magleby, 1983). A systematic analysis of the relative permeabilities has not been performed on muscle megachannels.

2. Voltage Dependence

Megachannels exhibit a characteristic voltage-dependent kinetic activity. With no potential difference across the membrane, megachannels are predominantly in the open state. When the membrane potential is stepped from 0 mV to either a positive or negative value, the open probability of these channels decreases with a time constant of about 1 sec to extremely small values (Fig. 7). It is this odd voltage dependence that has led to questions about the physiological relevance of the megachannel, since, under normal physiological conditions, one would expect the channel to remain closed.

III. NON-TRANSMITTER-ACTIVATED NEURONAL Cl CHANNELS

The surface membranes of neurons contain a variety of Cl-selective ion channels. By far the most widely studied have been those channels activated by GABA (for review, see Barker *et al.*, 1990). Non-transmitter-activated Cl-selective channels have been reported in only a few types of neurons such as tissue-cultured rat hippocampus (Franciolini and Nonner, 1987; Franciolini and Petris, 1988), spinal cord (Hughes *et al.*, 1987), and acutely dissociated rat cerebral cortex (Blatz, 1991).

A. Ion Selectivity

The most complete investigation of the permeability properties of neuronal Cl-selective ion channels was performed on tissue-cultured rat hippocampus neurons (Franciolini and Nonner, 1987). In this study it was found

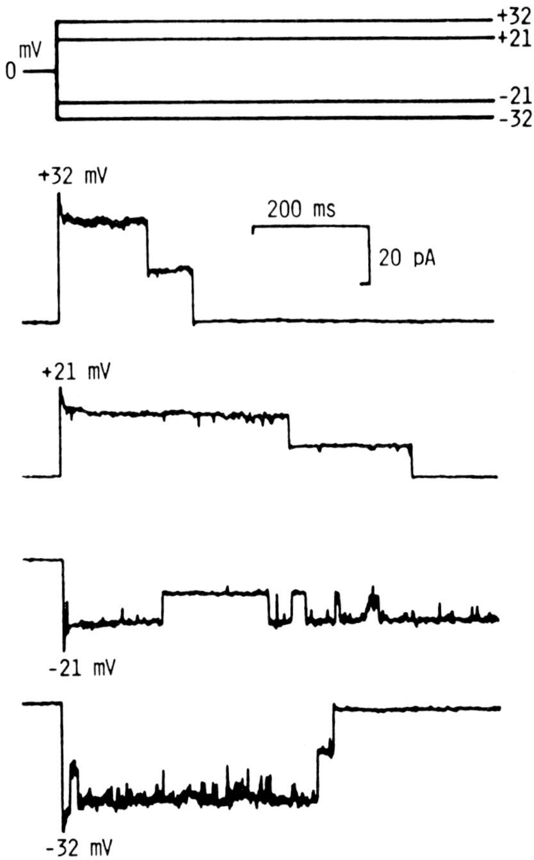

FIGURE 7 Large-conductance Cl channels (megachannels) recorded from inside-out membrane patches from rat skeletal muscle. Membrane patch was held at 0 mV and then stepped to the voltage indicated next to each trace. Internal and external bathing solutions contained 143 m*M* KCl. The initial transient current decline seen in each record during the first few milliseconds is a capacitive artifact. (Reproduced by copyright permission from the Biophysical Society, Blatz and Magleby, 1983).

that a variety of monovalent anions were permeable, as were monovalent cations such as K and Na. Given the relatively high permeability ratio for cation permeation through neuronal Cl channels it might be expected that ion currents carried by these permeant cations should be measured. This has not been the case. When currents are measured in the presence of asymmetric or symmetric permeant cations in the absence of permeant anions, no currents can be detected. This suggests that the permeation pathway of neuronal Cl channels is complex and may involve interactions between the channel pore and ions and between the ions themselves inside of the pore. Franciolini and Nonner (1987) proposed that cations binding to a site located within the membrane electrical field give rise to the measured cation permeability. When the cations leave the channel pore, they do so coupled to an anion, and therefore no net current would be measured.

B. Single-Channel Kinetics

Cl channels with properties very similar to those of the fast Cl channels found in tissue-cultured skeletal muscle have been reported in acutely dissociated rat cerebral cortical neurons (Blatz, 1991). The ion selectivity, voltage dependence, and kinetic activity of these neuronal fast Cl channels are nearly identical to their muscle counterparts. The stability of the gating of neuronal fast Cl channels makes them ideal objects for the study of single-channel kinetics. When dwell-time analysis was performed on neuronal fast Cl channels, it was found that the unconditional distributions of open and closed interval durations could be best fit by sums of two and six exponential components, respectively, suggesting that, during normal activity, these channels readily enter at least two open kinetic states and at least six closed states. The various modes of gating behavior that are characteristic of muscle fast Cl channels (Blatz and Magleby, 1986b) are also present in the neuronal channels (Blatz, 1991).

C. Block by Quaternary Ammonium Ions

Quaternary ammonium ions (QAs) have been used as probes of ion channels selective for K for many years (for review, see Stanfield, 1983). Even in light of the well-known nonselectivity of QA among cation channels, it is often reported that, since a particular process is reduced by QAs, then that process must depend on K channels. It has been learned that some Cl channels are also blocked by QAs (Lukás and Moczydlowski, 1990; Sanchez and Blatz, 1992), making reliance on QA block even more suspect.

The QA used most commonly in electrophysiology is tetraethylammonium ion (TEA). This compound blocks many K channels in the micro- to millimolar range (Stanfield, 1983). In many tissues, the TEA binding site is thought to be located inside the channel pores, within the electrical field of the membrane. Extracellularly applied TEA (TEA_o) blocks neuronal fast Cl channels in a similar fashion (Sanchez and Blatz, 1992). Figure 8 shows the reduction of fast Cl channel currents recorded at

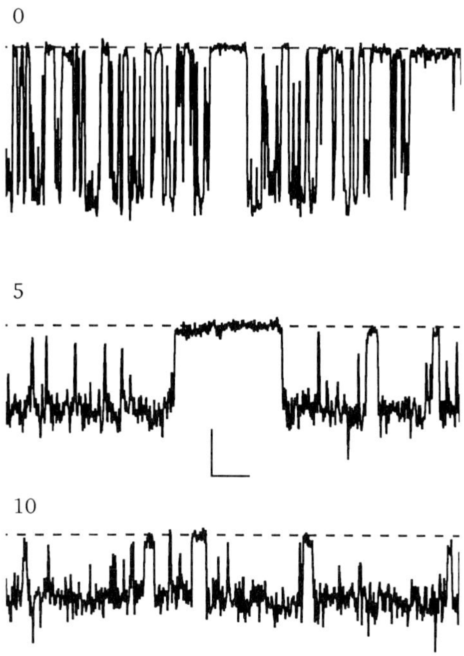

FIGURE 8 Fast Cl channel block by external TEA. Single Cl channel currents from three separate excised inside-out membrane patches from acutely dissociated rat cortical neurons. Concentrations of external TEA are indicated above each trace. Low-pass filtered at 1.4 kHz. Membrane potential of −40 mV. Horizontal and vertical calibrations are 10 msec and 3.0 pA, respectively. (Reproduced by copyright permission from Rockefeller University Press, Sanchez and Blatz, 1992).

−40 mV in the presence of 0, 5, and 10 m*M* TEA_o. TEA block appears as a reduction in single-channel current suggesting that TEA binds and unbinds to its blocking site with kinetics that are much more rapid than can be resolved in these experiments (6 kHz). That the blocking kinetics must be much more rapid than is resolvable is suggested by the fact that the variance of the open channel current is not increased by concentrations of TEA_o that reduce single-channel currents by 50% (Sanchez and Blatz, 1992). The dose–response relationship between the fractional current and [TEA_o] can be adequately fit assuming that two TEA molecules bind to the channel. Further analysis of the dose–response relationship revealed that it was impossible to differentiate between a blocking reaction involving two TEA molecules and a unimolecular reaction where a single TEA binding to the channel causes a reduction of current to 20% of the unblocked magnitude. The block of K channels by TEA is often dependent on membrane potential (Stanfield, 1938). This is also true for the TEA block of neuronal fast Cl channels as is shown in Fig. 9. Only the block by TEA_o is voltage dependent under the conditions of these experiments suggesting that the sites for intracellular and extracellular TEA block are different. Analysis of the voltage dependence of TEA_o block suggests that the blocking site is located within the electrical field of the membrane much as is imagined for K channel block.

Increasing the hydrocarbon chains of symmetric QAs often leads to altered K channel blocking kinetics. Similarly, tetrapropyl (TPA) and tetrabutyl (TBA)-ammonium ions block neuronal fast Cl channels, at significantly lower concentrations than those required with TEA (Sanchez and Blatz, 1991). In the presence of TPA, fast Cl channel currents exhibit significantly reduced mean open interval durations while the open current level remains at the control value. Using the methods that were used to characterize the normal gating properties of Cl channels we are currently

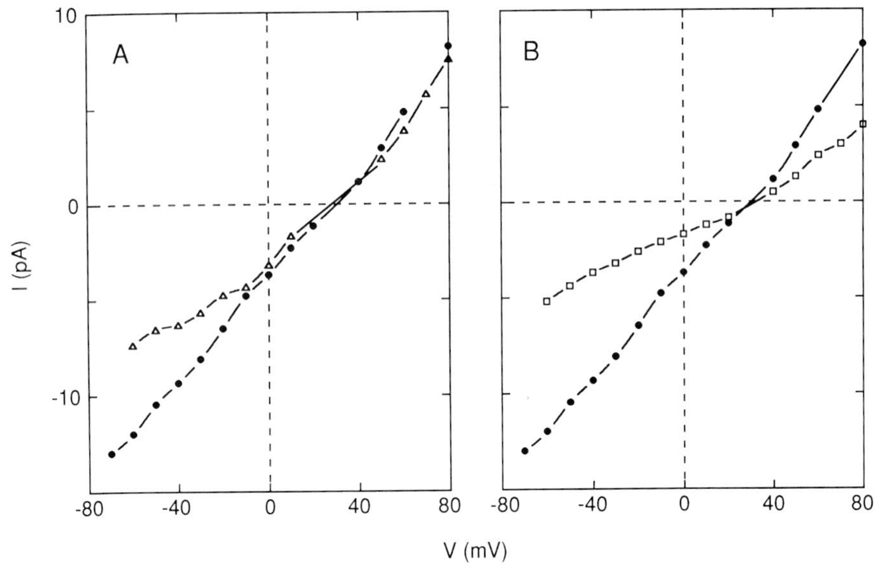

FIGURE 9 Voltage dependence of neuronal fast Cl channel block by TEA_o (A) and TEA_i (B). Steady-state current–voltage relationships were obtained from patch-clamp records by plotting open channel current level against holding potential. Filled circles represent control data and open symbols represent currents recorded in the presence of 10 m*M* $[TEA]_o$ (A) or 25 m*M* $[TEA]_i$ (B). Data shown in (A) were obtained from two membrane patches. Currents in (B) were recorded from a single patch. Bathing solutions for (A) and (B) were 1 *M* $[KCl]_i$, 140 m*M* $[Kcl]_o$. (Reproduced by permission from Rockefeller University Press, Sanchez and Blatz, 1992.)

attempting to determine the kinetic schemes underlying these blocking reactions.

IV. SUMMARY

The study of Cl channels is finally starting to approach the level of the study of the cation permeant channels. The advent of single-channel methods has led to the discovery of many different types of Cl channels in excitable tissues such as nerve and muscle. Several types of mammalian Cl channels have been cloned and their primary sequences determined. The way is now clear for the application of the techniques of molecular biology that have led to molecular models for cation channels to be applied to anion channels as well.

Acknowledgment

Supported by NIH Grant GM-39731.

References

Adrian, R. H., and Bryant, S. H. (1974). On the repetitive discharge in myotonic muscle fibres. *J. Physiol. (London)* **240,** 505–515.

Babel, D., Walter, G., Gotz, H., Thinnes, F. P., Jurgens, L., Konig, U., and Hilschmenn, N. (1991). Studies on human porin. VI. production and characterization of eight monoclonal mouse antibodies against the human VDAC "porin 31HL" and their application for histotopological studies in human skeletal muscle. *Biol. Chem. Hoppe-Seyler* **372,** 1027–1034.

Barker, J. L., Harrison, N. L., and Owen, D. G. (1990). Pharmacology and physiology of Cl^- conductances activated by GABA in cultured mammalian central neurons. *In* "Chloride Channels and Carriers in Nerve, Muscle, and Glial Cells" (F. J. Alvarez-Leefmans and J. M. Russell, eds.), pp. 273–298. Plenum, New York.

Barrett, J. N., Magleby, K. L., and Pallotta, B. S. (1982). Properties of single Ca-activated K channels in cultured rat muscle. *J. Physiol. (London)* **331,** 211–230.

Behrens, M. I., Oberhauser, A., Bezanilla, F., and Latorre, R. (1989). Batrachotoxin-modified sodium channels from squid optic nerve in planar bilayers. Ion conduction and gating properties. *J. Gen. Physiol.* **93,** 23–41.

Blatz, A. L. (1990). Chloride channels in skeletal muscle. *In* "Chloride Channels and Carriers in Nerve, Muscle, and Glial Cells" (F. J. Alvarez-Leefmans and J. M. Russell, eds.), pp. 407–420. Plenum, New York.

Blatz, A. L. (1991). Properties of single fast chloride channels from rat cerebral cortex neurons. *J. Physiol. (London)* **441,** 1–21.

Blatz, A. L., and Magleby, K. L. (1983). Voltage dependent Cl^- selective channels of large conductance in cultured rat muscle. *Biophys. J.* **43,** 237–241.

Blatz, A. L., and Magleby, K. L. (1984). Ion conductance and selectivity of single calcium-activated potassium channels in cultured rat muscle. *J. Gen. Physiol.* **84,** 1–23.

Blatz, A. L., and Magleby, K. L. (1985). Single chloride-selective channels active at resting membrane potentials in cultured rat skeletal muscle. *Biophys. J.* **47,** 119–123.

Blatz, A. L., and Magleby, K. L. (1986a). Correcting single channel data for missed events. *Biophys. J.* **49,** 967–980.
Blatz, A. L., and Magleby, K. L. (1986b). Quantitative description of three modes of activity of fast chloride channels from rat skeletal muscle. *J. Physiol. (London)* **378,** 141–174.
Blatz, A. L., and Magleby, K. L. (1989). Adjacent interval analysis distinguishes among gating mechanisms for the fast chloride channel from rat skeletal muscle. *J. Physiol. (London)* **410,** 561–585.
Bretag, A. H. (1987). Muscle chloride channels. *Physiol. Rev.* **67,** 618–724.
Colombini, M. (1979). A candidate for the permeability pathway of the outer mitochondrial membrane. *Nature (London)* **279,** 643–645.
Colquhoun, D., and Hawkes, A. G. (1983). The principles of stochastic interpretation of ion-channel mechanisms. *In* "Single Channel Recording" (B. Sakmann and E. Neher, eds.), pp. 135–176. Plenum, New York.
Colquhoun, D., and Sigworth, F. J. (1983). Fitting and statistical analysis of single-channel records. *In* "Single Channel Recording" (B. Sakmann and E. Neher, eds.), pp. 191–264. Plenum, New York.
Franciolini, F., and Nonner, W. (1987). Anion and cation permeability of a chloride channel in rat hippocampal neurons. *J. Gen. Physiol.* **90,** 453–478.
Franciolini, F., and Petris, A. (1988). Single chloride channels in cultured rat neurons. *Arch. Biochem. Biophys.* **261,** 97–102.
Gray, P. T. A., Bevan, S., and Ritchie, J. M. (1984). High conductance anion-selective channels in rat cultured Schwann cells. *Proc. R. Soc. London, Ser. B* **221,** 395–409.
Hamill, O. P., Marty, A., Neher, E., Sakmann, B., and Sigworth, F. J. (1981). Improved patch-clamp techniques for high-resolution current recording from cells and cell-free membrane patches. *Pfluegers Arch.* **391,** 85–100.
Hanrahan, J. W., Alles, W. P., and Lewis, S. A. (1985). Single anion-selective channels in basolateral membrane of a mammalian tight epithelium. *Proc. Natl. Acad. Sci. U.S.A.* **82,** 7791–7795.
Hille, B. (1984). "Ionic Channels of Excitable Membranes." Sinauer Assoc., Sunderland, MA.
Hodgkin, A. L., and Horowicz, P. (1959). The influence of potassium and chloride ions on the membrane potential of single muscle fibres. *J. Phyiol. (London)* **148,** 127–160.
Hodgkin, A. L., and Huxley, A. F. (1952). A quantitative description of membrane current and its application to conduction and excitation in nerve. *J. Physiol. (London)* **117,** 500–544.
Hughes, D., McBurney, R. N., Smith, S. M., and Zortec, R. (1987). Caesium ions activate chloride channels in rat cultured spinal cord neurones. *J. Phyiol. (London)* **392,** 231–251.
Hutter, O. F., and Noble, D. (1960). The chloride conductance of frog skeletal muscle. *J. Physiol. (London)* **151,** 89–102.
Keller, B. U., Hartshorne, R. P., Talvenheimo, J. A., Catterall, W. A., and Montal, M. (1986). Sodium channels in planar lipid bilayers: channel gating kinetics of purified sodium channels modified by Batrachotoxin. *J. Gen. Physiol.* **88,** 1–23.
Lindau, M., and Fernandez, J. M. (1986). A patch-clamp study of histamine-secreting cells. *J. Gen. Physiol.* **88,** 349–368.
Lukács, G. L., and Moczydlowski, E. (1990). A chloride channel from lobster walking leg nerves. Characterization of single-channel properties in planar bilayers. *J. Gen. Physiol.* **96,** 707–734.
Magleby, K. L., and Stevens, C. F. (1972). The effect of voltage on the time course of end-plate currents. *J. Physiol. (London)* **223,** 151–172.

Magleby, K. L., and Weiss, D. S. (1990). Estimating kinetic parameters for single channels with simulation. A general method that resolves the missed event problem and accounts for noise. *Biophys. J.* **58,** 1411–1426.

Palade, P. T., and Barchi, R. L. (1977). Characteristics of the chloride conductance in muscle fibers of the rat diaphragm. *J. Gen. Physiol.* **69,** 325–342.

Roux, B., and Sauve, R. (1985). A general solution to the time interval omission problem applied to single channel analysis. *Biophys. J.* **48,** 149–158.

Sanchez, D. Y., and Blatz, A. L. (1991). Tetraethylammonium and related quaternary ammonium ions block chloride channels in cerebral cortical neurons. *Soc. Neurosci. Abstr.* **17,** 1521 (abstr.).

Sanchez, D. Y., and Blatz, A. L. (1992). Voltage-dependent block of fast chloride channels from rat cortical neurons by external tetraethylammonium ion. *J. Gen. Physiol.* **100,** 217–231.

Schneider, G. T., Cook, D. I., Gage, P. W., and Young, J. A. (1985). Voltage sensitive, high conductance chloride channels in the luminal membrane of cultured pulmonary alveolar (type II) cells. *Pfluegers Arch.* **404,** 354–377.

Schwarze, W., and Kolb, H. A. (1984). Voltage-dependent kinetics of an anionic channel of large unit conductance in macrophages and myotube membranes. *Pfluegers Arch.* **402,** 281–291.

Stanfield, P. R. (1983). Tetraethylammonium ions and the potassium permeability of excitable cells. *Rev. Physiol. Biochem. Pharmacol.* **97,** 1–67.

Weiss, D. S. (1988). Membrane potential modulates the activation of GABA-gated channels. *J. Neurophysiol.* **59,** 514–527.

Weiss, D. S., and Magleby, K. L. (1990). Voltage dependence and stability of the gating kinetics of the fast chloride channel from rat skeletal muscle. *J. Physiol.* (*London*) **426,** 145–176.

Wilson, D. L., and Brown, A. M. (1985). Effect of limited interval resolution on single channel measurements with application to calcium channels. *IEEE Trans. Biomed. Eng.* **32,** 786–797.

CHAPTER 7

The CFTR Chloride Channel

Michael J. Welsh, Matthew P. Anderson, Devra P. Rich, Herbert A. Berger, and David N. Sheppard
Howard Hughes Medical Institute, Departments of Internal Medicine and Physiology and Biophysics, University of Iowa College of Medicine, Iowa City, Iowa 52242

Cystic fibrosis transmembrane conductance regulator (CFTR) is a regulated Cl^- channel located in the apical membrane of several Cl^- secretory epithelia. Mutations in the gene encoding CFTR cause cystic fibrosis (CF) (Boat *et al.*, 1989). CFTR was discovered and its primary structure was identified by positional cloning (Rommens *et al.*, 1989; Riordan *et al.*, 1989; Kerem *et al.*, 1989). Amino acid sequence analysis and comparison with other proteins suggested that CFTR consists of five domains (Riordan *et al.*, 1989) (Fig. 1). Beginning at the amino terminus, there is a putative membrane-spanning domain (MSD1), composed of six hydrophobic sequences predicted to cross the membrane as α-helices. MSD1 is followed by a nucleotide-binding domain (NBD1) in which there is sequence similarity with nucleotide binding domains from a number of other proteins. Then there is the R domain, which contains multiple phosphorylation sites for cAMP-dependent protein kinase (PKA) and protein kinase C (PKC). After the R domain, the protein reenters the membrane with a second membrane-spanning domain (MSD2) composed of six putative membrane-

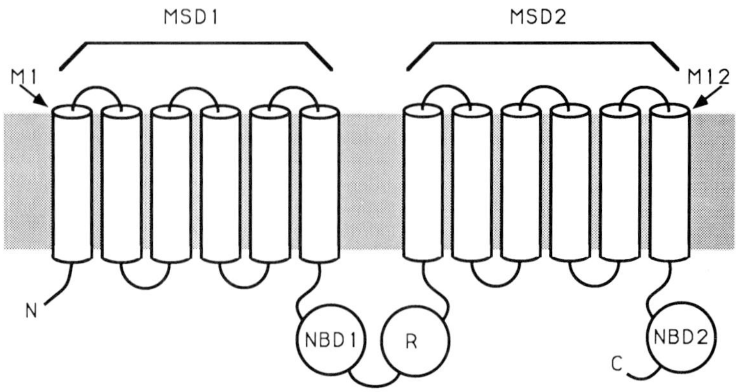

FIGURE 1 Model showing the proposed domain structure of CFTR. MSD refers to the membrane-spanning domains, NBD refers to the nucleotide-binding domains, and PKA refers to cAMP-dependent protein kinase. The membrane is represented by the shaded area. Some amino acids are indicated by residue number.

spanning sequences, followed by a second nucleotide binding domain (NBD2).

Sequence similarity in the NBDs and the topology of CFTR (with the exception of the R domain) suggest that CFTR belongs to a family of proteins called the traffic ATPases (Ames *et al.*, 1990) or the *A*TP binding cassette (ABC) transporters (Hyde *et al.*, 1990). Most members of this family are ATP-dependent transporters, including periplasmic permeases in prokaryotes and the multidrug-resistant P-glycoprotein (MDR) which is responsible for chemotherapeutic drug resistance in eukaryotes. The similarity of CFTR to other members of this family suggested that CFTR might be an active transporter. However, studies indicate that CFTR is a Cl^- channel, regulated by cAMP-dependent phosphorylation and by intracellular ATP. The cytoplasmic NBDs and R domain distinguish CFTR from the structure of previously described voltage- and ligand-gated ion channels. Thus, CFTR may represent the first identified member of a new family of ion channels. The observation that expression of MDR is associated with volume-regulated Cl^- channels (Valverde *et al.*, 1992; Gill *et al.*, 1992) indicates that MDR may also be a Cl^- channel and suggests that other members of this family may also be ion channels.

The conclusion that CFTR forms a Cl^- channel does not exclude the possibility that CFTR has additional, as yet undiscovered, functions (for example, see Egan *et al.*, 1992; Boucher *et al.*, 1986; Bradbury *et al.*, 1992; Cheng *et al.*, 1989; Tsui and Buchwald, 1991). Perhaps several members of the ABC transporter family will have more than one function.

I. TOPOLOGY AND LOCALIZATION OF CFTR

The general topology of CFTR shown in Fig. 1 is not known with certainty, but the location of several sites with reference to the plasma membrane has been assessed (Denning *et al.*, 1992a). The amino-terminus is likely to be intracellular because of the lack of a signal sequence (Riordan *et al.*, 1989). The first predicted extracellular loop is likely to be extracellular because it is recognized by antibodies directed to that epitope in nonpermeabilized cells (Denning *et al.*, 1992a,b) and because mutation of an arginine in that loop (R117) to histidine causes the channel to become dependent on extracellular pH (Sheppard *et al.*, 1993). NBD1 and the R domain are likely to be intracellular because they are regulated by kinases and nucleotides applied to the cytosolic surface of excised, inside-out membrane patches (see below) and because an antibody directed against the R domain only stains permeabilized cells (Denning *et al.*, 1992a). The fourth predicted extracellular loop is extracellular, because it contains sites that are glycosylated (Cheng *et al.*, 1990; Gregory *et al.*, 1991). NBD2 is intracellular because it interacts with intracellular nucleotides (see below). The carboxy-terminus is likely to be intracellular because it is recognized by antibodies only after cells have been permeabilized (Denning *et al.*, 1992a).

Although these observations support the model shown in Fig. 1, it is interesting that other members of the ABC transporter family contain fewer than 12 membrane-spanning sequences: MDR appears to have 6 membrane-spanning sequences in MSD1 and 4 membrane-spanning sequences in MSD2 (Zhang and Ling, 1991; Skach *et al.*, 1993). Additional studies are required to define the topology of CFTR with more precision.

II. LOCALIZATION OF CFTR

A number of antibodies have been used to immunolocalize CFTR. CFTR has been localized to the apical region of several epithelia, including small pancreatic ducts, intestinal epithelia, and several Cl^--secreting epithelial cell lines (Marino *et al.*, 1991; Cohn *et al.*, 1991; Crawford *et al.*, 1991; Denning *et al.*, 1992a,b; Engelhardt and Wilson, 1992; Kartner *et al.*, 1992; Trezise and Buchwald, 1991; Trezise *et al.*, 1992,1993). Evidence that CFTR is located within, rather than near, the apical membrane came from two types of studies. One set of studies showed that an antibody directed against an extracellular epitope labeled the apical membrane of unpermeabilized intestinal epithelial cell lines and primary cultures of airway epithelial cells (Denning *et al.*, 1992a,b). Another study showed that CFTR could be labeled with biotin from the extracellular surface (Prince *et al.*, 1993). CFTR has also been localized by *in situ* hybridization to several epithelia involved in CF (Trezise and Buchwald, 1991; Trezise *et al.*, 1992,1993; Engelhardt and Wilson, 1992).

The conclusion that CFTR is a regulated Cl^- channel and the observation that it is located in the apical membrane place it in a location where its activation by PKA-dependent phosphorylation would directly mediate Cl^- exit from the cell during transepithelial Cl^- secretion. Thus, it is possible to explain, at least at a superficial level, the observation that mutations in the gene for CFTR produce Cl^- impermeable epithelia in CF. But, it is also possible that CFTR is located beneath the apical membrane and functions on intracellular membranes (Barasch *et al.*, 1991; Lukács *et al.*, 1992; Bradbury *et al.*, 1992).

III. BIOPHYSICAL PROPERTIES OF THE CFTR Cl^- CHANNEL

CFTR cDNA has now been expressed in a variety of cell types by a number of different laboratories. In the first functional studies, CFTR cDNA was expressed in primary cultures of CF airway epithelial cells (Rich *et al.*, 1990) and in a CF pancreatic epithelial cell line (Drumm *et al.*, 1990). Expression of wild-type CFTR in these cells corrected the defect in cAMP-regulated Cl^- permeability.

Since then CFTR has been expressed in a number of cells that do not normally contain PKA-regulated Cl^- channels and express little or no endogenous CFTR. CFTR has been expressed in both mammalian and nonmammalian cells, including NIH 3T3 fibroblasts (Anderson *et al.*, 1991b,c; Berger *et al.*, 1991), Chinese hamster ovary cells (Anderson *et al.*, 1991b; Tabcharani *et al.*, 1991; Lukács *et al.*, 1992), airway epithelial cells (Egan *et al.*, 1992; Johnson *et al.*, 1992), HeLa cells (Anderson *et al.*, 1991b), mouse L cells (Rommens *et al.*, 1991; Dalemans *et al.*, 1992), Vero cells (Dalemans *et al.*, 1991), lymphocytes (Krauss *et al.*, 1992), Sf9 insect cell line (Kartner *et al.*, 1991; Bear *et al.*, 1992), and *Xenopus* oocytes (Bear *et al.*, 1991; Drumm *et al.*, 1991; Cunningham *et al.*, 1992). In each case, expression of CFTR generated a unique Cl^- current.

The whole-cell properties of recombinant CFTR include a linear current–voltage (*I–V*) relationship and minimal time-dependent voltage effects. The channels are selective for Cl^- over Na^+ and, in most reported cases, show an anion selectivity sequence of $Br^- > Cl^- > I^-$. Work from Hanrahan and co-workers suggests that I^- binds to and blocks the channel reducing its permeability (Tabcharani *et al.*, 1992). In any case, the anion selectivity sequence is a distinguishing feature of CFTR Cl^- currents, because it differs from the reported anion selectivity of some, but not all, other epithelial and nonepithelial Cl^- channels (Frizzell and Halm, 1990; Anderson *et al.*, 1992).

The biophysical properties of recombinant CFTR Cl^- channels are the same as those of endogenous CFTR Cl^- channels. Under baseline conditions, there is little, if any, Cl^- current in cells endogenously expressing CFTR. But following addition of cAMP agonists, there is a dramatic increase in Cl^- current in both whole-cell patch-clamp studies in individual epithelial cells (Tabcharani *et al.*, 1990; Cliff and Frizzell, 1990; Anderson and Welsh, 1991; Bear and Reyes, 1992; Haws *et al.*, 1992; Wagner *et al.*, 1992) and studies of the apical membrane of Cl^- secretory epithelia (Anderson and Welsh, 1991). Cl^- channel activation does not occur in response to an increase in intracellular Ca^{2+}.

The CFTR Cl^- channel has also been studied at the single-channel level using the cell-attached and excised patch-clamp techniques. Figure 2A shows an example of single-channel traces from a CFTR Cl^- channel in an excised, inside-out membrane patch; Fig. 2B shows the corresponding I–V relationship. The channel has a small conductance (approximately 6–10 pS, depending on the study conditions) and a linear I–V relationship and is selective for anions over cations. The anion selectivity is similar to that observed in whole-cell patch-clamp studies and in studies of the apical membrane.

CFTR has also been purified from Sf9 cells, reconstituted into proteoliposomes, and fused with planar lipid bilayers in the presence of PKA and ATP (Bear *et al.*, 1992). After fusion, low conductance Cl^--selective channels appeared which had a linear I–V relationship and showed no appreciable time-dependent voltage effects. These properties were the same as those observed in the native cell membrane and support the conclusion that CFTR is a Cl^- channel. Similar results, including regulation by PKA, have been obtained when membrane vesicles from cells expressing high levels of CFTR have been fused with bilayers (Tilly *et al.*, 1992; Bear *et al.*, 1992). The ability of purified CFTR to function in bilayers argues against the requirement for loosely associated factors for regulation or function of the channel.

Studies of CFTR containing site-directed mutations indicate that the MSDs may contribute to the channel pore. CFTR contains six positively charged amino acids within the putative membrane-spanning sequences. These residues are conserved across species, suggesting conservation of function. Four of the basic residues were individually converted to acidic

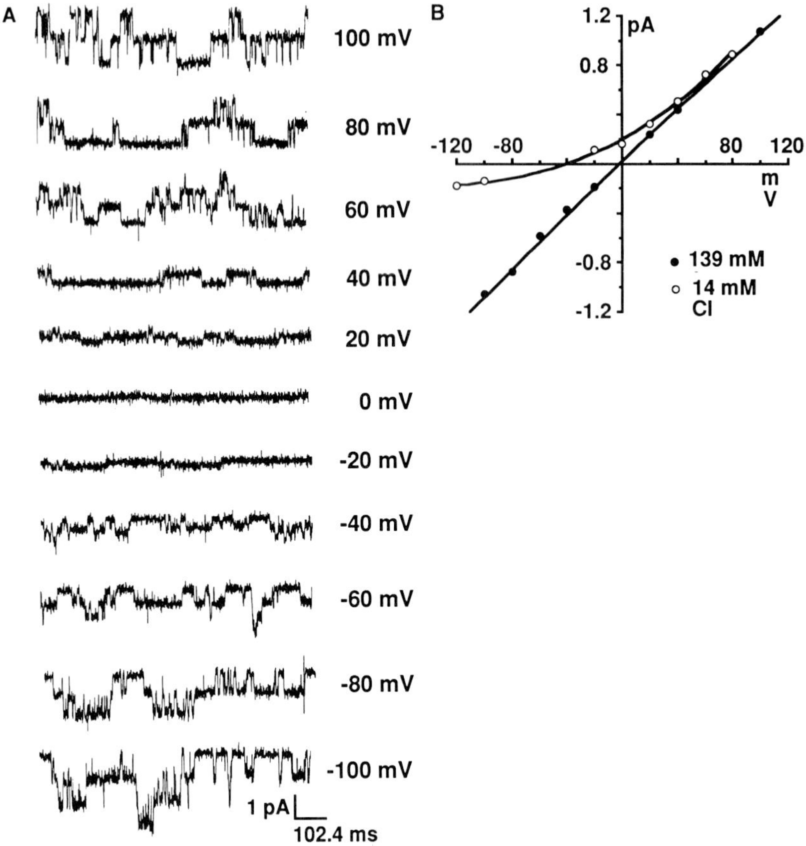

FIGURE 2 Single-channel current activated by PKA and ATP in excised, inside-out patch. Traces in (A) were obtained from a 3T3 fibroblast stably expressing CFTR. Voltages are indicated on the right. *I–V* relationship is shown in (B). Pipette contained 140 m*M* Cl^-; the bath Cl^- concentration was 139 mM in (A) and as indicated in (B). Reproduced from Berger et al. (1991) by copyright permission of the American Society for Clinical Investigation.

residues and the CFTR variants were studied using the whole-cell patch-clamp technique (Anderson *et al.*, 1991c). Each of the mutant forms of CFTR generated Cl^- channel currents. Many properties of these channels were unchanged by the mutations: in each mutant, PKA-dependent channel regulation was intact, selectivity for Cl^- over Na^+ was unaltered, and no time-dependent voltage effects were introduced. Thus the mutations did not cause a general disruption of channel structure. Yet two mutations,

K95D and K335E,[1] each altered anion selectivity. With both mutations, the permeability sequence was converted from $Br^- > Cl^- > I^-$ to $I^- > Br^- > Cl^-$. The two mutations were not, however, equivalent: K335E altered the conductivity sequence more dramatically than did K95D. Mutation of two other basic residues, R347E and R1030E, did not change the selectivity sequence. It is interesting that, in the topological model of CFTR (Fig. 1), both K95 and K335 are predicted to lie toward the outer portion of the membrane-spanning sequence, whereas R347 and R1030 are predicted to lie toward the inner half of the channel.

Mutation of other amino acids located in MSD1 can also change the conductive properties of CFTR. Three CF-associated mutations (R117H, R334W, and R347P) replace positively charged amino acids with neutral residues. Each variant retains normal regulation by PKA and ATP (Sheppard *et al.*, 1993). Moreover, each variant was selective for Cl^- over Na^+, each had a normal anion-selectivity sequence, there were no time-dependent voltage effects, and the I–V relationships were linear. Nevertheless, the whole-cell currents were quantitatively reduced as compared to wild-type CFTR. The reduction in current resulted from decreases in both single-channel conductance and open-channel probability. Under the conditions of the study, wild-type CFTR had a slope conductance of 7.8 pS, R117H had a conductance of 6.76 pS, and R347P had a conductance of 2.34 pS. Discrete channel openings could never be resolved with R334W, suggesting that its conductance was markedly reduced. The kinetics of the R117H variant were also altered, such that channel-open time appeared to be decreased and, as a result, open-state probability (P_o) was reduced to less than one-third that of wild-type CFTR. The gating of the R347P channel appeared to be like that of the wild type. The finding that three different mutations (R334W, R347P, and K335E) altered specific properties of the conduction mechanism suggests that the putative sixth membrane-spanning sequence (M6) contributes to the pore of the CFTR Cl^- channel. The observations that mutations in M1 (K95D) and at the external surface of M2 (R117H) alter specific pore properties suggest that those sequences may also contribute to the formation of the channel pore. However, these data alone are insufficient to specifically identify the pore and additional studies will be required.

[1] Mutants are named to include the amino acid residue number preceded by the wild-type amino acid, and followed by the amino acid to which the residue was changed, using the single-letter amino acid code. Thus K95D means that lysine residue 95 was changed to aspartate.

Because, as we describe below, the NBDs regulate the opening and closing of the CFTR Cl^- channel, it is interesting to speculate that they might also contribute to formation of the channel pore (Arispe *et al.*, 1992). That suggestion is supported by the work of Ames and co-workers (Baichwal *et al.*, 1993; Shyamala *et al.*, 1991) on the histidine transport system, another member of the ABC transporter family. HisP, which is analogous to the NBDs of CFTR, interacts with HisQ and HisM, two membrane proteins that are analogous to the MSDs of CFTR. Both genetic and biochemical studies suggest that HisP may span the bilayer. It is possible that such an interaction between analogous domains in CFTR may provide a mechanism for coupling the activity of the NBDs with ion flow.

IV. PHOSPHORYLATION-DEPENDENT REGULATION OF CFTR

CFTR is regulated by phosphorylation of the R domain. First, addition of cAMP agonists increases the apical membrane Cl^- permeability of normal, but not CF, epithelia (Knowles *et al.*, 1983; Widdicombe *et al.*, 1985), and addition of cAMP agonists activates CFTR Cl^- channels in heterologous cells expressing recombinant CFTR (see above). Second, addition of the catalytic subunit of PKA to the intracellular surface of excised, inside-out membrane patches activates CFTR Cl^- channels (Tabcharani *et al.*, 1991; Berger *et al.*, 1991). Third, phosphorylation of CFTR was increased *in vivo* by addition of cAMP agonists to cells incubated with $^{32}P_i$ and *in vitro* by addition of PKA plus [γ-^{32}P]ATP (Cheng *et al.*, 1991; Picciotto *et al.*, 1992). Fourth, mutation of the phosphorylation sites within the R domain alters channel regulation (see below).

In addition to regulation by PKA, PKC can also phosphorylate and activate CFTR Cl^- channels (Tabcharani *et al.*, 1991; Berger *et al.*, 1993; Picciotto *et al.*, 1992). However, the multifunctional Ca^{2+}/calmodulin-dependent protein kinase failed to either activate or phosphorylate CFTR Cl^- channels, suggesting that this enzyme has no direct effect on CFTR (Berger *et al.*, 1993). Cyclic GMP-dependent protein kinase (cGK), purified from bovine lung, phosphorylated CFTR *in vitro*, but failed to activate CFTR Cl^- channels (Berger *et al.*, 1993). These results suggest that, if cGK phosphorylates CFTR *in vivo*, it does so at sites not involved in CFTR Cl^- channel activation. There is, however, some controversy about the effect of cGK and cGMP. Cyclic GMP failed to increase apical membrane Cl^- permeability in human airway epithelia in one study (Berger *et al.*, 1993), but another study suggested that in T84 cells, cGK could activate CFTR (Lin *et al.*, 1992). A third study suggested there was no role for cGK in regulation of Cl^- secretion in T84 cells (Forte *et al.*, 1992). Additional studies may be required to evaluate the role of cGK.

Stimulation of Cl^- secretion is reversible: the rate of secretion returns to basal values once the agonist is removed, and readdition of the agonist restimulates Cl^- secretion. This result indicates that phosphorylation of CFTR is reversible. CFTR that has been activated by PKA-dependent phosphorylation is dephosphorylated and inactivated by protein phosphatase 2A (PP2A) (Berger *et al.*, 1993). The effect of PP2A was blocked by the phosphatase inhibitor calyculin A. Neither protein phosphatase 1 nor protein phosphatase 2B inactivated or dephosphorylated the channel. One study suggested that alkaline phosphatase might inactivate PKA-phosphorylated CFTR Cl^- channels (Tabcharani *et al.*, 1991). However, another study suggested that such inactivation might result from a phosphatase-dependent reduction in the ATP concentration of the bathing solution without dephosphorylating the channel (Berger *et al.*, 1993). However, it remains possible that different isoforms of alkaline phosphatase might regulate CFTR. Although PP2A can dephosphorylate CFTR *in vitro*, it is not clear whether PP2A is the predominant phosphatase that is responsible for regulation of CFTR *in vivo*. Future studies may be required to identify and characterize epithelial-specific phosphatases.

The R domain contains a number of potential phosphorylation sites for PKA (Riordan *et al.*, 1989). PKA favors the consensus phosphorylation sequence R–R/K–*X*–S*/T* (where the phosphoacceptor site is indicated by an asterisk and *X* indicates any amino acid). There are 10 "classic" PKA consensus sequences within CFTR: eight serines plus one threonine within the R domain and one serine just prior to NBD1. To identify the specific sites of phosphorylation, each of the 10 potential PKA phosphorylation sites was mutated individually to alanine. Each of the mutant CFTRs was phosphorylated *in vitro* and digested with trypsin, and the phosphopeptides were separated by two-dimensional electrophoresis (Cheng *et al.*, 1991). Each phosphopeptide was assigned to a specific serine by comparison of the wild-type and mutant fingerprints. Of the 10 potential phosphorylation sites, 7 were identified as phosphoacceptors *in vitro* and all were located within the R domain. Neither the threonine located within the R domain (T788) nor the serine located outside the R domain (S422) were phosphorylated *in vitro* by PKA. Of the 8 remaining potential phosphorylation sites, only 4 appeared to be phosphoacceptors *in vivo:* upon stimulation with cAMP, serines 660, 737, 795, and 813 were phosphorylated. All are located in the R domain. In the T84 intestinal epithelial cell line, an additional serine within the R domain (S700) was reported to be phosphorylated *in vivo* after stimulation by cAMP agonists (Picciotto *et al.*, 1992).

To investigate the functional consequences of phosphorylation, each of the mutant CFTRs (containing a serine to alanine mutation at one of the PKA phosphorylation sites) was expressed in heterologous cells and the response to cAMP agonists was assessed. Mutation of any single serine did not prevent cAMP from opening CFTR Cl^- channels, as assessed by the 6-methoxy-(3′-sulfopropyl)-quinolinium (SPQ)-halide efflux assay (Cheng *et al.*, 1991). Because mutation of any one serine to alanine did not prevent PKA-dependent activation of the mutant CFTR Cl^- channels, multiple serines were mutated simultaneously to alanine. When all four of the serines that are phosphorylated *in vivo* were mutated to alanine, CFTR Cl^- channel activity was substantially reduced. However, studies suggest that CFTR can still respond to cAMP agonists even when all four of the *in vivo* phosphorylation sites are mutated, albeit with a substantially lower level of channel activity.

The functional consequences of deleting the R domain were also examined (Rich *et al.*, 1991,1993). Expression of CFTR in which amino acids 708–835 were deleted (CFTRΔR) generated Cl^- channels that had the same biophysical properties as wild-type CFTR with one important exception: they were open even without an increase in cAMP. Addition of cAMP agonists further activated CFTRΔR Cl^- channels. These channels retain S660, one of the previously identified *in vivo* phosphorylation sites. Mutation of S660 to alanine in the combined mutant, CFTRΔRS660A, generated only constitutively active channels; addition of cAMP agonists caused no further increase in activity (Cheng *et al.*, 1991).

These data suggest that the R domain serves a regulatory role by keeping the channel closed. When the R domain is phosphorylated or when part of it is deleted, the channel opens. How could phosphorylation of the R domain open the channel? Phosphorylation might cause a conformational change in the R domain. Alternatively, phosphorylation-induced changes in charge might produce electrostatic forces that alter the interaction between the R domain and another part of the protein. This latter suggestion is supported by observations that mutation of the PKA consensus serines to negatively charged amino acids (such as aspartate) generate a channel that is open even without an increase in cAMP.

It is interesting that regulation of CFTR by phosphorylation is so complex. It appears that phosphorylation is degenerate, in that phosphorylation occurs at multiple serines and phosphorylation of any combination of serines is sufficient to activate the channel. What is the reason for this complexity? It is possible that phosphorylation of multiple sites could provide a backup mechanism to ensure that phosphorylation activates the channel. Alternatively, it could slow the rate of complete dephosphorylation and thereby delay closing after removal of kinase. Finally, it is possible that graded levels of phosphorylation might produce graded levels of CFTR Cl^- channel activity.

V. NUCLEOTIDE-DEPENDENT REGULATION OF CFTR Cl^- CHANNELS

As indicated above, CFTR contains NBDs that have amino acid sequence similarity with members of the ABC transporter family (Riordan *et al.*, 1989; Ames *et al.*, 1990; Hyde *et al.*, 1990). In most members of this family, the NBDs appear to be the site of ATP hydrolysis; the energy released during ATP hydrolysis is used to actively transport substrate across the cell membrane. Because CFTR functions as a Cl^- channel, rather than a Cl^- pump, the hypothesis that ATP might regulate the CFTR Cl^- channel was tested (Anderson *et al.*, 1991a). In excised, inside-out patches of membrane from cells expressing recombinant CFTR, CFTR Cl^- channels were activated by addition of PKA and ATP. However, once the ATP was removed, current returned to basal values (Fig. 3). Readdition of ATP alone reactivated the channels. The results indicate that ATP regulates the channel, but only when it has first been phosphorylated by PKA. A series of studies showed that PKA-dependent phosphorylation is relatively irreversible in those excised, inside-out membrane patches. The data also suggest that the effect of ATP does not occur through reversible phosphorylation of the CFTR Cl^- channel. Additional studies showed that as the concentration of Mg-ATP increased, the open-state probability (P_0) of the phosphorylated channels increased. But the data appeared complex; they could not be fit by simple kinetic models, suggesting that there may be more than one site at which ATP interacts: those sites may be NBD1 and NBD2.

To determine whether ATP interacts directly with the NBDs in CFTR, the effect of Mg-ATP on CFTR containing site-directed mutations in the NBDs was examined (Anderson and Welsh, 1992). Each NBD contains a Walker A motif and a Walker B motif (Walker *et al.*, 1982) which are conserved in the ABC transporter family and in many other proteins that hydrolyze ATP. The Walker A motif is G–*X*–*X*–G–*X*–G–K, where *X* is any amino acid. In the Walker A motif, lysine is thought to interact with either the α or γ phosphate of ATP. Mutation of the Walker A lysine in NBD1 (K464A) or in NBD2 (K1250M) resulted in an altered relationship between (Mg-ATP concentration and channel activity. In both CFTR variants in which the Walker A lysine was mutated, Mg-ATP was less

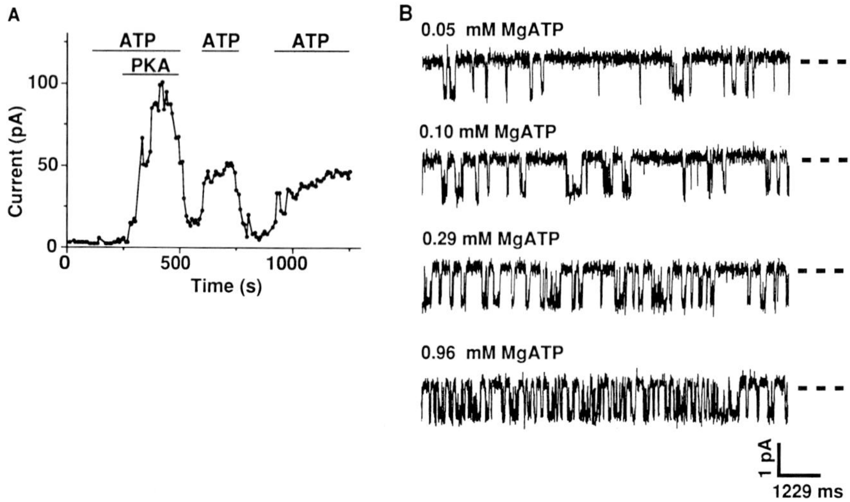

FIGURE 3 Effect of ATP on CFTR Cl^- channel activity. (A) Data points are current at −40 mV from an excised, inside-out patch containing large numbers of channels. ATP (1 m*M*) and PKA (75 n*M*) were added during the times indicated. (B) Single-channel traces of a CFTR Cl^- channel of increasing concentrations of Mg-ATP applied to the cytosolic surface. From Anderson *et al.*, (1991a) and Anderson and Welsh (1992), with permission. © AAAS.

potent at stimulating channel activity. When the conserved aspartate in the Walker B motif (hhhhD, where h indicates a hydrophobic amino acid) of NBD2 was mutated (D1370M), the potency of Mg-ATP was also reduced. The Walker A lysine is followed by two hydroxyl amino acids in most members of the ABC transporter family. In NBD1 of CFTR, the sequence is K–T–S. When the order of the two hydroxyls was switched to K–S–T, the Mg-ATP dose–response curve was shifted to the left. This result suggests that Mg-ATP is more potent at stimulating CFTR-TS465/6ST than wild-type CFTR.

Biochemical studies have shown that nucleotides interact with a peptide from NBD1 (Thomas *et al.*, 1991,1992), a fusion protein that includes NBD1 (Hartman *et al.*, 1992), and with intact membrane-associated CFTR (Travis *et al.*, 1993). These results plus the functional studies indicate that ATP interacts directly with CFTR and, more specifically, that it interacts with both NBDs. The data also suggest that an interaction of Mg-ATP with both NBDs is required for maximal channel activity.

In some members of the ABC transporter family, the NBDs hydrolyze ATP. To assess this possibility for CFTR, the effect of nonhydrolyzable analogues was examined (Anderson *et al.*, 1991a; Carson and Welsh, 1993). Nonhydrolyzable ATP analogues, such as adenosine 5′-(β,γ-methyline) triphosphate (AMP-PCP) and adenosine 5′-(β,γ-imido) trisphosphate (AMP-PNP), failed to activate CFTR Cl^- channels. Other nucleotides, such as ADP and AMP, also failed to activate CFTR. However, the requirement for nucleoside triphosphates was not highly specific. At a concentration of 1 m*M*, the nucleotide specificity sequence was ATP > AMP-CPP > GTP > ITP ≅ UTP > CTP. This broad nucleotide specificity contrasts with the high specificity for ATP that is observed in a number of kinases, but is similar to that reported for other members of the ABC transporter family.

Although AMP-PNP failed to activate CFTR, it also failed to inhibit the effect of ATP (Carson and Welsh, 1993). This result suggests that although AMP-PNP shares considerable structural similarity with ATP, it does not appreciably interact with the channel even when used at relatively high concentration ratios (0.3 m*M* ATP and 10 m*M* AMP-PNP). Mg^{2+} is a cofactor that is required in ATP hydrolysis reactions in other proteins. Mg^{2+} was also required for nucleotide-dependent regulation of CFTR. This result further suggests that ATP hydrolysis may be required to open CFTR Cl^- channels.

Although studies in excised patches suggest that nonhydrolyzable analogues of ATP cannot activate the channel, studies of CFTR in sweat gland ducts and in T84 cells in which the basolateral membrane has been permeabilized suggest that nonhydrolyzable analogues can stimulate apical membrane CFTR Cl^- channels (Quinton and Reddy, 1992; Bell and Quinton, 1993). After permeabilizing the basolateral membrane, both cAMP and ATP were required to activate apical membrane CFTR Cl^- channels. However, in the presence of cAMP plus low concentrations of ATP (0.1 m*M*), addition of either AMP-PMP or ATPγS stimulated secretion. Those data were interpreted to mean that nonhydrolyzable analogues may regulate CFTR via a nonhydrolytic interaction of ATP with the NBDs. At present it is not clear how to reconcile the observations in excised patches of membrane (AMP-PNP does not stimulate current in the presence of half-maximal concentrations of ATP) and in the intact cell. However, insights into the differences in these studies are likely to yield important new insights into the function and regulation of CFTR Cl^- channels.

ADP is a product of ATP hydrolysis and is another abundant intracellular nucleotide whose concentration is influenced by the metabolic status of the cell. Although ADP alone did not stimulate channel activity, increasing concentrations of ADP progressively inhibited CFTR Cl^- channel current

(Anderson *et al.*, 1991a). ADP proved to be a competitive antagonist of ATP, suggesting that ATP and ADP interact with the same site. Analogues of ADP also inhibited CFTR Cl^- channel activity: ADP > GDP ≅ IDP > UDP > CDP > ADPβS.

What is the physiologic relevance of CFTR regulation by Mg-ATP? One possibility is that nucleotide regulation provides a way of matching the rate of transepithelial Cl^- secretion with the availability of cellular ATP. Because apical membrane Cl^- channels are a key site in the regulation of transepithelial Cl^- transport, they may represent an advantageous point at which to couple cellular ATP and ADP levels to transport. A decrease in cellular ATP (or increase ADP) would close apical Cl^- channels, thereby decreasing the rate of Cl^- secretion and consequently, the metabolic demand on the cells (Anderson *et al.*, 1991a; Quinton and Reddy, 1992; Bell and Quinton, 1993).

VI. CONCLUSION

CFTR appears to be the first identified member of a novel class of Cl^- channels. Regulation of CFTR is complex and involves both phosphorylation and an interaction with intracellular nucleotides. The combination of a molecular biologic approach (using site-directed mutagenesis and expression of CFTR in heterologous cells), a biochemical approach (using CFTR-specific antibodies), and an electrophysiologic approach (using the patch-clamp technique) is beginning to provide essential information about the structure and function of the various domains of this interesting channel.

Acknowledgments

This work was supported in part by the Howard Hughes Medical Institute, The National Heart Lung and Blood Institute (HL42385, HL29851), and the Cystic Fibrosis Foundation. We also gratefully acknowledge our laboratory colleagues and our collaborators, Drs. Alan E. Smith, Richard J. Gregory, Seng H. Cheng, and their colleagues at Genzyme Corporation.

References

Ames, G. F.-L., Mimura, C. S., and Shyamala, V. (1990). Bacterial periplasmic permeases belong to a family of transport proteins operating from *Escherichia coli* to human: Traffic ATPases. *FEMS Microbiol. Rev.* **75,** 429–446.

Anderson, M. P., and Welsh, M. J. (1991). Calcium and cAMP activate different chloride channels in the apical membrane of normal and cystic fibrosis epithelia. *Proc. Natl. Acad. Sci. U.S.A.* **88,** 6003–6007.

Anderson, M. P., and Welsh, M. J. (1992). Regulation by ATP and ADP of CFTR chloride channels that contain mutant nucleotide-binding domains. *Science* **257,** 1701–1704.

Anderson, M. P., Berger, H. A., Rich, D. P., Gregory, R. J., Smith, A. E., and Welsh M. J. (1991a). Nucleoside triphosphates are required to open the CFTR chloride channel. *Cell (Cambridge, Mass.)* **67,** 775–784.

Anderson, M. P., Rich, D. P., Gregory, R. J., Smith, A. E., and Welsh, M. J. (1991b). Generation of cAMP-activated chloride currents by expression of CFTR. *Science* **251,** 679–682.

Anderson, M. P., Gregory R. J., Thompson, S., Souza, D. W., Paul, S., Mulligan, R. C., Smith, A. E., and Welsh, M. J. (1991c). Demonstration that CFTR is a chloride channel by alteration of its anion selectivity. *Science* **253,** 202–205.

Anderson, M. P., Sheppard, D. N., Berger, H. A., and Welsh, M. J. (1992). Chloride channels in the apical membrane of normal and cystic fibrosis airway and intestinal epithelia. *Am. J. Physiol.* **263,** L1–L14.

Arispe, N., Rojas, E., Hartman, J., Sorscher, E. J., and Pollard H. B. (1992). Intrinsic anion channel activity of the recombinant first nucleotide-binding fold domain of the cystic fibrosis transmembrane regulator protein. *Proc. Natl. Acad. Sci. U.S.A.* **89,** 1539–1543.

Baichwal, V., Liu, D., and Ames, G. F. (1993). The ATP-binding component of a prokaryotic traffic ATPase is exposed to the periplasmic (external) surface. *Proc. Natl. Acad. Sci. U.S.A.* **90,** 620–624.

Barasch, J., Kiss, B., Prince, A., Saiman, L., Gruenert, D., and Al-Awqati, Q. (1991). Defective acidification of intracellular organelles in cystic fibrosis. *Nature (London)* **352,** 70–73.

Bear, C. E., and Reyes, E. F. (1992). cAMP-activated chloride conductance in the colonic cell line, Caco-2. *Am. J. Physiol.* **262,** C251–C256.

Bear, C. E., Duguay, F., Naismith, A. L., Kartner, N., Hanrahan, J. W., and Riordan, J. R. (1991). Cl^- channel activity in *Xenopus* oocytes expressing the cystic fibrosis gene. *J. Biol. Chem.* **266,** 19142–19145.

Bear, C. E., Li, C., Kartner, N., Bridges, R. J., Jensen, T. J., Ramjeesingh, M., and Riordan, J. R. (1992). Purification and functional reconstitution of the cystic fibrosis transmembrane conductance regulator (CFTR). *Cell (Cambridge, Mass.)* **68,** 809–818.

Bell, C. L., and Quinton, P. M. (1993). Regulation of CFTR Cl^- conductance in secretion by cellular energy levels. *Am. J. Physiol.* **264,** C925–C931.

Berger, H. A., Anderson, M. P., Gregory, R. J., Thompson, S., Howard, P. W., Maurer, R. A., Mulligan, R., Smith, A. E., and Welsh, M. J. (1991). Identification and regulation of the cystic fibrosis transmembrane conductance regulator-generated chloride channel. *J. Clin. Invest.* **88,** 1422–1431.

Berger, H. A., Travis, S. M., and Welsh, M. J. (1993). Regulation of the cystic fibrosis transmembrane conductance regulator Cl^- channel by specific protein kinases and protein phosphatases. *J. Biol. Chem.* **268,** 2037–2047.

Boat, T. F., Welsh, M. J., and Beaudet, A. L. (1989). Cystic fibrosis. *In* "The Metabolic Basis of Inherited Disease" (C. R. Scriver, A. L. Beaudet, W. S. Sly, and D. Valle, eds.), 6th ed., Vol. II, pp 2649–2680. McGraw-Hill, New York.

Boucher, R. C., Stutts, M. J., Knowles, M. R., Cantley, L., and Gatzy, J. T. (1986). Na^+ transport in cystic fibrosis respiratory epithelia. Abnormal basal rate and response to adenylate cyclase activation. *J. Clin. Invest.* **78,** 1245–1252.

Bradbury, N. A., Jilling, T., Berta, G., Sorscher, E. J., Bridges, R. J., and Kirk, K. L. (1992). Regulation of plasma membrane recycling by CFTR. *Science* **256,** 530–532.

Carson, M. R., and Welsh, M. J. (1993). 5′-Adenylylimidodphosphate (AMP-PNP) does not activate CFTR chloride channels in cell-free patches of membrane. *Am. J. Physiol.* **265,** L27–32.

Cheng, P. W., Boat, T. F., Cranfill, K., Yankaskas, J. R., and Boucher, R. C. (1989). Increased sulfation of glycoconjugates by cultured nasal epithelial cells from patients with cystic fibrosis. *J. Clin. Invest.* **84,** 68–72.

Cheng, S. H., Gregory, R. J., Marshall, J., Paul, S., Souza, D. W., White, G. A., O'Riordan, C. R., and Smith, A. E. (1990). Defective intracellular transport and processing of CFTR is the molecular basis of most cystic fibrosis. *Cell (Cambridge, Mass.)* **63,** 827–834.

Cheng, S. H., Rich, D. P., Marshall, J., Gregory, R. J., Welsh, M. J., and Smith, A. E. (1991). Phosphorylation of the R domain by cAMP-dependent protein kinase regulates the CFTR chloride channel. *Cell (Cambridge, Mass.)* **66,** 1027–1036.

Cliff, W. H., and Frizzell, R. A. (1990). Separate Cl^- conductances activated by cAMP and Ca^{2+} in Cl^--secreting epithelial cells. *Proc. Natl. Acad. Sci. U.S.A.* **87,** 4956–4960.

Cohn, J. A., Melhus, O., Page, L. J., Dittrich, K. L., and Vigna, S. R. (1991). CFTR: Development of high-affinity antibodies and localization in sweat gland. *Biochem. Biophys. Res. Commun.* **181,** 36–43.

Crawford, I., Maloney, P. C., Zeitlin, P. L., Guggino, W. B., Hyde, S. C., Turley, H., Gatter, K. C., Harris, A., and Higgins, C. F. (1991). Immunocytochemical localization of the cystic fibrosis gene product CFTR. *Proc. Natl. Acad. Sci. U.S.A.* **88,** 9262–9266.

Cunningham, S. A., Worrell, R. T., Benos, D. J., and Frizzell, R. A. (1992). cAMP-stimulated ion currents in *Xenopus* oocytes expressing CFTR cRNA. *Am. J. Physiol.* **262,** C783–C788.

Dalemans, W., Barbry, P., Champigny, G., Jallat, S., Dott, K., Dreyer, D., Crystal, R. G., Pavirani, A., Lecocq, J. P., and Lazdunski, M. (1991). Altered chloride ion channel kinetics associated with the Δ F508 cystic fibrosis mutation. *Nature (London)* **354,** 526–528.

Dalemans, W., Hinnrasky, J., Slos, P., Dreyer, D., Fuchey, C., Pavirani, A., and Puchelle, E. (1992). Immunocytochemical analysis reveals differences between the subcellular localization of normal and Δ Phe508 recombinant cystic fibrosis transmembrane conductance regulator. *Exp. Cell. Res.* **201,** 235–240.

Denning, G. M., Ostedgaard, L. S., Cheng, S. H., Smith, A. E., and Welsh, M. J. (1992a). Localization of cystic fibrosis transmembrane conductance regulator in chloride secretory epithelia. *J. Clin. Invest.* **89,** 339–349.

Denning, G. M., Ostedgaard, L. S., and Welsh, M. J. (1992b). Abnormal localization of cystic fibrosis transmembrane conductance regulator in primary cultures of cystic fibrosis airway epithelia. *J. Cell. Biol.* **118,** 551–559.

Drumm, M. L., Pope, H. A., Cliff, W. H., Rommens, J. M., Marvin, S. A., and Tsui, L.-C., Collins, F. S., Frizzell, R. A., and Wilson, J. M. (1990). Correction of the cystic fibrosis defect in vitro by retrovirus-mediated gene transfer. *Cell (Cambridge, Mass.)* **62,** 1227–1233.

Drumm, M. L., Wilkinson, D. J., Smit, L. S., Worrell, R. T., Strong, T. V., Frizzell, R. A., Dawson, D. C., and Collins, F. S. (1991). Chloride conductance expressed by Δ F508 and other mutant CFTRs in *Xenopus* oocytes. *Science* **254,** 1797–1799.

Egan, M., Flotte, T., Afione, S., Solow, R., Zeitlin, P. L., Carter, B. J., and Guggino, W. B. (1992). Defective regulation of outwardly rectifying Cl^- channels by protein kinase A corrected by insertion of CFTR. *Nature (London)* **358,** 581–584.

Engelhardt, J. F., and Wilson, J. M. (1992). Submucosal glands are the predominant site of CFTR expression in the human bronchus. *Nat. Genet.* **2,** 240–248.

Forte, L. R., Thorne, P. K., Eber, S. L., Krause, W. J., Freeman, R. H., Francis, S. H., and Corbin, J. D. (1992). Stimulation of intestinal Cl^- transport by heat-stable enterotoxin:

Activation of cAMP-dependent protein kinase by cGMP. *Am. J. Physiol.* **263,** C607–C615.

Frizzell, R. A., and Halm, D. R. (1990). Chloride channels in epithelial cells. *Curr. Top. Membr. Transp.* **37,** 247–282.

Gill, D. R., Hyde, S. C., Higgins, C. F., Valverde, M. A., Mintenig, G. M., and Sepulveda, F. V. (1992). Separation of drug transport and chloride channel functions of the human multidrug resistance P-glycoprotein. *Cell (Cambridge, Mass.)* **71,** 23–32.

Gregory, R. J., Rich, D. P., Cheng, S. H., Souza, D. W., Paul, S., Manavalan, P., Anderson, M. P., Welsh, M. J., and Smith, A. E. (1991). Maturation and function of cystic fibrosis transmembrane conductance regulator variants bearing mutations in putative nucleotide-binding domains 1 and 2. *Mol. Cell. Biol.* **11,** 3886–3893.

Hartman, J., Huang, Z., Rado, T. A., Peng, S., Jilling, T., Muccio, D. D., and Sorscher, E. J. (1992). Recombinant synthesis, purification, and nucleotide binding characteristics of the first nucleotide binding domain of the cystic fibrosis gene product. *J. Biol. Chem.* **267,** 6455–6458.

Haws, C., Krouse, M. E., Xia, Y., Gruenert, D. C., and Wine, J. J. (1992). CFTR channels in immortalized human airway cells. *Am. J. Physiol.* **263,** L692–L707.

Hyde, S. C., Emsley, P., Hartshorn, M. J., Mimmack, M. M., Gileadi, U., Pearce, S. R., Gallagher, M. P., Gill, D. R., Hubbard, R. E., and Higgins, C. F. (1990). Structural model of ATP-binding proteins associated with cystic fibrosis, multidrug resistance and bacterial transport. *Nature (London)* **346,** 362–365.

Johnson, L. G., Olsen, J. C., Sarkadi, B., Moore, K. L., Swanström, R., and Boucher, R. C. (1992). Efficiency of gene transfer for restoration of normal airway epithelial function in cystic fibrosis. *Nat. Genet.* **2,** 21–25.

Kartner, N., Hanrahan, J. W., Jensen, T. J., Naismith, A. L., Sun, S., Ackerley, C. A., Reyes, E. F., Tsui, L.-C., Rommens, J. M., Bear, C. E., and Riordan, J. R. (1991). Expression of the cystic fibrosis gene in non-epithelial invertebrate cells produces a regulated anion conductance. *Cell (Cambridge, Mass.)* **64,** 681–691.

Kartner, N., Augustinas, O., Jensen, T. J., Naismith, A. L., and Riordan, J. R. (1992). Mislocalization of ΔF508 CFTR in cystic fibrosis sweat gland. *Nat. Genet.* **1,** 321–327.

Kerem, B.-S., Rommens, J. M., Buchanan, J. A., Markiewicz, D., Cox, T. K., Chakravarti, A., Buchwald, M., and Tsui, L.-C. (1989). Identification of the cystic fibrosis gene: Genetic analysis. *Science* **245,** 1073–1080.

Knowles, M., Gatzy, J., and Boucher, R. (1983). Relative ion permeability of normal and cystic fibrosis nasal epithelium. *J. Clin. Invest.* **71,** 1410–1417.

Krauss, R. D., Bubien, J. K., Drumm, M. L., Zheng, T., Peiper, S. C., Collins, F. S., Kirk, K. L., Frizzell, R. A., and Rado, T. A. (1992). Transfection of wild-type CFTR into cystic fibrosis lymphocytes restores chloride conductance at G1 of the cell cycle. *EMBO J.* **11,** 875–883.

Lin, M., Nairn, A. C., and Guggino, S. E. (1992). cGMP-dependent protein kinase regulation of a chloride channel in T84 cells. *Am. J. Physiol.* **262,** C1304–C1312.

Lukács, G. L., Chang, X. B., Kartner, N., Rotstein, O. D., Riordan, J. R., and Grinstein, S. (1992). The cystic fibrosis transmembrane regulator is present and functional in endosomes. Role as a determinant of endosomal pH. *J. Biol. Chem.* **267,** 14568–14572.

Marino, C. R., Matovcik, L. M., Gorelick, F. S., and Cohn, J. A. (1991). Localization of the cystic fibrosis transmembrane conductance regulator in pancreas. *J. Clin. Invest.* **88,** 712–716.

Picciotto, M. R., Cohn, J. A., Bertuzzi, G., Greengard, P., and Nairn, A. C. (1992). Phosphorylation of cystic fibrosis transmembrane conductance regulator. *J. Biol. Chem.* **267,** 12742–12752.

Prince, L. S., Tousson, A., and Marchase, R. B. (1993). Cell surface labeling of CFTR in T84 cells. *Am. J. Physiol.* **264,** C491–C498.

Quinton, P. M., and Reddy, M. M. (1992). Control of CFTR chloride conductance by ATP levels through non-hydrolytic binding. *Nature (London)* **360,** 79–81.

Rich, D. P., Anderson, M. P., Gregory, R. J., Cheng, S. H., Paul, S., Jefferson, D. M., McCann, J. D., Klinger, K. W., Smith, A. E., and Welsh, M. J. (1990). Expression of cystic fibrosis transmembrane conductance regulator corrects defective chloride channel regulation in cystic fibrosis airway epithelial cells. *Nature (London)* **347,** 358–363.

Rich, D. P., Gregory, R. J., Anderson, M. P., Manavalan, P., Smith, A. E., and Welsh, M. J. (1991). Effect of deleting the R domain on CFTR-generated chloride channels. *Science* **253,** 205–207.

Rich, D. P., Gregory, R. J., Cheng, S. H., Smith, A. E., and Welsh, M. J. (1993). Effect of deletion mutations on the function of CFTR chloride channels. *Recept. Channels* **1,** 221–232.

Riordan, J. R., Rommens, J. M., Kerem, B.-S., Alon, N., Rozmahel, R., Grzelczak, Z., Zielenski, J., Lok, S., Plavsic, N., Chou, J.-L., Drumm, M. L., Iannuzzi, M. C., Collins, F. S., and Tsui, L.-C. (1989). Identification of the cystic fibrosis gene: Cloning and characterization of complementary DNA. *Science* **245,** 1066–1073.

Rommens, J. M., Iannuzzi, M. C., Kerem, B.-S., Drumm, M. L., Melmer, G., Dean, M., Rozmahel, R., Cole, J. L., Kennedy, D., Hidaka, N., Zsiga, M., Buchwald, M., Riordan, J. R., Tsui, L.-C., and Collins, F. S. (1989). Identification of the cystic fibrosis gene: Chromosome walking and jumping. *Science* **245,** 1059–1065.

Rommens, J. M., Dho, S., Bear, C. E., Kartner, N., Kennedy, D., Riordan, J. R., Tsui, L.-C., and Foskett, J. K. (1991). cAMP-inducible chloride conductance in mouse fibroblast lines stably expressing the human cystic fibrosis transmembrane conductance regulator. *Proc. Natl. Acad. Sci. U.S.A.* **88,** 7500–7504.

Sheppard D. N., Rich, D. P., Ostedgaard, L. S., Gregory, R. J., Smith, A. E., and Welsh, M. J. (1993). Mutations in CFTR associated with mild disease from Cl^- channels with altered pore properties. *Nature (London)* **362,** 160–164.

Shyamala, V., Baichwal, V., Beall, E., and Ames, G. F. (1991). Structure-function analysis of the histidine permease and comparison with cystic fibrosis mutations. *J. Biol. Chem.* **266,** 18714–18719.

Skach, W. R., Calayag, M. C., and Lingappa, V. R. (1993). Evidence for an alternate model of human P-glycoprotein structure and biogenesis. *J. Biol. Chem.* **268,** 6903–6908.

Tabcharani, J. A., Low, W., Elie, D., and Hanrahan, J. W. (1990). Low-conductance chloride channel activated by cAMP in the epithelial cell line T84. *FEBS Lett.* **270,** 157–164.

Tabcharani, J. A., Chang, X.-B., Riordan, J. R., and Hanrahan, J. W. (1991). Phosphorylation-regulated Cl^- channel in CHO cells stably expressing the cystic fibrosis gene. *Nature (London)* **352,** 628–631.

Tabcharani, J. A., Chang, X.-B., Riordan, J. R., and Hanrahan, J. W. (1992). The cystic fibrosis transmembrane conductance regulator chloride channel. Iodide block and permeation. *Biophys. J.* **62,** 1–4.

Thomas, P. J., Shenbagamurthi, P., Ysern, X., and Pedersen, P. L. (1991). Cystic fibrosis transmembrane conductance regulator: Nucleotide binding to a synthetic peptide. *Science* **251,** 555–557.

Thomas, P. J., Shenbagamurthi, P., Sondek, J., Hullihen, J. M., and Pedersen, P. L. (1992). The cystic fibrosis transmembrane conductance regulator. Effects of the most common cystic fibrosis-causing mutation on the secondary structure and stability of a synthetic peptide. *J. Biol. Chem.* **267,** 5727–5730.

Tilly, B. C., Winter, M., Ostedgaard, L. S., O'Riordan, C., Smith, A. E., and Welsh, M. J. (1992). cAMP-dependent protein kinase activation of CFTR chloride channels in planar lipid bilayers. *J. Biol. Chem.* **267,** 9470–9473.

Travis, S. M., Carson, M. R., Ries, D. R., and Welsh, M. J. (1993). Interaction of nucleotides with membrane-associated cystic fibrosis transmembrane conductance regulator. *J. Biol. Chem.* **268,** 15336–15339.

Trezise, A. E. O. and Buchwald, M. (1991). *In vivo* cell-specific expression of the cystic fibrosis transmembrane conductance regulator. *Nature (London)* **353,** 434–437.

Trezise, A. E. O., Romano, P. R., Gill, D. R., Hyde, S. C., Sepulveda, F. V., Buchwald, M., and Higgins, C. F. (1992). The multidrug resistance and cystic fibrosis genes have complementary patterns of epithelial expression. *EMBO J* **11,** 4291–4303.

Trezise, A. E. O., Linder, C. C., Grieger, D., Thompson, E. W., Meunier, H., Griswold, M. D., and Buchwald, M. (1993). CFTR expression is regulated during both the cycle of the seminiferous epithelium and the oestrous cycle of rodents. *Nat. Genet.* **3,** 157–164.

Tsui, L.-C., and Buchwald, M. (1991). Biochemical and molecular genetics of cystic fibrosis. *In Adv. Hum. Genet.* **20,** 153–266.

Valverde, M. A., Diaz, M., Sepulveda, F. V., Gill, D. R., Hyde, S. C., and Higgins, C. F. (1992). Volume-regulated chloride channels associated with the human multidrug-resistance P-glycoprotein. *Nature (London)* **355,** 830–833.

Wagner, J. A., McDonald, T. V., Nghiem, P. T., Lowe, A. W., Schulman, H., Gruenert, D. C., Stryer, L., and Gardner, P. (1992). Antisense oligodeoxynucleotides to the cystic fibrosis transmembrane conductance regulator inhibit cAMP-activated but not calcium-activated chloride currents. *Proc. Natl. Acad. Sci. U.S.A.* **89,** 6785–6789.

Walker, J. E., Saraste, M., Runswick, M. J., and Gay, N. J. (1982). Distantly related sequences in the α- and β-subunits of ATP synthase, myosin, kinases and other ATP-requiring enzymes and a common nucleotide binding fold. *EMBO J.* **1,** 945–951.

Widdicombe, J. H., Welsh, M. J., and Finkbeiner, W. E. (1985). Cystic fibrosis decreases the apical membrane chloride permeability of monolayers cultured from cells of tracheal epithelium. *Proc. Natl. Acad. Sci. U.S.A.* **82,** 6167–6171.

Zhang, J. T., and Ling, V. (1991). Study of membrane orientation and glycosylated extracellular loops of mouse P-glycoprotein by *in vitro* translation. *J. Biol. Chem.* **266,** 18224–18232.

CHAPTER 8

Chloride Conductances of Salt-Secreting Epithelial Cells

Raymond A. Frizzell and Andrew P. Morris*
Department of Physiology and Biophysics, University of Alabama at Birmingham, Birmingham, Alabama 35294; and *Departments of Physiology and Cell Biology, and Gastroenterology, University of Texas Health Science Center at Houston, Houston, Texas 77030

I. INTRODUCTION

Chloride conductance pathways at the apical membranes of epithelial cells play a well-defined role in determining the rate of salt secretion across the epithelia that line the intestines, airways, and exocrine glands (Halm and Frizzell, 1990). Salt secretion is regulated on the seconds–

minutes time scale by a variety of endocrine substances, neurotransmitters, and immune cell products whose receptors lie predominantly at the plasma-facing, basolateral membranes. Secretion can be activated also by bacterial enterotoxins acting from the lumen, for example, those inducing intestinal secretion and diarrhea. There has been much interest in apical membrane receptors for regulatory substances released from immune cells (Madara *et al.*, 1992; Castro *et al.*, 1987).

It is now generally accepted that the rate-determining step in transepithelial salt transport is the apical Cl conductance, G_{Cl}. The effects of secretory agonists are generally mediated by changes in the intracellular concentrations of two principal second-messenger pathways: cAMP and Ca. Thus, agonists such as VIP, prostaglandins, and isoproterenol activate adenylate cyclase-mediated cAMP accumulation. The resulting increase in protein kinase A (PKA) activity stimulates Cl secretory mechanisms by phosphorylation. Another class of agonists includes acetylcholine, histamine, and bradykinin, which stimulate inositol phosphate production and a cellular Ca rise. In addition to these pathways, activation of protein kinase C (PKC) stimulates airway and intestinal secretion (Rao and DeJonge, 1990; Musch *et al.*, 1990; Fondacaro and Henderson, 1985; Barthelson *et al.*, 1987) and increases in cellular cGMP can activate Cl secretion in intestine (DeJonge and Rao, 1990). Their effects on the conductance pathways of secretory cells mimic the activation by cAMP and, as with cAMP-mediated agonists, are defective in cystic fibrosis (CF) intestine. The influence of different second-messenger pathways varies in different salt-secreting epithelial cells. For example, in the airways, cAMP-mediated agonists are primarily responsible for sustained stimulation of Cl secretion, whereas in sweat gland secretory coil, cholinergic agonists are predominately active. In other tissues, such as the intestines and several exocrine glands, both cAMP- and Ca-mediated secretory events are physiologically important. Moreover, the rate of secretion achieved when both classes of agonists act together is much greater than the sum of their individual activities, i.e., there is potentiation among agonists (see below).

A. Cellular Mechanisms for Cl Secretion

The cellular model for salt secretion that has dominated this field for 15 years (Frizzell *et al.*, 1979) is shown in Fig. 1. Chloride enters secretory cells via a Na/K/Cl co-transport process that is sensitive to inhibition by loop diuretics, e.g., furosemide and bumetanide. Sodium entering with Cl returns to the serosal interstitial space via the ouabain-sensitive Na/K pump. In tissues where net K secretion does not accompany Cl secretion,

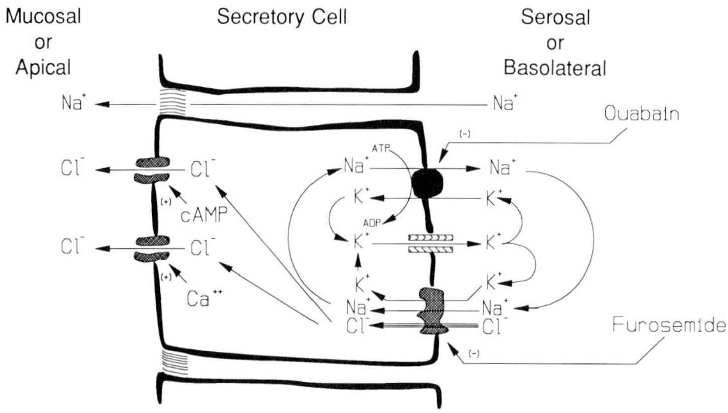

FIGURE 1 Model for Cl secretion across secretory epithelia. The basolateral membranes contain the Na/K-ATPase, K channels, and the Na/K/2Cl co-transporter. The sensitivity of these transport processes to ouabain, K channel blockers, and furosemide, respectively, account for their inhibitory effects on Cl secretion. At the apical membrane are shown two Cl channels, one activated by cAMP-dependent phosphorylation, the other by a cellular Ca rise. Further details can be found in the text.

the K that enters via the pump is entirely recycled across the basolateral membrane through K conductance pathways. As a result of these processes, Cl is accumulated within secretory cells to levels that exceed the thermodynamic activity of Cl in the external solutions. Thus, Cl ions can exit across the apical membrane by diffusion in response to agonists that increase apical Cl conductance or enhance the driving force for apical Cl exit.

As discussed above, the conductance pathways in secretory cells are activated principally by increases in cellular cAMP or Ca. Together, the apical Cl and basolateral K conductance pathways carry a transepithelial current during secretory stimulation that causes the lumen to become electrically negative with respect to the serosal solution. This transepithelial voltage arises from a depolarization of the apical membrane potential produced by activation of the apical membrane Cl conductance and from the hyperpolarizing effect of the basolateral membrane K conductance on the basolateral membrane potential. This generates a lumen-negative transepithelial voltage that is proportional to the rate of Cl secretion, and this voltage provides the driving force for transepithelial Na secretion. Sodium traverses the paracellular pathways inasmuch as the driving force at the apical membrane favors Na entry rather than Na exit. As a result of these cellular transport mechanisms, salt accumulates in the lumen, providing an osmotic driving force for net water secretion. The secretory products of most tissues of this type are approximately isotonic to plasma.

The activities of several of the transport mechanisms shown in Fig. 1 are subject to regulation by intracellular second messengers or cellular homeostatic mechanisms (Rao and DeJonge, 1990; DeJonge and Rao, 1990). However, the principal focus of this article is on the regulated Cl conductance pathways that contribute to salt secretion across epithelial cells. These pathways have received considerable attention in recent years because of their role in the etiology of cystic fibrosis (CF). CF affects the transport properties of virtually all salt-secreting epithelial cells (Quinton, 1990); a loss of apical membrane Cl permeability is a common phenotypic feature of this disease. This review has two primary goals. The first is to provide an update on the properties of Cl conductances in salt-secreting epithelial cells and to resolve, insofar as possible, their single-channel basis and mechanisms of regulation. A second goal is to consider the implications for cystic fibrosis research of the presence of different Cl conductances in these epithelia.

B. Levels of Resolution

A variety of methods have been used to quantify epithelial Cl secretion and assess the properties of the regulated Cl conductances of secretory cells. Early measurements performed on intestinal epithelial sheets identified a transepithelial component of the short-circuit current (I_{sc}) across the intestine that was equal to the net Cl flux from serosa-to-mucosa (Field *et al.*, 1968). Ion substitution experiments showed that Cl replacement abolished net Cl flow and the I_{sc}, arguing that Cl was the actively transported species (Field *et al.*, 1972). Stimulation of both current and transepithelial conductance by secretory agonists led to the concept of a secretory Cl conductance. The dependence of transepithelial Cl transport on serosal Na and its inhibition by loop diuretics and ouabain implicated the basolateral transport mechanisms shown in Fig. 1 in providing for net Cl entry across the basolateral membranes and cellular Cl accumulation (Silva *et al.*, 1977; Heintze *et al.*, 1983). Studies conducted in airway epithelial sheets and cultured intestinal cell monolayers provided confirmation for the basic transport features of this model (Welsh *et al.*, 1982, 1983). Voltage and ion-sensing microelectrode studies provided additional support for the roles of these specific transport events in establishing the driving forces necessary for Cl secretion and implicated the secretory Cl conductance at the apical membrane as the primary regulated transport event.

Secretagogue responses such as that illustrated in Fig. 2 were identified in canine tracheal apical membranes (Welsh *et al.*, 1983). Stimulation by

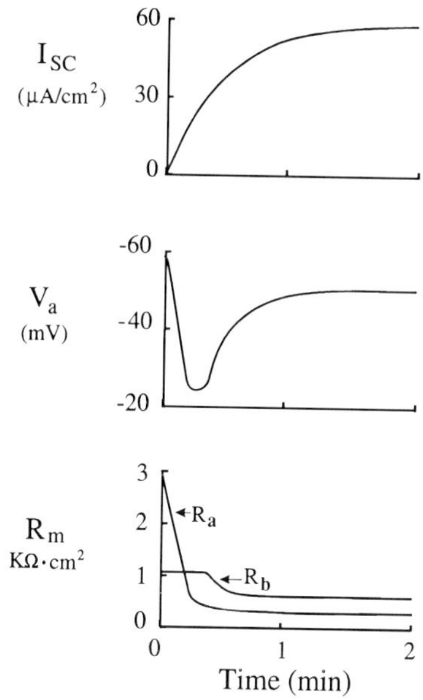

FIGURE 2 Time course of stimulation of Cl secretion as reflected by changes in the Cl secretory current (I_{sc}), the apical membrane voltage (V_a), and the apical and basolateral membrane resistances (R_a and R_b). Further discussion of these changes can be found in the text; data adapted from Welsh *et al.* (1983).

a cAMP-mediated agonist, in this case epinephrine, produces a time-dependent increase in a short-circuit current (I_{sc}), reflecting the rate of Cl secretion, and a distinct pattern of changes in the apical membrane potential (V_a) and the apical and basolateral membrane resistances (R_a and R_b). The initial event in stimulation of transepithelial current flow is a marked decrease in the electrical resistance of the apical membrane, which required the presence of Cl in the bathing media. This increase in apical Cl conductance is the first and presumably rate-determining step in stimulation of transepithelial Cl transport. As a result of this marked increase in apical Cl conductance, the electrical potential difference across the apical membrane, V_a, approaches the equilibrium potential for Cl across this barrier (−30 mV). With Cl at equilibrium, no further secretion would occur; however, a secondary change in basolateral membrane K conductance (shown by the delayed decrease in R_b) hyperpolarizes the voltages at both the basolateral and the apical membranes (electrical coupling between the opposing membranes being provided by current flow through the paracellular pathway), and this cellular hyperpolarization establishes the electrochemical driving force for apical Cl exit. The essential role of the basolateral membrane K conductance change is illustrated by the abolition of net Cl secretion elicited by K channel blockers or by depolarization of the membrane potentials by elevated serosal solution K concentration (Smith and Frizzell, 1984).

Importantly, ion substitution experiments conducted during microelectrode recordings showed unequivocally the locations of the Cl and K conductances at the apical and basolateral membranes, respectively. However, further resolution and characterization of the secretory Cl conductance by traditional methods, such as the use of ion channel blockers, did not provide much additional insight into the number or nature of these Cl conductance pathways due to the lack of sufficient specificity and the voltage dependency of several pharmacologic compounds in blocking Cl conductance pathways.

The major limitations imposed by microelectrode studies, i.e., lack of control of driving forces and failure to isolate the individual conductance pathways, were overcome by the application of patch-clamp techniques to secretory epithelial cells (Frizzell *et al.*, 1986; Welsh and Liedtke, 1986). Cell-attached, single-channel, and whole-cell current recordings have provided the opportunity to identify Cl channels activated by secretagogues that are known to perturb specific signal transduction pathways. Continued activity of these channels in excised membrane patches or after incorporation into planar lipid bilayers has allowed explicit characterization of their biophysical properties, including single-channel conductance, kinetics, voltage-dependence, ion selectivity, and blocker pharmacology. Identification of the single-channel events responsible for the secretory Cl currents evoked by different classes of agonists is still proceeding. However, the search has so far identified several of the players. It has emphasized also the importance of attempting to keep track of the biophysical fingerprints of these channels at all levels of resolution, from the identification of single-channel events to the measurement of transepithelial Cl fluxes.

II. SECRETAGOGUE-ACTIVATED Cl CONDUCTANCES

Transepithelial measurements of Cl secretion provided suggestive evidence for the presence of more than one apical membrane Cl conductance pathway. Different agonists produce secretory responses that differ in time course (sustained vs transient) (Dharmsathaphorn *et al.*, 1984). Those agonists that produce similar secretory responses are thought to perturb the same signaling and transport mechanisms because their responses are not additive. On the other hand, the responses to agonists that produce different patterns of secretory response are more than additive, they are synergistic. Thus, the combined presence of cAMP- and Ca-mediated secretory agonists (e.g., VIP plus carbachol) does not merely produce an additive secretory response, but shows potentiation, a phenomenon in which the secretion rate is greater than the sum of the individual agonist effects. Potentiation is due, in part, to the tendency for cAMP-mediated agonists to stimulate primarily the apical membrane Cl conductance, whereas Ca-dependent agonists act primarily on the basolateral membrane K conductance (see below, and Cartwright *et al.*, 1985). Thus, their combined effects potentiate due to the activation of different transport components of the secretory pathway; however, the ability of each agonist to elicit secretion on its own also raised the possibility that Ca and cAMP activate different apical Cl conductances. The presence of distinct Cl conductances activated by cAMP and Ca has been demonstrated in airway and intestinal cells (Cliff and Frizzell, 1990; Anderson and Welsh, 1991) and in nonepithelial cells as well (McDonald *et al.*, 1992).

A. The cAMP-Activated Cl Conductance: G_{Cl}^{cAMP}

A variety of agonists utilize increases in intracellular cAMP as the transducer for their secretory effects, and they include VIP, prostaglandins, β-adrenergic agonists, and adenosine. Stimulation via this mechanism usually elicits a sustained level of salt secretion, whereas the stimulation produced by Ca-dependent agonists is transient (see below).

1. Properties of the Macroscopic Whole-Cell Current

a. Time and Voltage Dependence. The Cl conductance pathway activated by cAMP analogues, forskolin, or agonists that raise cellular cAMP displays distinct biophysical properties that distinguish it from other cellular Cl conductances (Cliff and Frizzell, 1990). Figure 3A shows the Cl currents activated by cAMP during whole-cell voltage clamp. The cAMP-activated currents are time independent, showing no significant activation or inactivation at holding potentials of ± 100 mV (Cliff and Frizzell, 1990; Wagner *et al.*, 1991; McDonald *et al.*, 1992; Anderson and Welsh, 1991). The corresponding current–voltage relation, constructed from currents taken at any time following the initial capacitative current transient, is linear. The current magnitude and time course are independent of prepulse voltage history. This is an ohmic pathway, and the underlying single-channel events are expected to conform to this behavior.

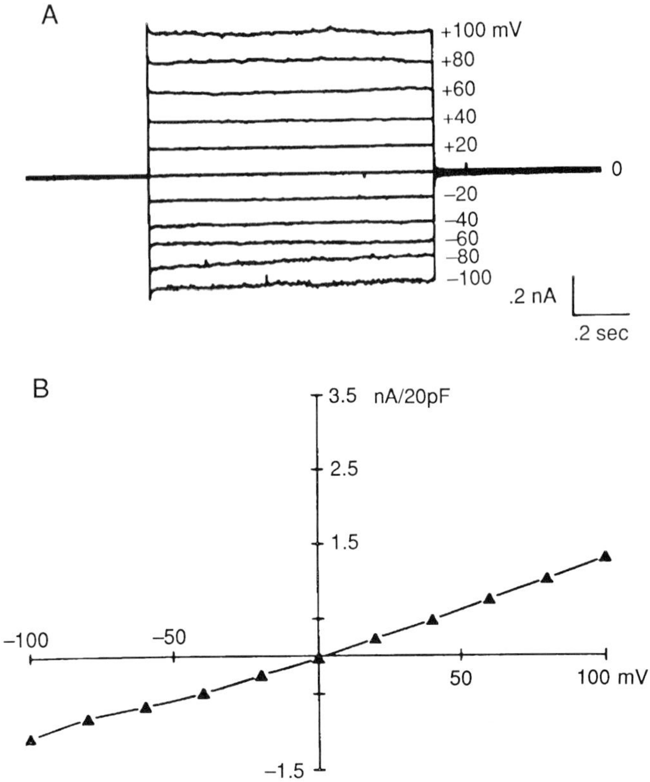

FIGURE 3 Properties of G_{Cl}^{cAMP}. (A) cAMP-activated whole-cell currents in a T84 cell. Overlay of currents recorded during voltage pulses to ±100 mV during steady-state stimulation by 5 μM forskolin. (B) Instantaneous *I–V* relation for the currents shown in A.

b. Ion Selectivity. The halide selectivity determined during ion replacement under whole-cell patch-clamp conforms to Eiseman sequence III for the halide anions: Br > Cl > I > F (Cliff and Frizzell, 1990; Anderson and Welsh, 1991). P_{Cl}/P_{I} is ~ 2.5 at standard physiologic concentrations of these anions. The low apparent iodide permeability is in part due to channel block by I, which apparently can be relieved at high I concentrations (Tabcharani *et al.*, 1992). At high salt, the selectivity reverts to the Eiseman I sequence (I > Br > Cl > F), which is similar to that found in other epithelial anion conductance pathways (see below).

c. Blocker Sensitivity. The blocker sensitivity of this conductance pathway has not yet been examined in great detail. Gray *et al.* (1990)

showed block of the single channels corresponding to G_{Cl}^{cAMP} (see below) by 5-nitro-2-(3-phenylpropylamino) benzoic acid (NPPB) during cell-attached recording from human and rat pancreatic duct cells in primary culture. They found no effect of 4,4′-diisothiocyanostilbene-2,2′-disulfonic acid (DIDS) on the same channel. This conductance pathway can generally be distinguished from the other Cl conductances by its lack of sensitivity to the disulfonic stilbenes (Anderson and Welsh, 1991). Both inward and outward currents through G_{Cl}^{cAMP} are unaffected by DIDS or DNDS at concentrations as high as 100 μM.

Two nonsteroidal anti-inflammatory agents, niflumic acid and flufenamic acid, have been shown to inhibit Cl secretion across cultured airway cell monolayers (Chao and Mochizuki, 1992) with IC_{50} values in the low micromolar range. Their effects on the single Cl channels corresponding to G_{Cl}^{cAMP} have not been examined.

Reagents that block or open ATP-sensitive K channels are blocker of G_{Cl}^{cAMP} (Sheppard and Welsh, 1992). Sulfonylureas inhibit whole-cell Cl currents stimulated by cAMP; the most potent was glibenclamide with an IC_{50} of 20 μM. Potassium channel openers also blocked G_{Cl}^{cAMP}. The single-channel events opened by cAMP-dependent phosphorylation are ATP sensitive (see below) so that similar mechanisms may govern the interaction of the nucleotide binding domains of these channels with these compounds. Detailed single-channel studies will be required to assess whether their binding sites correspond with those for nucleotides (DeRoos *et al.*, 1993). These agents appear to act as fast open channel blockers at the single-channel level and their apparent specificity for G_{Cl}^{cAMP} suggests that they may be useful for the development of antidiarrheal drugs.

2. Single-Channel Basis of G_{Cl}^{cAMP}

The single-channel responsible for the cAMP-activated Cl conductance was first identified in cell-attached patches using primary cultures of pancreatic duct cells (Gray *et al.*, 1988). Similar channels have been observed to be activated by cAMP in a variety of systems in which the cystic fibrosis gene is expressed endogenously, or exogenously in both epithelial or nonepithelial cells (Tabcharani *et al.*, 1990, 1991; Cliff *et al.*, 1992; Pollard *et al.*, 1991; Champigny *et al.*, 1990; Levesque *et al.*, 1992; Nagel *et al.*, 1992; Duszyk *et al.*, 1992). Despite the range of single-channel conductances observed in different expression systems (ca. 4–14 pS), a consistent finding is this channel's linear *I*–V relation and its activation by cAMP. These small conductance Cl channels share similarities in single-channel conductance, kinetics, and lack of sensitivity to blockade by the stilbene disulfonates and thus resemble the properties of the whole-cell currents (Cliff and Frizzell, 1990; Tabcharani *et al.*, 1990). The anion selectivity of these channels also corresponds to that observed for the whole-cell current, in particular, the low permeability to iodide relative to Cl (Gray *et al.*, 1990; Berger *et al.*, 1991).

There is now little or no doubt that this small linear Cl channel is the cystic fibrosis transmembrane conductance regulator (CFTR). Evidence favoring this view is derived from three types of observations: (a) expression of CFTR cDNA in CF cells or cells that normally do not express CFTR results in the appearance of low-conductance Cl channel activity activated by cAMP and inhibited by NPPB (Rich *et al.*, 1990; Drumm *et al.*, 1990, 1991; Anderson *et al.*, 1991b; Kartner *et al.*, 1991; Bear *et al.*, 1991; Cunningham *et al.*, 1992), (b) introduction of mutations into CFTR alters its halide selectivity or the channel's gating properties (Anderson *et al.*, 1991c; Dalemans *et al.*, 1991; Denning *et al.*, 1992b), and, most convincing, (c) reconstitution of purified CFTR protein into planar lipid bilayers produces a channel with properties like those of the low conductance, linear I–V Cl channel in CFTR-expressing cells, and its activity in the bilayer requires phosphorylation by protein kinase A (Bear *et al.*, 1992).

In cell-attached patches, CFTR Cl channels are activated by increases in cellular cAMP. However, the details of channel regulation have been further defined using excised inside-out membrane patches. CFTR Cl channels can be activated by PKA and inactivated by protein phosphatase, indicating that reversible protein phosphorylation is required for channel opening (Tabcharani *et al.*, 1991; Berger *et al.*, 1993). Further studies have identified two distinct steps in this activation sequence which conform to the predicted presence, from structural modeling and homology, of CFTR domains for regulation by phosphorylation and for ATP binding and/or hydrolysis. The central portion of this molecule consists of a highly charged region of ~250 amino acids that comprises the R (regulatory) domain. This region possesses several putative sites for phosphorylation by protein kinases (Riordan *et al.*, 1989). Site-directed mutagenesis of four serines in the R domain has shown that their phosphorylation is crucial for regulated Cl transport. *In vivo* phosphorylation studies have implicated these residues (S660, S737, S795, and S813) as sites of phosphorylation by protein kinase A (Rich *et al.*, 1991; Cheng *et al.*, 1991). However, studies suggest that other phosphorylation sites, perhaps sites that are not traditional consensus sequences for PKA-mediated phosphorylation, may also be important in CFTR activation (J. W. Hanrahan, personal communication). Our knowledge of the extent of channel activity relative to the phosphorylation status of the protein is far from complete.

The conceptual model upon which most studies are currently based involves a phosphorylation-induced conformational change in the R domain that exposes a conduction pathway for transmembrane Cl flow. This is conceptually similar to the "ball and chain" model for regulation of K channel gating (Hoshi *et al.*, 1990). The idea that the R domain occludes the conduction pathway is supported by studies in which part of the R domain was removed by deletional mutation, producing a mutant channel with constitutive activity. Studies modeled after those on K channel regulation, which have employed synthetic peptides to mimic the channel gating process (Zagotta *et al.*, 1990), will be difficult for CFTR because of the size and complexity of regulatory sites in the R domain.

Subsequent to phosphorylation-induced CFTR channel activation, removal of PKA and ATP produces a cessation of channel activity which can be restored by adding back ATP alone (Anderson *et al.*, 1991a). Thus, two steps appear to be involved in CFTR conduction: R domain phosphorylation by PKA + ATP, followed by ATP binding to one or both NBFs; the latter being necessary to maintain channel activity. These results provide evidence that ATP itself directly regulates CFTR gating, as for other ATP-dependent ion channels. These two regulatory steps fit well with the predicted structure of CFTR, involving separate regulatory and nucleotide binding domains. Many CFTR mutations involve the nucleotide binding domains, confirming the importance of ATP binding for the normal function of CFTR.

There is also evidence that ATP binds to CFTR. Synthetic peptides and purified recombinant proteins corresponding to the first nucleotide binding domain bind ATP and ATP derivatives (Thomas *et al.*, 1991; Hartman *et al.*, 1992). These findings on ATP binding and channel gating appear comparable to those obtained from other transport ATPase family members, which bind and hydrolyze ATP to drive solute transport across the plasma membrane. However, the issue of ATP hydrolysis by CFTR, at least in the context of Cl channel function, is not resolved at the present time. The initial studies of Anderson *et al.* (1991a) demonstrated that in contrast to ATP and other nucleotide triphosphates, nonhydrolyzable ATP analogues did not support CFTR Cl channel activity. Conflicting results have been obtained for the apical membrane Cl conductance of sweat duct and T84 cells. In epithelia where the basolateral membrane was permeabilized with α-toxin, Quinton and colleagues (Quinton and Reddy, 1992; Bell and Quinton, 1993) found that some nonhydrolyzable ATP analogues could keep the apical Cl conductance open. In excised membrane patches, Schultz *et al.*(1993) found that the lack of CFTR channel activation by nonhydrolyzable ATP analogues could be explained by a failure of these compounds to interact with CFTR. Unlike ADP, these substances did not compete with ATP for channel activation, and, therefore, their inability to gate CFTR cannot be used to infer that CFTR hydrolyzes ATP. The results of the Quinton and colleagues are still difficult to reconcile with these findings. For example, they could maintain apical Cl conductance with ATPγS, whereas this compound was not effective in the single-channel studies of CFTR. There are some potential alternate explanations for the studies in permeabilized epithelia or epithelial monolayers, however. The ATP concentration dependence of conductance activation occurs at very high ATP concentrations (in the millimolar range compared to a micromolar EC_{50} observed in the excised patch experiments). The possibility of displacement of bound or compartmentalized ATP by ATP analogues should be evaluated. Second, it is not certain that the conductance pathway opened by ATPγS, for example, is due to CFTR. Nucleotides activate alternate Cl conductance pathways in other epithelia (Stutts *et al.*, 1992).

The results of recent studies show that ATP increases the open probability of CFTR Cl channels as a simple Michaelis–Menten function of [ATP] in excised membrane patches (Venglarik *et al.*, 1994). The effect of ATP on P_O was half-maximal at 24 μM ATP and P_O reached a maximum value of 0.44 at saturating [ATP]. The relation of P_O to ATP showed a Hill coefficient of unity, suggesting that there is no cooperativity among the two potential sites (NBFs) for ATP binding. The results of current fluctuation analysis of multiple single-channel currents yielded identical results and also provided the on and off rate coefficients for ATP association. These values agreed well with those derived from the single-channel analysis and were most consistent with a noncooperative three-state model for CFTR activation by ATP subsequent to phosphorylation. Accordingly, ATP binding to the channel places it in another closed conformational state, which can then open spontaneously. Thus, there are two ATP bound states: one open and one closed. The ATP bound state is not simply open, because P_O would be 1.0 at high [ATP], and it is less than half this value. This model of CFTR gating by ATP should serve as a basis to further define the CFTR channel effects of pharmacological modulators and of CFTR mutations.

A rapid loss of channel activity after patch excision is sometimes a feature of this small conductance Cl channel. This inactivation of channels in excised patches has been interpreted as suggesting that there are substances in the cell required for channel activity and that these factors are retained during whole-cell recordings, but are rapidly lost after patch excision (Kunzelmann *et al.*, 1991b; Krick *et al.*, 1991). It seems more likely that this "rundown" phenomenon represents a loss of channel activation by phosphorylation, which is likely to be mediated by the presence of membrane-associated phosphatase(s) in some cell types (Tabcharani *et al.*, 1991; Berger *et al.*, 1991). Following channel rundown, exposure of excised patches to the catalytic subunit of PKA and ATP restores channel activity to patches excised from CFTR-expressing CHO cells.

3. Occurrence

This conductance pathway is found at the apical membranes of cAMP responsive secretory epithelia, including intestinal (Tabcharani *et al.*, 1990), airway (Gabriel *et al.*, 1993), pancreatic (Gray *et al.*, 1988), choroid plexus (Christensen *et al.*, 1989), thyroid (Champigny *et al.*, 1990), renal (Marunaka and Eaton, 1990), and epididymal (Pollard *et al.*, 1991) cells. Currents with these properties are present also in lymphocytes (Krauss *et al.*, 1992) and cardiac cells (Levesque *et al.*, 1992; Nagel *et al.*, 1992) following activation by cAMP analogues. These currents, in both epithelial and nonepithelial cell types, correlate with the expression of the CF gene product, CFTR. The presence of a cAMP-activated Cl conductance at the apical membranes of other epithelia, including sweat gland reabsorptive duct (Cohn *et al.*,1991; Kartner *et al.*, 1992), pancreatic and biliary duct cells (Marino *et al.*, 1991; Crawford *et al.*, 1991), is inferred from CFTR localization studies. Nevertheless, the single-channel currents themselves have not been monitored directly in all of these systems.

4. Cellular Location

A variety of immunolocalization studies have demonstrated that CFTR is present at the apical membranes of secretory epithelial cells, corresponding to its role in Cl secretion (Fig. 1). CFTR has also been localized to the apical and basolateral membranes of sweat duct cells, with the densest staining at the apical border (Cohn *et al.*, 1991; Kartner *et al.*, 1992). Other studies have demonstrated that CFTR can be detected at the apical membranes of the T84 colonic cell line, using both antibody and protein-labeling immunoprecipitation methods (Denning *et al.*, 1992a; Prince *et al.*, 1993). Studies conducted with Brefeldin A (BFA), an inhibitor of the outward migration of integral membrane glycoproteins, confirm this correlation between apical CFTR location and the cAMP-activated Cl conductance response (Morris *et al.*, 1993). BFA inhibits the cAMP-activated short-circuit current across HT-29 colon cell monolayers with a half-time of ~10 hr. With a similar time course, BFA causes the loss of apical domain staining of CFTR, and the protein accumulates in intracellular vesicles, unable to target to the apical domain. BFA prevents the outward migration of glycoproteins along the protein secretory pathway and thus interferes with the apical targeting of CFTR. The loss of cAMP-activated apical Cl conductance reflects the time course of CFTR retrieval from the apical membrane domain, and these results are in agreement with the time course of transport downregulation following inhibition of CFTR transcription (Trapnell *et al.*, 1991).

Other evidence suggests that membrane-trafficking events, including both endocytosis and exocytosis, are regulated by cAMP in CFTR-expressing epithelia (Sorscher *et al.*, 1992; Bradbury *et al.*, 1992a,b). The exogenous expression of CFTR brings these membrane-trafficking events under regulation by cAMP, where no such regulation existed prior to CFTR expression (Bradbury *et al.*, 1992a). Thus, CFTR confers cAMP-dependent regulation on these membrane-trafficking events, and this raises the possibility that CFTR itself targets into the surface membrane during stimulation of Cl secretion. Accordingly, there may be two mechanisms for increasing the apical membrane Cl conductance of secretory epithelia in response to cAMP: (a) direct regulation, via phosophorylation, of membrane-resident CFTR and (b) an increase in the amount of CFTR in the membrane brought about by both a decrease in CFTR retrieval from, and an increase in CFTR insertion into, the apical membrane.

Although the extent of Cl conductance stimulation generally correlates with the level of CFTR expression, there are exceptions to this paradigm. One of these occurs during cellular differentiation in colonic epithelial cells. Subclones of the HT-29 human colon cell lines exist in a variety of levels of differentiation. The parental cells are undifferentiated, pleuripotent cells that do not polarize to form apical and basolateral membranes, and no increase in Cl conductance is observed in response to elevated cellular cAMP (Morris *et al.*, 1992). In contrast, the differentiated Cl.19A cell line forms a resistive monolayer when grown on permeable supports and secretes Cl by the mechanism shown in Fig. 1 in response to elevated cAMP. Despite these qualitative differences in cAMP-dependent transport properties, these cell lines show identical quantities of CFTR expression at both the mRNA and protein levels (Morris *et al.*, 1992). The results of recent studies (Morris *et al.*, 1994) have shown that the basis for this difference in cAMP-activated Cl conductance is in the differentiation-dependent cellular location of CFTR. Undifferentiated colonic cells retain the protein in a perinuclear location as visualized by immunofluorescence. This site of CFTR retention must be at the level of the trans-Golgi network or sorting endosomes since the CFTR protein is fully glycosylated. In contrast, differentiated HT-29 cells target CFTR to the polarized apical membrane domain. The signals responsible for retention of CFTR in the undifferentiated cells appear to be epithelial cell specific, inasmuch as the recombinant protein, when expressed in heterologous systems (e.g. fibroblasts), targets out to the plasma membrane and is cAMP responsive (Cheng *et al.*, 1990; Yang *et al.*, 1993). Thus, the signals which permit apical targeting of CFTR in epithelial cells appear to be expressed in concert with those that generate cellular polarization, the formation of tight junctions, and discrete apical and basolateral membrane domains. Other apical membrane proteins target in a polarization-dependent fashion (Rodriguez-Boulan and Nelson, 1989), and CFTR apparently falls into this category.

5. Regulation of G_{Cl}^{cAMP}

The changes in the electrical potential profile of secretory cells that are evoked by an increase in cellular cAMP show that the primary event in stimulation of Cl secretion is an increase in apical membrane Cl conductance. This is usually followed or accompanied by an increase in basolateral membrane K conductance (Welsh *et al.*, 1982). In intestinal and airway cells, this secondary K conductance change is thought to be important in enhancing the driving force for Cl exit by depolarizing the cellular electrical potential profile, particularly the apical membrane voltage (Welsh *et al.*, 1983; Smith and Frizzell, 1984). The secondary K conductance change does not occur in all systems. Both microelectrode and whole-cell patch-clamp studies of HT-29 cells have revealed evidence of the Cl conductance activation without a corresponding K conductance change (Bajnath *et al.*, 1991). Interestingly, the undifferentiated nonpolarized HT-29 cells show evidence of a cAMP-mediated K conductance change which appears to be absent in the differentiated Cl.19A subclone. The increase in both apical Cl and basolateral K conductance observed in airway epithelia is found in both native tissue and primary cultures of the epithelial cells. However, evidence suggests that cAMP can evoke an increase in intracellular Ca (McCann *et al.*, 1989a) so that in some cell types the increase in basolateral K conductance may be induced by a cAMP-induced cellular Ca rise.

Activation of protein kinase C by phorbol esters induces Cl secretion across both intestinal and airway epithelia (Chang *et al.*, 1985; Fondacaro and Henderson, 1985; Barthelson *et al.*, 1987). However, much of the Cl secretion elicited by phorbol esters could be blocked by indomethacin (Musch *et al.*, 1990), suggesting that the activation of PKC is also accompanied by prostaglandin formation and the stimulation of cAMP production. Studies have shown clearly that Cl secretion can be induced by the phorbol ester PMA in the presence of indomethacin (Barthelson *et al.*, 1987) and, in HT-29 Cl.19A cells, stimulation by phorbol ester or carbachol results in translocation of PKC from cytosolic to membrane fractions (Van den Berghe *et al.*, 1992). In some cell culture systems, conflicting results have been obtained for phorbol ester stimulation of Cl secretion and this may be related to the expression of different PKC isoforms in different cell lines or under different culture conditions.

A number of studies have implicated activation of the same Cl conductance by PKC and PKA. In a study of CFTR phosphorylation by forskolin and phorbol ester, Picciotto *et al.* (1992) found multiple phosphorylation sites arising from cAMP stimulation, but only a subset of these were cophosphorylated by PKC. In addition, PKC was found to phosphorylate a residue in the R-domain that was not PKA sensitive. The activation of CFTR Cl channels has been found to follow a pattern which could be explained on this basis (Tabcharani *et al.*, 1991). Channel activity stimulated by PKC alone was much smaller than that observed with PKA, but together their effects on P_O potentiated one another. Thus, phosphorylation at the PKC sites may have only a marginal effect on Cl channel activity, but when combined with PKA, the additional PKC-induced phosphorylation may greatly potentiate the activity of CFTR (Tabcharani *et al.*, 1991). In other studies, synergism between PKA and PKC in excised patches has not been found (Berger *et al.*, 1993). The differences in these findings could stem from the study of endogenous vs exogenously expressed CFTR or from the use of epithelial vs nonepithelial cells.

cGMP also activates Cl secretion and has been most extensively studied in colonic epithelial cell lines (e.g., T84) (DeJonge and Rao, 1990). An interesting aspect of this activation process is that these cells express no cGMP-sensitive protein kinase activity, and therefore the effects of elevated cGMP are thought to reflect activation of protein kinase A. The cGMP levels achieved in response to stimulation are sufficient to activate PKA. Nevertheless, it has been shown that *in vitro* phosphorylation by protein kinase G is capable of activating a low-conductance Cl channel in membrane patches excised from T84 cells (Lin *et al.*, 1992). In intestinal cells, DeJonge *et al.* (1993) have shown that the activation of a specific PKG isoform (type II) is associated with Cl conductance activation. Therefore, in systems where protein kinase G is expressed, there is every reason to believe that CFTR would be a phosphorylation substrate for this kinase.

B. The Ca-Activated Cl Conductance: G_{Cl}^{Ca}

A second class of secretory agonists uses increases in intracellular Ca to stimulate Cl secretion, and they include muscarinic agonists, histamine, bradykinin, and neurotensin. In general, these agents stimulate a transient secretory response, in contrast to the sustained activation induced by cAMP-mediated agonists. The transport and regulatory mechanisms responsible for Ca-mediated Cl secretion, and for its transient nature, are beginning to be identified.

1. Properties of the Macroscopic Whole-Cell Current

a. Time and Voltage Dependence. During both depolarizing and hyperpolarizing voltage pulses, the currents activated by increased cell Ca show time-dependent behavior (Cliff and Frizzell, 1990; Morris *et al.*, 1990; Anderson and Welsh, 1991; McDonald *et al.*, 1992). With an initial holding voltage of 0 mV, the currents observed at depolarizing voltages activate, whereas those recorded at hyperpolarizing voltages inactivate (Fig. 4A). This behavior clearly differs from that associated with cAMP activation where the currents are time independent; however, in some cell types (e.g., T84 cells) the time dependence of G_{Cl}^{Ca} wanes with time after stimulation and more closely resembles the currents activated by cAMP (W. H. Cliff and R. A. Frizzell, personal observation). Since total current is the product of the single-channel conductance, channel number, and channel open probability (P_O), it is useful to minimize the time-dependent inactivation of G_{Cl}^{Ca} by using a conditioning prepulse that maximizes P_O (in this case, a large depolarizing voltage). The instantaneous currents can then be recorded at various test voltages before time-dependent changes in current intervene. Following a conditioning prepulse to +100 mV (Fig. 4B), the currents recorded at progressively more depolarized voltages inactivate, and the steady-state current–voltage relation is outwardly rectified. However, the instantaneous currents show a near-linear current–voltage relation, which should reflect the current–voltage relation of the underlying single-channel events. Thus, the *I–V* of the macroscopic currents traversing G_{Cl}^{Ca} is markedly dependent on the voltage prehistory of this conductance pathway. Fig. 4C illustrates this phenomenon in terms of the current–voltage relations shown for two different holding voltages, 0 and +100 mV. Currents determined from an initial holding potential of 0 mV show marked outward rectification because G_{Cl}^{Ca} is partially inactivated at this voltage. Further depolarization activates additional current, while hyperpolarization produces voltage-dependent inhibition of current flow. On the other hand, the instantaneous currents recorded from a holding potential of +100 mV show a nearly linear *I–V* relation, which turns out to be similar to the *I–V* of the underlying single-channel events (see below).

b. Ion Selectivity. The Na/Cl selectivity of the Ca-activated Cl conductance is approximately 0.14 (Anderson and Welsh, 1991). Inasmuch as the data shown in Fig. 4 were obtained in the absence of permeant cations, the current properties shown reflect those of the G_{Cl}^{Ca}. The halide selectivity sequence obeys Eiseman sequence I selectivity:

elicited by cAMP-dependent mechanisms is absent (Frizzell *et al.*, 1986; Welsh and Liedtke, 1986), and (d) the Ca-activated conductance is present in undifferentiated intestinal cells where no cAMP-activated channels are found due to restrictions on CFTR targeting into the plasma membrane (Morris *et al.*, 1992).

Studies have identified a Ca-activated Cl channel in HT-29 cells, during both cell-attached and excised patch recording (Morris and Frizzell, 1993a,b). The properties of these channels differ from those associated with CFTR. Cl channels were activated when membrane patches were excised into solutions containing high (1 μM) Ca concentrations. With equal Cl concentrations bathing both membrane surfaces, the I–V relation of the single-channel events was slightly outwardly rectified (as are the instantaneous whole-cell currents, see above). The single-channel chord conductance averaged 13 pS at -90 mV and 16 pS at $+90$ mV. As with the Ca-activated whole-cell currents, the single-channel events showed time-dependent activation at depolarizing membrane voltages. During cell-attached recording, Cl channel activity was rarely observed at resting Ca levels, but channels with the same properties as those found in excised patches were activated in response to the Ca ionophore, ionomycin, or the Ca-dependent agonist, neurotensin. When cell-attached patch-clamp measurements were combined with simultaneous intracellular Ca determinations, Cl channel activation correlated in time with the Ca rise induced by ionomycin. With the Ca-mediated agonist, neurotensin, Cl channel activation preceded the generalized cellular Ca rise, as reported previously for the whole-cell Cl current evoked by neurotensin (Morris *et al.*, 1990). These simultaneous measurements indicate that Ca-dependent agonists release Ca from peripheral stores in the vicinity of the plasma membrane, whereas the Ca rise produced by ionophore is more homogeneous. The Ca dependency, voltage and time dependence, and kinetics of the ~15-pS Cl channel indicate that it is the basis of the Ca-activated whole-cell Cl current.

When patches containing this 15-pS Cl channel are excised from the cell, its activity decreases, or runs down (Morris and Frizzell, 1993b). Cell-attached channel activity recorded from detergent-permeabilized cells shows a similar phenomenon. Thus, it appears that an essential cellular regulatory factor(s) may be lost upon patch excision or membrane permeabilization. In some studies, this may occur as a result of cell dialysis during conventional whole-cell recording, leading to a loss of function of these Ca-dependent Cl channels (Devor *et al.*, 1993). After run-down of the 15-pS channel, a large conductance multistate Cl-selective channel (maxi-Cl channel) is observed. This channel is Ca insensitive. Maximal single gating events of this channel yield current transitions corresponding to a conductance of approximately 200–300 pS. However, this maxichannel shows complex subconductance behavior, in which the smallest subconductance state is similar in magnitude to the conductance of the Ca-dependent Cl channel (~15 pS). In addition, the voltage sensitivity and halide selectivity of this maxi-Cl channel are also similar to those of the 15-pS channel and the Ca-dependent whole-cell Cl currents. Activity of this maxi-Cl channel has not been observed in intact cells, nor were the large and small channels observed simultaneously. Its substate behavior with similar properties suggests that this maxichannel may be a functionally altered remnant of the Ca-sensitive 15-pS Cl channel, but cloning and expression studies will probably be necessary to verify this idea.

The molecular basis of G_{Cl}^{Ca} is unknown, but there are some interesting leads in this direction. The properties of G_{Cl}^{Ca} do not correspond to the properties of already cloned Cl channels when these are expressed in heterologous systems (Frizzell and Cliff, 1992). However, Ran and Benos (1991) have reconstituted a Cl channel from bovine tracheal apical membranes whose properties are interesting with respect to G_{Cl}^{Ca}. The single-channel conductance, anion selectivity, and stilbene sensitivity of this channel in planar lipid bilayers are in rough agreement with those of the Ca-activated Cl channel discussed above (Ran *et al.*, 1992). Its activity appears to depend on disulfide bond interactions and can be abolished by dithiotheitol (DTT). Interestingly, DTT blocks the ^{125}I efflux from T84 monolayers stimulated by Ca ionophore, but not that evoked by forskolin (C. M. Fuller and D. J. Benos, personal communication). An antibody has been raised to the 38-kDa protein that was used in the reconstitution studies, and it should be useful in attempts to further characterize this channel's structural properties.

3. Regulation

Activation of the 15-pS channel by Ca-mediated agonists does not require bath Ca; however, the agonist effects are abolished by preloading cells with BAPTA (Morris and Frizzell, 1993b). Thus, agonist-induced activation of this Cl channel involves mobilization of Ca from intracellular stores. Several studies have documented a stimulation of inositol polyphosphate hydrolysis and cellular Ca mobilization that mediates the actions of primary Ca-dependent agonists such as carbachol and neurotensin (Turner *et al.*, 1990; Morris *et al.*, 1990; Devor *et al.*, 1993).

The activation of G_{Cl}^{Ca} in HT-29 cells by neurotensin could be abolished by loading the cells with a synthetic peptide corresponding to the calmodulin binding domain of the Ca/calmodulin-dependent protein kinase II (Morris and Frizzell, 1993b). This competitive antagonist of calmodulin-mediated signaling inhibited Cl channel activation without a corresponding effect on either agonist-induced Ca mobilization or activation of K conductance pathways. These results suggest that calmodulin confers Ca sensitivity to the 15-pS Cl channel.

Inhibitors of calmodulin's interaction with its binding proteins, including trifluoperazine, calmidizolium, and sphingosine, also reversibly abolished the Ca-stimulated Cl currents (Worrell and Frizzell, 1991). The addition of pseudosubstrate inhibitor peptides for various kinases to whole-cell patch pipettes has demonstrated that the Ca-activated Cl conductance is not altered by inclusion of the peptide inhibitors of protein kinases A or C, but that the peptide inhibitor of Ca/calmodulin-dependent protein kinase II (CaMKII283-302) specifically abolishes the activation of G_{Cl}^{Ca} in both T84 cells and lymphocytes (Worrell and Frizzell, 1991; Wagner *et al.*, 1991). Truncated peptides lacking essential elements of the autoinhibitory domain or the Ca binding region were ineffective. These findings suggest that the Ca/calmodulin-dependent protein kinase II activates G_{Cl}^{Ca} in both epithelial and nonepithelial cells. At present, it is not clear whether this represents a direct effect of the kinase on the 15-pS Cl channel or whether other proteins are also involved in transducing this signal.

4. Occurrence and Location

A Ca-activated Cl conductance with the properties cited above is found in a variety of secretory epithelial cells, including colonic and pancreatic cell lines and primary cultures of airway, nasal, and sweat gland epithelial cells (Cliff and Frizzell, 1990; Wagner *et al.*, 1991; Anderson and Welsh, 1991). It is also present in certain nonepithelial cells, e.g., lymphocytes, where activation of a Cl conductance with similar conductance and kinetic properties has been demonstrated in response to Ca ionophores (McDonald *et al.*, 1992). It has also been suggested that the Ca-mediated activation of Cl conductance in the G_1 phase of the cell cycle is a feature of lymphocyte proliferation; several steps in this process are Ca dependent (Bubien *et al.*, 1993).

The cellular location of G_{Cl}^{Ca} in epithelial cells and its role in Ca-mediated Cl secretion has been the focus of some studies. According to the model of Dharmsathaphorn and colleagues (Cartwright *et al.*, 1985), Ca-mediated agonists stimulate Cl secretion across the apical membrane through an already open Cl conductance. An increased rate of Cl exit across the apical membrane is driven by hyperpolarization of the cellular electrical potential profile produced by a primary Ca-dependent activation of the basolateral K conductance. This idea has gained additional support from several observations. Anderson and Welsh (1991) showed that when the basolateral membranes of confluent T84 cells were permeabilized by nystatin in the presence of a transepithelial Cl gradient, no Ca-stimulated Cl current was detected. However, a transapical Cl current was stimulated by cAMP. They concluded that the Ca-stimulated Cl conductance detected in whole-cell recordings is present in subconfluent cells, but is not expressed once the cells polarize. Accordingly, the Ca-dependent secretory response would depend on basal activity of the G_{Cl}^{cAMP} and K channel activation. Morris *et al.* (1993) lent further support to this idea in experiments with BFA, an inhibitor of glycoprotein targeting to plasma membranes. They showed that BFA abolished the cAMP-activated Cl secretory response across HT-29 monolayers with a time course similar to that for the retrieval of CFTR from the apical membrane domain. The Ca-activated Cl secretory response was small and unaffected by BFA. When a Ca-mediated agonist was added during stimulation by cAMP, the additional secretory response to Ca was large, but like the cAMP-activated secretory response, the secretion evoked by Ca with cAMP present was abolished by BFA. This suggests that the apical Cl conductance pathway stimulated by cAMP is dominant under conditions where Ca and cAMP are acting together. Thus in many epithelia, and especially in cells of intestinal origin, CFTR may play a major role in establishing the apical Cl conductance for Ca-dependent Cl secretion. Indeed, this would account for the observation that both cAMP and Ca-mediated Cl secretion are defective in the intestinal epithelia from cystic fibrosis patients (Berschneider *et al.*, 1988; Goldstein *et al.*, 1988). Since CFTR accounts for G_{Cl}^{cAMP}, basal Cl conductance activity of CFTR must play a major role in intestinal secretion activated by Ca.

Other studies suggest that a Ca-stimulated Cl conductance is expressed in the apical membranes of intestinal cells. Evidence from microelectrode studies has shown that Ca-stimulated Cl conductances are present at both the apical and basolateral membranes of polarized HT-29 cells (Bajnath *et al.*, 1992). Moreover, when these cells were treated with nystatin, the Cl conductance response to ionomycin at the apical membrane was lost, but the response to forskolin was still present. These results suggest that the experimental conditions used to monitor transapical Cl currents, permeabilization of the basolateral membranes with nystatin, may interfere with the Cl conductance response to Ca. Additional evidence of an apical G_{Cl}^{Ca} in intestine was provided by Morris *et al.* (1992) who found that the 15-pS Cl channel in the apical membranes of confluent monolayers of HT-29 cells was activated by ionomycin and neurotensin during cell-attached recording. Their data show a higher density of the 15-pS Ca-activated Cl channel in undifferentiated, unpolarized HT-29 cells than in the polarized subclone. Thus, the Ca-activated channel targets to the plasma membrane in the absence of cellular polarization, and its targeting and/or expression level may fall as intestinal cells take on more differentiated functions. This pattern is the converse of that associated with apical membrane CFTR expression. It should probably not be surprising to find cells at intermediate levels of differentiation where both G_{Cl}^{cAMP} and G_{Cl}^{Ca} are present in the apical membrane. The HT-29 (Cl.19A) cells appear to express both pathways when polarized; T84 cells may not.

In contrast to the results from intestinal epithelia, studies of airway epithelia have indicated that Ca-mediated Cl secretion is not affected in cystic fibrosis (Boucher *et al.*, 1989; Willumsen and Boucher, 1989). Both transepithelial transport and microelectrode studies have shown that Ca ionophores and primary Ca agonists increase the Cl conductance of the apical membrane. It will be interesting to determine whether the 15-pS Cl channel identified in intestinal cell lines is expressed also in the apical membranes of airway cells. It is possible that the cellular targeting controls on the Ca-dependent Cl channel differ in airway and intestinal cells, so that Ca-dependent Cl secretion is less dependent on CFTR function at the apical membranes of airway cells. The pharmacologic manipulation of G_{Cl}^{Ca} in airway cells represents a potential therapeutic approach to providing a missing Cl secretory conductance in the airways (Knowles *et al.*, 1991; Stutts *et al.*, 1992).

C. Combined Effects of cAMP and Ca

The combined effects of Ca and cAMP on whole-cell Cl currents are additive (Cliff and Frizzell, 1990; Anderson and Welsh, 1991). The currents stimulated by cAMP show characteristic time-independent, ohmic properties, and Ca stimulates an additional Cl current component having the time-dependent properties of activation at depolarizing voltages that are characteristic of G_{Cl}^{Ca}. The additivity of these currents fits with the idea that separate conductance pathways and different single channels are responsible for the effects of cAMP and Ca.

As noted earlier, the combined effects of Ca and cAMP-mediated agonists on transepithelial Cl transport are more than additive. When both cAMP and Ca are elevated, a potentiated or synergistic response occurs in which the rate of Cl secretion exceeds the sum of the individual responses. Potentiation of the secretory response has been observed in a variety of epithelial tissues and cultured cell monolayers (Cartwright *et al.*, 1985, Yajima *et al.*, 1988; Barrett and Dharmsathaphorn, 1990; Simmons, 1992). As mentioned previously, the basis of potentiation between cAMP and Ca-dependent agonists lies in the primary activation of different conductance pathways by these second messengers. cAMP acts primarily on apical Cl conductance whereas Ca acts primarily to increase the basolateral K conductance and hyperpolarize the apical membrane potential.

Studies have utilized the sustained vs transient nature of the transepithelial response to secretagogues to categorize them as cAMP or Ca mediated (Madara *et al.*, 1992). Combining secretagogue additions to probe for potentiation with agonists that are known to operate through cAMP or Ca can provide a preliminary categorization of the second-messenger and conductance pathways activated by an agonist whose mechanism of action is unknown. Whole-cell current measurements suggest that this potentiated response is not due to interactions between G_{Cl}^{cAMP} and G_{Cl}^{Ca}. Rather, it derives from Ca activation of the basolateral membrane K conductance and an increase in the driving force for apical Cl exit. A Ca-dependent K channel in the basolateral membranes of airway and intestinal cells that appears to be responsible for this response has been identified (McCann *et al.*, 1990; Devor and Frizzell, 1993). This inwardly rectified K channel shows a similar pharmacological profile to that determined previously from measurements of transepithelial Cl secretion (Barrett and Dharmsathaphorn, 1990). During cell-attached recording from T84 cells, it was activated by the Ca-dependent agonists carbachol, ionomycin, and taurodeoxycholate (Devor and Frizzell, 1993).

III. THE VOLUME-SENSITIVE Cl CONDUCTANCE: G_{Cl}^{Vol}

Following an initial cell swelling, cells exposed to hypotonic media generally undergo a regulated return toward their original size (Grinstein and Foskett, 1990). This regulatory volume decrease (RVD) is caused by the activation of conductance pathways for Cl and/or K, which elicits KCl exit. Water follows osmotically to effect a reduction in cell volume. In both lymphocytes and epithelial cells, cell swelling and Cl conductance activation occur during conventional whole-cell patch clamp (Cahalan and Lewis, 1988; Worrell *et al.*, 1989). The Cl conductance activation can be enormous as cells continue to swell under these conditions. Both the volume increase and the Cl conductance activation can be reversed by imposing a suitable osmotic pressure difference between the pipette and bath solutions. In some cells (e.g., T84), the imposition of an osmotic gradient between the bath and the pipette is necessary to prevent cell swelling and provide a low basal cellular conductance upon which the effects of secretory agonists (as above) can be discerned (Cliff *et al.*, 1991).

A. Properties of the Whole-Cell Current

The swelling-activated Cl conductance shows unique biophysical properties that permit its distinction from the agonist-activated Cl conductances (Worrell *et al.*, 1989; Wagner *et al.*, 1991; McDonald *et al.*, 1992). The Cl currents evoked by cell swelling are outwardly rectified, and during both hyperpolarizing and depolarizing voltage pulses, they exhibit time dependence. This property differs from G_{Cl}^{Ca}, in that the currents inactivate during depolarizing voltage pulses and activate with hyperpolarization of the membrane potential (Fig. 5A). Accordingly, and as with G_{Cl}^{Ca}, the shape of the current–voltage relation is influenced by the degree to which this conductance pathway is activated prior to voltage clamp. Nevertheless, even when G_{Cl}^{Vol} is maximally activated by conditioning prepulses to large negative voltages, the instantaneous currents recorded at progressively more depolarized voltages are outwardly rectified (Fig. 5B). The source and potential function of this time dependence of the currents are not known, but it appears to be an inherent and stable feature of this conductance pathway at all times subsequent to activation.

1. Anion Selectivity

The properties illustrated in Fig. 5 are characteristic of the Cl conductance activated by cell swelling inasmuch as the currents are recorded in the absence of permeant cations. As with G_{Cl}^{Ca}, the halide-selectivity sequence obeys the Eiseman I series, I > Br > Cl > F, and as for the other anion conductances, differences in anion permeability are not large (Worrell *et al.*, 1989).

2. Blockers

The swelling-induced conductance is sensitive to the disulfonic stilbenes, DIDS and DNDS, which block these currents with IC_{50} values of approximately 5 μM (R. T. Worrell and R. A. Frizzell, personal communication; Solc and Wine, 1991). Interestingly, DIDS added to the bath under whole-cell recording conditions blocks primarily the outward currents (Cl entry) with little effect on inward Cl currents (Cl exit). Over brief periods of exposure (1–2 min), DIDS was a reversible inhibitor; nevertheless, the sensitivity of these currents to DNDS suggests that the disulfonic stilbene effect is due to specific channel block. Other studies (McCann *et al.*, 1989b) showed that the G_{Cl}^{Vol} can be inhibited by diphenylamine-2-carboxylic acid (DPC) and 5-nitro-2-(3-phenylamine) benzoate (NPPB). The pattern of block is similar to that observed for the outwardly rectified Cl channel (ORCC). Evidence that the ORCC may account for the swelling-activated Cl conductance will be presented below.

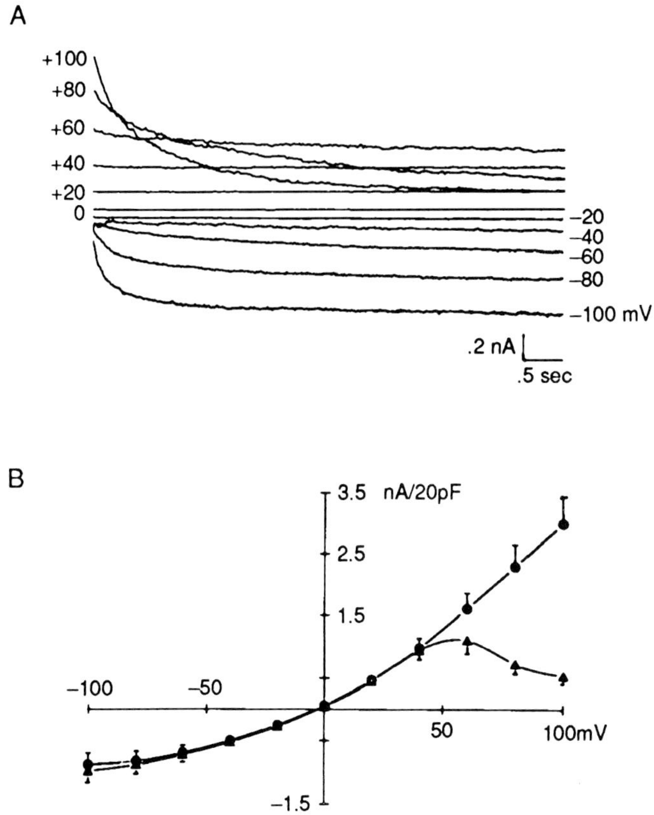

FIGURE 5 Properties of G_{Cl}^{Vol}. (A) Currents recorded during steady-state cell swelling at membrane voltages between ±100 mV. (B) Instantaneous (circles) and steady-state (triangles) *I–V* relations of G_{Cl}^{Vol}. In this experiment, a holding potential of −100 mV was used to activate these currents prior to the test pulse.

B. Occurrence and Location

The swelling-activated Cl conductance has been observed in a variety of epithelial and nonepithelial cells. In epithelia, it was first identified in the colonic tumor cell line, T84, but it is also present in other intestinal cells (Hazama and Okada, 1988) in the pancreatic cell lines, Panc-1 and CFPAC, and in primary cultures of airway and sweat gland cells. There is no discernible difference in G_{Cl}^{Vol} recorded from CF and non-CF cells (Solc and Wine, 1991; Wagner *et al.*, 1991; McDonald *et al.*, 1992; McCann *et al.*, 1989b; R. T. Worrell and R. A. Frizzell, personal communication). In nonepithelial cells, the properties of this conductance have been most completely characterized in human T lymphocytes (Cahalan and Lewis, 1988; McDonald *et al.*, 1992) and Xenopus oocytes (Parker and Miledi, 1988). The current–voltage relations and kinetics of the swelling-activated Cl conductance are qualitatively similar in all systems examined (Fig. 5).

The location of this swelling-induced conductance in epithelial cells is somewhat uncertain. Studies by McCann *et al.* (1989b), using tracheal cells in primary culture, suggest that at least some of these channels are located in the apical membrane. In these studies, tracheal cell monolayers were mounted in Ussing chambers and swelling was elicited by exposure to hypotonic media. The development of a short-circuit current suggested stimulation of an apical membrane Cl conductance. ORCC have been recorded from the upward-facing membranes of T84 cells in islands. In addition, a swelling-induced K conductance lies in the basolateral membrane and its activation could lead to Cl secretion via already active apical Cl conductance pathways (e.g., as in the response to Ca-dependent agonists). The localization question should be reexamined in confluent cell monolayers on permeable supports or using transepithelial current measurements in monolayers where the apical or basolateral membranes are permeabilized with nystatin or amphotericin. The activation of single-channel events or transepithelial currents with the appropriate voltage-dependent kinetics in response to cell swelling would clearly indicate an apical location. However, inasmuch as these channels are also present in nonepithelial cells, it seems likely that they reside, at least in part, in the basolateral membranes.

C. Regulation

The intracellular mediator(s) of G_{Cl}^{Vol} are poorly characterized and may vary among different cell types. For example, in intestinal 407 cells, Cl conductance activation during swelling requires Ca, but the G_{Cl}^{Vol} is insensitive to intracellular Ca concentration (Hazama and Okada, 1988). Similar results are obtained from T84 cells in which the development of G_{Cl}^{Vol} is still present at high EGTA concentrations in pipette and bath (calc. cell Ca 25 nM, Worrell *et al.*, 1989). In HT-29 cells, simultaneous whole-cell conductance and Ca measurements showed no increase in cell Ca as Cl currents developed during cell swelling (A. P. Morris and R. A. Frizzell, unpublished observation). Arachidonic acid metabolites appear to be important in activating the swelling conductance in some cell types, e.g., lymphocytes (Lambert, 1987). RVD was accelerated by exposure of lymphocytes to leukotriene D_4 and inhibited on exposure to arachidonic acid (Lambert, 1989). Preliminary experiments showed inhibition of the Cl conductance response by NDGA in T84 cells (R. T. Worrell and R. A. Frizzell, unpublished observation). In the presence of the lipoxygenase inhibitor, cells swell in hypotonic media, but Cl currents do not develop. These findings suggest that an arachidonic acid metabolite, perhaps a leukotriene, mediates the activation of G_{Cl}^{Vol}; its generation could be controlled by membrane deformation, or in systems where cell Ca rises in response to swelling, by Ca-induced activation of phospholipase A_2. It would be interesting to attempt to uncouple the conductance response from the volume stimulus with PLA_2 inhibitors.

D. Single-Channel Basis

The single-channel event responsible for the swelling-induced Cl current appears to be a mid-sized Cl channel of ~50 pS (at 0 mV), which shows voltage-dependent kinetic properties similar to those of the whole-cell swelling-induced current (Worrell *et al.*, 1989; Solc and Wine, 1991; Banderali and Roy, 1992). This channel is active at hyperpolarizing voltages, and test pulses to depolarizing voltages produce step-wise inactivation of single-channel currents (Worrell *et al.*, 1989). Summation of many such single-channel records yields an ensemble current whose inactivation time constant matches that of the swelling-induced whole-cell current. Single-channel events with these properties have been observed during both cell-attached and excised inside-out recordings from swollen cells but their activity drops after seal formation (run-down) (Solc and Wine, 1991; Banderali and Roy, 1992).

A member of the transport ATPase family related to CFTR is the P-glycoprotein, the product of the multiple drug resistance gene, mdr1. Valverde *et al.* (1992) found that the swelling-induced whole-cell Cl currents in fibroblasts expressing mdr1 were much larger than those in non-transfected cells. They suggested that mdr1 encodes the channel responsible for G_{Cl}^{Vol}. A compelling feature of their findings was the inhibition of swelling-induced currents by P-glycoprotein transport substrates. These authors have suggested (Valverde *et al.*, 1992) that P-glycoprotein is obligatorily bifunctional and that its drug transport or channel functions may be separated by hydrolyzable vs nonhydrolyzable ATP analogues. Whereas drug transport requires ATP hydrolysis, channel function apparently can be supported by nucleotide binding. Resolution of whether P-glycoprotein is responsible for G_{Cl}^{Vol} will require more detailed studies, as was true of CFTR. In particular, single-channel measurements would permit direct assessment of the relation to ORCC, more detailed studies of drug interactions, and structure–function studies. Reconstitution of the purified protein would be the ultimate goal.

IV. THE OUTWARD RECTIFIER

In recent years, much attention has been focused on the properties of a ~40-pS outwardly rectified Cl-selective channel that is present in a variety of cell types. The distinctive biophysical features of this channel and its relatively large unitary currents facilitated its identification in trachea (Welsh and Liedtke, 1986; Frizzell *et al.*, 1986), sweat gland (Wine *et al.*, 1991), colon (Halm *et al.*, 1988), and pancreatic cells (Schoumacher *et al.*, 1990). A channel with similar properties has been detected in human skin fibroblasts and lymphocytes (Bear, 1988; Chen *et al.*, 1989). Its biophysical properties have been described in detail (Halm and Frizzell, 1992).

A. Ion Selectivity

The selectivity for Cl over Na found in several cell types varies between 10 and 100 for P_{Cl}/P_{Na}. The relative halide anion permeabilities of the outward rectifier (Halm and Frizzell, 1992), as determined from bi-ionic reversal potential measurements, is I (1.7)/Br (1.4)/Cl (1.0)/F ≅ HCO_3 (0.4). As determined from conductance measurements, I, Br, and Cl are equally effective current carriers for anion efflux (the secretory current). The conductance ratios for HCO_3 and F (relative to Cl) are ~0.4 (Tabcharani *et al.*, 1989; Kunzelmann *et al.*, 1991a). The open probability of the outward rectifier increases with membrane depolarization. Thus, P_O increases as the cytoplasmic side is made more positive and may increase as much as fivefold between −80 and +80 mV (Halm *et al.*, 1988).

B. Blockers

Compounds of three chemical classes have been shown to block this Cl channel. Derivatives of DPC were characterized for their inhibition of basolateral Cl conductance in renal tubules by Greger and colleagues (Wangemann *et al.*, 1986). Their structure–activity studies led to the identification of NPPB, the most potent blocker in this assay. A second class of compounds was developed by Landry *et al.* (1987) based on the initial work of Cragoe with indanyloxy alkanoic acids (Cragoe *et al.*, 1986). They identified IAA-94 (2-[(2-cyclopenytyl-6,7-dichloro-2,3-dihydio-2-methyl-1-oxo-1H-inden-5-yl)oxy]acetic acid) as the most potent inhibitor of C1 uptake into renal membrane vesicles. The third class of blockers is the disulfonic stilbenes, traditionally used to inhibit the anion exchanger in red cells. Bridges *et al.* (1989) and Becq *et al.* (1992) found that the ORCC was blocked by disulfonic stilbenes. DNDS (4,4′-dinitro stilbene-2,2′-disulfonic acid) reversibly blocked the outwardly rectified Cl channel from rat colon incorporated into planar lipid bilayers with the characteristics of an open channel blocker. Finally, a channel with properties similar to those of the ORCC may be blocked by a scorpion venom toxin (DeBin and Strichartz, 1991).

Bridges and co-workers (Singh *et al.*, 1991) have now compared the potencies of the lead compounds from these three classes of anion channel antagonists in blocking the ORCC from rat colon in planar lipid bilayers and characterized their mechanism of block. NPPB and IAA-94 were equipotent when added to either side of the bilayer, and this probably reflects their high lipid solubilities. DNDS was about 10-times more potent as a blocker of the ORCC when added to the outside of the channel. NPPB caused a decrease in channel conductance (fast block), whereas the other compounds caused a flickery block (increased brief closures) that decreased P_0. Perhaps most interestingly, no compound by itself could be used to infer a role for the ORCC in contributing to a macroscopic Cl current. The relative blocker potency order and sidedness of block would be needed to implicate the outward rectifier as contributing to a whole-cell or transepithelial Cl current. Kinetic analysis suggested that these compounds act as open channel blockers. While decreasing P_0 within open bursts of channel activity, these compounds increased burst duration. This may account for their relatively weak effects on transepithelial Cl secretory currents. Caution also is warranted in interpreting the effects of these compounds on macroscopic currents, since all of these agents influence other transport and cellullar processes. At present, highly selective reagents that would permit distinction of different Cl channels or conductance pathways based on the actions of Cl channel blockers alone have not been identified. Nevertheless, G_{Cl}^{Vol} and G_{Cl}^{Ca} exhibit sensitivity to the disulfonic stilbenes while G_{Cl}^{cAMP} does not. Accordingly, the absence of an effect of the reversible blocker, DNDS, could be useful together with other evidence to infer that CFTR contributes to a whole-cell or transepithelial secretory Cl current. Newly developed and more potent relatives of these compounds, the calixarenes, should be even more useful in this regard (Singh *et al.*, 1993).

C. Regulation

In most studies, secretory agonist-induced activation of the outward rectifier during cell-attached recording has not been detected (see, however, the report of Henderson *et al.*, 1992). It is now generally thought that the outward rectifier is active in excised membrane patches (however, see Volume-Activated Cl Channels; Solc and Wine, 1991). Most studies of the outward rectifier and its regulation have been performed in patches excised from activated cells.

Early conclusions regarding the cystic fibrosis phenotype involved studies of the regulation of the outward rectifier, both in (presumably) cell-attached and in excised membrane patches. A consistent finding in four different laboratories (Frizzell *et al.*, 1986; Welsh and Liedtke, 1986; Hwang *et al.*, 1989; Chen *et al.*, 1989) was the observation that PKA plus ATP activated the outward rectifier in excised patches from normal cells but not in patches derived from CF epithelial cells. Nevertheless, the low incidence of this channel in cell-attached patches has led to uncertainty regarding its role in contributing to secretagogue-mediated increases in apical Cl conductance. The lack of an effect of cAMP during cell-attached recording led to the idea that the outward rectifier might be tonically inhibited by a cytosolic factor and to the proposal that abnormal control over the cellular concentration of this factor might prevent activation of the channel in CF cells. Two groups have reported candidate cytosolic inhibitors of ORCC (Kunzelmann *et al.*, 1991b; Krick *et al.*, 1991).

In addition, the cloning of the CF gene and its expression in both epithelial and nonepithelial cells focused attention on the low-conductance linear I–V channel observed in cells that endogenously and exogenously express CFTR, as discussed previously (see G_{Cl}^{cAMP}). Yet, two reports confirm the regulatory differences in outward rectifier regulation observed in comparisons of normal and CF cells. These findings have again raised the idea that this channel is, in some way, under regulation by the CF gene product. In a study that repeats the protocols of experiments performed previously on primary airway cell cultures, Egan *et al.* (1992) found that PKA activated the outward rectifier in membrane patches excised from normal airway cell lines but not in patches excised from CF cell lines. The outward rectifier was not observed during cell-attached recordings; the small linear channel (CFTR) was present in cell-attached recordings from normal, but not CF, cells. Because of the low levels of CFTR expressed in these cell lines, the authors concluded that the outward rectifier was not related to the low-conductance CFTR Cl channel, but was, in some way, regulated by CFTR in membrane patches where they coexisted.

More recently, studies by Gabriel *et al.* (1993) have added additional intrigue to this puzzle. These investigators studied primary airway cell cultures from normal and CFTR "knockout" transgenic mice. These animals express no detectable CFTR mRNA or protein. The investigators excised membrane patches from both normal and CF cells into solutions containing PKA + ATP. Under these conditions, outward rectifiers activated in patches derived from normal airway cells, but not in patches from CF cells. However, in the CF patches, the outward rectifier was present and could be activated by exposure to large depolarizing voltages (as in previous studies). Interestingly, this difference in regulation by PKA was also apparent in membrane patches that showed no detectable low conductance Cl channel activity. This suggests that the CFTR Cl channel does not have to be present within the same membrane patch in order for the difference in PKA-dependent regulation of the ORCC to exist. These results are consistent with the idea that the CF gene product can regulate the activity of other proteins. Theories based on functional activity of CFTR in intracellular membrane compartments, where its activity could modify the processing and/or trafficking of other proteins, have been proposed (Barasch *et al.*, 1991; Bradbury *et al.*, 1992a). According to this hypothesis, the interactions between CFTR and other proteins may be indirect (e.g., as in the studies of Gabriel *et al.*, 1993) and defects in this activity could affect a variety of other cellular processes which are thought to be manifestations of the CF phenotype (e.g., increased airway in Na absorption and changes in the composition of macromolecular secretory products). Further discussion along these lines, however, is beyond the scope of this review.

V. SUMMARY

From this discussion, it is evident that Cl conductances in secretory cells have been identified at a variety of levels of resolution. The role of secondary-active Cl transport in salt secretion (Fig. 1) was identified from transepithelial measurements and the location of Cl conductances at the apical membranes was resolved using microelectrode recordings. The introduction of single-channel recording techniques generated candidates for the apical membrane secretory Cl channel. However, the leap from the Ussing chamber to the patch pipette left in its wake considerable uncertainty as to the single-channel basis of the regulated Cl conductance pathways of secretory epithelial cells. Ultimately, whole-cell recording techniques bridged this gap and identified three regulated Cl current components, two of which are controlled by secretory agonists, the other by cell volume. The primary signal transduction events that control the activities of the secretory Cl conductances have emerged, but our understanding of their detailed modulation is incomplete. The mechanisms that mediate the swelling-induced Cl conductance appear to be complex and, perhaps cell specific. The distinct biophysical properties of these conductance pathways, summarized in Fig. 6, together with the effects of blockers, suggest that different single Cl channels underlie each conductance

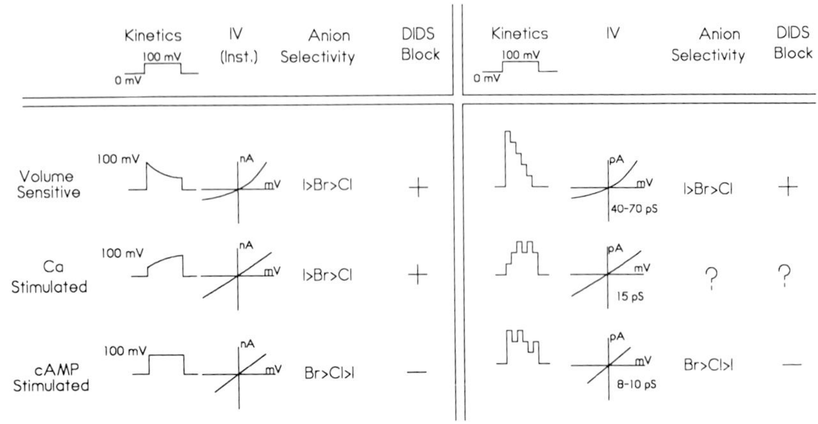

FIGURE 6 Summary of the whole-cell (left) and corresponding single-channel (right) properties of G_{Cl}^{Vol}, G_{Cl}^{Ca}, and G_{Cl}^{cAMP}.

pathway, and over the past several years, candidates for the single-channel basis of each conductance have emerged. Only in the case of CFTR, however, has the structural identity of a specific Cl conductance been verified by expression.

Clearly, there remain many unknowns. Chief among these is the molecular basis of the Ca and volume-activated Cl conductances. Structural information on these channel proteins will provide the framework for an evaluation of their apical vs basolateral locations in epithelial cells and a description of their detailed mechanisms of regulation. The rapid growth of our understanding of CFTR, which has emerged since cloning of the CF gene, will provide a conceptual blueprint for similar studies of the Ca and volume-sensitive Cl channels.

Acknowledgments

The authors are supported by NIH DK31091, DK38518, and DK41330. We thank Jan Tidwell for typing the manuscript, Edward Walthall for graphics, and Roger Worrell for help with the figures.

References

Anderson, M. P., and Welsh, M. J. (1991). Calcium and cAMP activate different chloride channels in the apical membrane of normal and cystic fibrosis epithelia. *Proc. Natl. Acad. Sci. U.S.A.* **88,** 6003–6007.

Anderson, M. P., Berger, H. A., Rich, D. P., Gregory, R. J., Smith, A. E., and Welsh, M. J. (1991a). Nucleoside triphosphates are required to open the CFTR chloride channel. *Cell (Cambridge, Mass.)* **67,** 775–784.

Anderson, M. P., Rich, D. P., Gregory, R. J., Smith, A. E., and Welsh, M. J. (1991b). Generation of cAMP-activated chloride currents by expression of CFTR. *Science* **251,** 679–682.

Anderson, M. P., Gregory, R. J., Thompson, S., Souza, D. W., Paul, S., Mulligan, R. C., Smith, A. E., and Welsh, W. J. (1991c). Demonstration that CFTR is a chloride channel by alteration of its anion selectivity. *Science* **253,** 202–204.

Bajnath, R. B., Augeron, C., Laboisse, C. L., Bijman, J., DeJonge, H. R., and Groot, J. A. (1991). Electrophysiological studies of forskolin-induced changes in ion transport in the human colon carcinoma cell line HT-29cl.19A: Lack of evidence for a cAMP-activated basolateral K^+ conductance. *J. Membr. Biol.* **122,** 239–250.

Bajnath, R. B., Dekker, K., Vaandrager, A. B., DeJonge, H. R., and Groot, J. A. (1992). Biphasic increase of apical Cl^- conductance by muscarinic stimulation of HT-29 Cl.19A human colon carcinoma cell line: Evidence for activation of different Cl^- conductances by carbachol and forskolin. *J. Membr. Biol.* **127,** 81–94.

Banderali, U., and Roy, G. (1992). Activation of K^+ and Cl^- channels in MDCK cells during volume regulation in hypotonic media. *J. Membr. Biol.* **126,** 219–234.

Barasch, J., Kiss, B., Prince, A., Saiman, L., Gruenert, D., and Al-Awqati, Q. (1991). Defective acidification of intracellular organelles in cystic fibrosis. *Nature (London)* **352,** 70–73.

Barrett, K. E., and Dharmsathaphorn, K. (1990). Mechanisms of chloride secretion in a colonic epithelial cell line. *In* "Textbook of Secretory Diarrhea" (E. Lebenthal and M. E. Duffey, eds.), pp. 59–66. Raven Press, New York.

Barthelson, R. A., Jacoby, D. B., and Widdicombe, J. H. (1987). Regulation of chloride secretion in dog tracheal epithelium by protein kinase C. *Am. J. Physiol.* **253,** C802–C808.

Bear, C. E. (1988). Phosphorylation-activated chloride channels in human skin fibroblasts. *FEBS Lett.* **237,** 145–149.

Bear, C. E., Duguay, F., Naismith, A. L., Kartner, N., Hanrahan, J. W., and Riordan, J. R. (1991). Cl^- channel activity in *Xenopus* oocytes expressing the cystic fibrosis gene. *J. Biol. Chem.* **266,** 19142–19145.

Bear, C. E., Li, C., Kartner, N., Bridges, R. J., Jensen, T. J., Ramjeesingh, M., and Riordan, J. R. (1992). Purification and functional reconstitution of the cystic fibrosis transmembrane conductance regulator (CFTR). *Cell (Cambridge, Mass.)* **68,** 809–818.

Becq, F., Fanjul, M., Mahieu, I., Berger, Z., Gola, M., and Hollande, E. (1992). Anion channels in a human pancreatic cancer cell line (Capan-1) of ductal origin. *Pfluegers Arch.* **420,** 46–53.

Bell, C. L., and Quinton, P. M. (1993). Regulation of CFTR Cl^- conductance in secretion by cellular energy levels. *Am. J. Physiol.* **264,** C925–C931.

Berger, H. A., Anderson, M. P., Gregory, R. J., Thompson, S., Howard, P. W., Maurer, R. A., Mulligan, R., Smith, A. E., and Welsh, M. J. (1991). Identification and regulation of the cystic fibrosis transmembrane conductance regulator-generated chloride channel. *J. Clin. Invest.* **88,** 1422–1431.

Berger, H. A., Travis, S. M., and Welsh, M. J. (1993). Regulation of the cystic fibrosis transmembrane conductance regulator Cl^- channel by specific protein kinases and protein phosphatases. *J. Biol. Chem.* **268**(3), 2037–2047.

Berschneider, H. M., Knowles, M. R., Azizkhan, R. G., and Boucher, R. C., Tobey, N. A., Orlando, R. C., and Powell, D. W. (1988). Altered intestinal chloride transport in cystic fibrosis. *FASEB J.* **2,** 2625–2629.

Boucher, R. C., Cheng, E. H. C., Paradiso, A. M., Stutts, M. J., Knowles, M. R., and Earp, H. S. (1989). Chloride secretory response of cystic fibrosis human airway epithelia: Preservation of calcium but not protein kinase C- and A-dependent mechanisms. *J. Clin. Invest.* **84,** 1424–1431.

Bradbury, N. A., Jilling, T., Berta, G., Sorscher, E. J., Bridges, R. J., and Kirk, K. L. (1992a). Regulation of plasma membrane recycling by CFTR. *Science* **256,** 530–532.

Bradbury, N. A., Jilling, T., Kirk, K. L., and Bridges, R. J. (1992b). Regulated endocytosis in a chloride secretory epithelial cell line. *Am. J. Physiol.* **262,** C752–C759.

Bridges, R. J., Worrell, R. T., Frizzell, R. A., and Benos, D. J. (1989). Stilbene disulfonate blockade of colonic secretory Cl^- channels in planar lipid bilayers. *Am. J. Physiol.* **256,** C902–C912.

Bubien, J. K., Zhou, L.-J., Bell, P. D., Frizzell, R. A., and Tedder, T. F. (1993). Transfection of the CD20 cell surface molecule into ectopic cell types generates a Ca^{2+} conductance found constitutively in B lymphocytes. *J. Cell Biol.* **121**(5), 1121–1132.

Cahalan, M. D., and Lewis, R. S. (1988). Role of potassium and chloride channels in volume regulation by T lymphocytes. *In* "Cell Physiology of Blood" (R. Gunn and J. Parker, eds.), pp. 281–301. Rockefeller Univ. Press, New York.

Cartwright, C. A., McRoberts, J. A., Mandel, K. G., and Dharmsathaphorn, K. (1985). Synergistic action of cyclic adenosine monophosphate- and calcium-mediated secretion in a colonic epithelial cell line. *J. Clin. Invest.* **76,** 1837–1842.

Castro, G. A., Harari, Y., and Russel, D. (1987). Mediators of anaphylaxis-induced ion transport changes in small intestine. *Am. J. Physiol.* **253,** G540–G548.

Champigny, G., Verrier, B., Gerard, C., Mauchamp, J., and Lazdunski, M. (1990). Small-conductance chloride channels in the apical membrane of thyroid cells. *FEBS Lett.* **259,** 263–268.

Chang, E. B., Wang, N. S., and Rao, M. C. (1985). Phorbol ester stimulation of active anion secretion in intestine. *Am. J. Physiol.* **249,** C356–C361.

Chao, A. C., and Mochizuki, H. (1992). Niflumic and flufenamic acids are potent inhibitors of chloride secretion in mammalian airway. *Life Sci.* **51,** 1453–1457.

Chen, J. H., Schulman, H., and Gardner, P. (1989). A cAMP-regulated chloride channel in lymphocytes that is affected in cystic fibrosis. *Science* **243,** 657–660.

Cheng, S. H., Gregory, R. J., Marshall, J., Paul, S., Souza, D. W., White, G. A., O'Riordan, C. R., and Smith, A. E. (1990). Defective intracellular transport and processing of CFTR is the molecular basis of most cystic fibrosis. *Cell (Cambridge, Mass.)* **63,** 827–834.

Cheng, S. H., Rich, D. P., Marshall, J., Gregory, R. J., Welsh, M. J., and Smith, A. E. (1991). Phosphorylation of the R-domain by cAMP dependent protein kinase regulates the CFTR chloride channel. *Cell (Cambridge, Mass.)* **66,** 1027–1036.

Christensen, O., Simon, M., and Randlev, T. (1989). Anion channels in a leaky epithelium. A patch clamp study of *Necturus* choroid plexus. *Pfluegers Arch.* **415,** 37–46.

Cliff, W. H., and Frizzell, R. A. (1990). Separate Cl^- conductances activated by cAMP and Ca^{2+} in Cl secretory epithelial cells. *Proc. Natl. Acad. Sci. U.S.A.* **87,** 4956–4960.

Cliff, W. H., Worrell, R. T., Morris, A. P., and Frizzell, R. A. (1991). Conductance pathways involved in chloride secretion and their regulation. *Adv. Exp. Med. Biol.* **290,** 197–208.

Cliff, W. H., Schoumacher, R. A., and Frizzell, R. A. (1992). cAMP-activated Cl channels in CFTR-transfected CF pancreatic epithelial cells. *Am. J. Physiol.* **262,** C1154–C1160.

Cohn, J. A., Melhus, O., Page, L. J., Dittrich, K. L., and Vigna, S. L. (1991). CFTR: Development of high affinity antibodies and localization in sweat gland. *Biochem. Biophys. Res. Commun.* **181,** 36–43.

Cragoe, E. J., Jr., Woltersdorf, O. W., Jr., Gould, N. P., Pietruszkiewicz, A. M., Ziegler, C., Sakurai, Y., Stokker, G. E., Anderson, P. S., Bourke, R. S., Kimelberg, H. K., Nelson, L. R., Barron, K. D., Rose, J. R., Szarowski, D., Popp, A. J., and Waldman, J. B. (1986). Agents for the treatment of brain edema. 2.[2,3,9,9a-tetrahydro-3-oxo-9a-substituted-1H-fluoren-7-yl)oxy] alkanoic acids and some of their analogues. *J. Med. Chem.* **29,** 825–841.

Crawford, I., Maloney, P. C., Zeitlin, P. L., Guggino, W. B., Hyde, S. C., Turly, H.,

Gatter, K. C., Harris, A., and Higgins, C. F. (1991). Immunocytochemical localization of the cystic fibrosis gene product, CFTR. *Proc. Natl. Acad. Sci. U.S.A.* **88,** 9262–9266.

Cunningham, S. A., Worrell, R. T., Benos, D. J., and Frizzell, R. A. (1992). cAMP stimulated ion currents in *Xenopus* oocytes expressing CFTR cRNA. *Am. J. Physiol.* **262,** C783–C788.

Dalemans, W., Barbry, P., Champigny, G., Jallat, S., Dott, K., Dreyer, D., Crystal, R. G., Pavirani, A., Lecocq, J.-P., and Lazdunski, M. (1991). Altered chloride ion channel kinetics associated with the ΔF508 cystic fibrosis mutation. *Nature (London)* **354,** 526–528.

DeBin, J. A., and Strichartz, G. R. (1991). Chloride channel inhibition by the venom of the scorpion *Leiurus quinquestriatus*. *Toxicon* **29,** 1403–1408.

DeJonge, H. R., and Rao, M. C. (1990). Cyclic nucleotide dependent kinases. *In* "Textbook of Secretory Diarrhea" (E. Lebenthal and M. E. Duffey, eds.), pp. 191–207. Raven Press, New York.

DeJonge, H. R., Scholte, B. J., French, P. J., Bijman, J., and Nairn, A. C. (1993). Phosphorylation of CFTR by different isotypes of cyclic GMP-dependent protein kinase. *Pediatr. Pulm. Suppl.* (abstr.) **9,** 212.

Denning, G. M., Ostedgaard, L. S., Cheng, S. H., Smith, A. E., and Welsh, M. J. (1992a). Localization of cystic fibrosis transmembrane conductance regulator in chloride secretory epithelia. *J. Clin. Invest.* **89,** 339–349.

Denning, G. M., Anderson, M. P., Amara, J. F., Marshall, J., Smith, A. E., and Welsh, M. J. (1992b). Processing of mutant cystic fibrosis transmembrane conductance regulator is temperature sensitive. *Nature (London)* **358,** 761–764.

DeRoos, A. D. G., Schultz, B. D., Venglarik, C. J., Singh, A. K., Frizzell, R. A., and Bridges, R. J. (1993). Glybenclamide blockade of CFTR. *Pediatr. Pulm. Suppl.* (abstr.) **9,** 213.

Devor, D. C., and Duffey, M. E. (1992). Carbachol induces a K^+, Cl^-, and non-selective cation conductance in the T84 cell line: A perforated patch-clamp study. *Am. J. Physiol.* **265,** C780–C787.

Devor, D. C., and Frizzell, R. A. (1993). Calcium-mediated agonists activate an inwardly rectified K^+ channel in colonic secretory cells. *Am. J. Physiol.* **265,** C1271–C1280.

Devor, D. C., Sekar, M. C., Frizzell, R. A., and Duffey, M. E. (1993). Taurodeoxycholate activates potassium and chloride conductances via an IP_3-mediated release of calcium from intracellular stores in a colonic cell line (T84). *J. Clin. Invest.* **92,** 2173–2181.

Dharmsathaphorn, K., McRoberts, J. A., Mandel, K. G., Tisdale, L. D., and Masui, H. (1984). A human colonic cell line that maintains vectorial electrolyte transport. *Am. J. Physiol.* **246,** G204–G208.

Drumm, M. L., Pope, H. A., Cliff, W. H., Rommens, J. M., Marvin, S. A., Tsui, L.-C., Collins, F. S., Frizzell, R. A., and Wilson, J. M. (1990). Correction of the cystic fibrosis defect *in vitro* by retrovirus-mediated gene transfer. *Cell (Cambridge, Mass.)* **62,** 1227–1233.

Drumm, M. L., Wilkinson, D. J., Smit, L. S., Worrell, R. T., Strong, T. V., Frizzell, R. A., Dawson, D. C., and Collins, F. S. (1991). Chloride conductance expressed by ΔF508 and other mutant CFTRs in *Xenopus* oocytes. *Science* **254,** 1797–1799.

Duszyk, M., French, A. S., and Man, S. F. P. (1992). Noise analysis and single-channel observations of 4 pS chloride channels in human airway epithelia. *Biophys. J.* **61,** 583–587.

Egan, M., Flotte, T., Afione, S., Solow, R., Zeitlin, P. L., Carter, B. J., and Guggino, W. B. (1992). Defective regulation of outwardly rectifying Cl^- channels by protein kinase A corrected by insertion of CFTR. *Nature (London)* **358,** 581–584.

Field, M., Plotkin, G. R., and Silen, W. (1968). Effects of vasopressin, theophylline and cyclic adenosine monophosphate on short circuit current across isolated rabbit ileal mucosa. *Nature (London)* **217,** 469–471.

Field, M., Fromm, D., Al-Awqati, Q., and Greenough, W. B. (1972). II. Effect of cholera toxin on ion transport across isolated ileal mucosa. *J. Clin. Invest.* **51,** 796–804.

Fondacaro, J. D., and Henderson, L. S. (1985). Evidence for protein kinase C as a regulator of intestinal electrolyte transport. *Am. J. Physiol.* **12,** G422–G426.

Frizzell, R. A., and Cliff, W. H. (1992). No common motif. *Curr. Biol.* **2**(6), 285–287.

Frizzell, R. A., Field, M., and Schultz, S. G. (1979). Sodium-coupled chloride transport by epithelial tissues. *Am. J. Physiol.* **236,** F1–F8.

Frizzell, R. A, Rechkemmer, G., and Shoemaker, R. L. (1986). Altered regulation of airway epithelial cell chloride channels in cystic fibrosis. *Science* **233,** 558–560.

Gabriel, S. E., Clarke, L. L., Boucher, R. C., and Stutts, M. J. (1993). CFTR and outward rectifying chloride channels are distinct proteins with a regulatory relationship. *Nature (London)* **363,** 263–266.

Goldstein, J. L., Nash, N. T., Al-Bazzaz, F., Layden, T. J., and Rao, M. C. (1988). Rectum has abnormal ion transport but normal cAMP-binding proteins in cystic fibrosis. *Am. J. Physiol.* **254,** C719–C724.

Gray, M. A., Greenwell, J. R., and Argent, B. E. (1988). Secretin-regulated chloride channel on the apical plasma membrane of pancreatic duct cells. *J. Membr. Biol.* **105,** 131–142.

Gray, M. A., Pollard, C. E., Harris, A., Coleman, L., Grenwell, J. R., and Argent, B. E. (1990). Anion selectivity and block of the small-conductance chloride channel on pancreatic duct cells. *Am. J. Physiol.* **259,** C752–C761.

Grinstein, S., and Foskett, J. K. (1990). Ionic mechanisms of cell volume regulation in leukocytes. *Annu. Rev. Physiol.* **52,** 399–414.

Halm, D. R., and Frizzell, R. A. (1990). Chloride channels in epithelial cells. *Curr. Top. Membr. Transp.* **37,** 247–282.

Halm, D. R., and Frizzell, R. A. (1992). Anion permeation in an apical membrane chloride channel of a secretory epithelial cell. *J. Gen. Physiol.* **99**(3), 339–366.

Halm, D. R., Rechkemmer, G. R., Schoumacher, R. A., and Frizzell, R. A. (1988). Apical membrane chloride channels in a colonic cell line activated by secretory agonists. *Am. J. Physiol.* **254,** C505–C511.

Hartman, J., Huang, Z., Rado, T. A., Peng, S., Jilling, T., Muccio, D. D., and Sorscher, E. J. (1992). Recombinant synthesis, purification and nucleotide binding characteristics of the first nucleotide binding domain of the cystic fibrosis gene product. *J. Biol. Chem.* **267,** 6455–6458.

Hazama, A., and Okada, Y. (1988). Ca-sensitivity of volume-regulatory K and Cl channels in cultured human epithelial cells. *J. Physiol. (London)* **402,** 687–702.

Heintze, K., Stewart, C. P., and Frizzell, R. A. (1983). Sodium-dependent chloride secretion across rabbit descending colon. *Am. J. Physiol.* **244,** G357–G365.

Henderson, R. M., Ashford, M. L., MacVinish, L. J., and Cuthbert, A. W. (1992). Chloride channels and anion fluxes in a human colonic epithelium (HCA-7). *Br. J. Pharmacol.* **106**(1), 109–114.

Hoshi, T., Zagotta, W. N., and Aldrich, R. W. (1990). Biophysical and molecular mechanisms of Shaker potassium channel inactivation. *Science* **250,** 533–538.

Hwang, T.-C., Lu, L., Zeitlin, P. L., Gruenert, D. C., Huganir, R., and Guggino, W. B. (1989). Cl^- channels in CF: Lack of activation by protien kinase C and cAMP-dependent protein kinase. *Science* **244,** 1351–1353.

Kartner, N., Hanrahan, J. W., Jensen, T. J., Naismith, A. L., Sun, S., Ackerly, C. A., Reyes, E. F., Tsui, L.-C., Rommens, J. M., Bear, C. E., and Riordan, J. R. (1991).

Expression of the cystic fibrosis gene in non-epithelial invertebrate cells produces a regulated anion conductance. *Cell (Cambridge, Mass.)* **64,** 681–691.

Kartner, N., Augustinas, O., Jensen, T. J., Naismith, A. L., and Riordan, J. R. (1992). Mislocalization of ΔF508 CFTR in cystic fibrosis sweat gland. *Nat. Genet.* **1,** 321–327.

Knowles, M. J., Clarke, L. L., and Boucher, R. C. (1991). Activation by extracellular nucleotides of chloride secretion in the airway epithelia of patients with cystic fibrosis. *N. Engl. J. Med.* **325,** 533–538.

Krauss, R. D., Bubien, J. K., Drumm, M. L., Zheng, T., Peiper, S. C., Collins, F. S., Kirk, K. L., Frizzell, R. A., and Rado, R. A. (1992). Transfection of wild-type CFTR into cystic fibrosis lymphocytes restores chloride conductance G_i of the cell cycle. *EMBO J.* **11**(3), 875–883.

Krick, W., Disser, J., Hazama, A., Burckhardt, G., and Fromter, E. (1991). Evidence for a cytosolic inhibitor of epithelial chloride channels. *Pfluegers Arch.* **418,** 491–499.

Kunzelmann, K., Gerlach, L., Frobe, U., and Greger, R. (1991a). Bicarbonate permeability of epithelial chloride channels. *Pfluegers Arch.* **417,** 616–621.

Kunzelmann, K., Tilmann, M., Hansen, C. P., and Greger, R. (1991b). Inhibition of epithelial chloride channels by cytosol. *Pfluegers Arch.* **418,** 479–490.

Lambert, I. H. (1987). Effect of arachidonic acid, fatty acids, prostaglandins, and leukotrienes on volume regulation in Ehrlich ascites tumor cells. *J. Membr. Biol.* **98,** 207–221.

Lambert, I. H. (1989). Leukotriene-D_4 induced cell shrinkage in Ehrlich ascites tumor cells. *J. Membr. Biol.* **108,** 165–176.

Landry, D. W., Reitman, M., Cragoe, E. J., Jr., and Al-Awqati, Q. (1987). Epithelial chloride channel development of inhibitory ligands. *J. Gen. Physiol.* **90,** 779–798.

Levesque, P. C., Hart, P. J., Hume, J. R., Kenyon, J. L., and Horowitz, B. (1992). Expression of cystic fibrosis transmembrane regulator Cl^+ channels in heart. *Circ. Res.* **71,** 1002–1007.

Lin, M., Nairn, A. C., and Guggino, S. E. (1992). cGMP-dependent regulation of a chloride channel in T84 cells. *Am. J. Physiol.* **262,** C1304–C1312.

Madara, J. L., Parkos, C., Colgan, S., McLeod, R. J., Nash, S., Mathews, J., Delp, C., and Lencer, W. (1992). Cl secretion in a model intestinal epithelium induced by a neutrophil-derived secretagogue. *J. Clin. Invest.* **89,** 1938–1944.

Marino, C. R., Matovcik, L. M., Gorelick, F. S., and Cohn, J. A. (1991). Localization of the cystic fibrosis transmembrane conductance regulator in pancreas. *J. Clin. Invest.* **88,** 712–716.

Marunaka, Y., and Eaton, D. C. (1990). Chloride channels in the apical membrane of a distal nephron A6 cell line. *Am. J. Physiol.* **258,** C352–C368.

McCann, J. D., Bhalla, R. C., and Welsh, M. J. (1989a). Release of intracellular calcium by two different second messengers in airway epithelium. *Am. J. Physiol.* **257,** L116–L124.

McCann, J. D., Li, M., and Welsh, M. J. (1989b). Identification and regulation of whole cell chloride currents in airway epithelium. *J. Gen. Physiol.* **94,** 1015–1036.

McCann, J. D., Matsuda, J., Garcia, M., Kaczorowski, G., and Welsh, M. J. (1990). Basolateral K^+ channels in airway epithelia. I. Regulation by Ca^{2+} and block by charybdotoxin. *Am. J. Physiol.* **258,** L334–L342.

McDonald, T. V., Nghiem, P. T., Gardner, P., and Martens, C. L. (1992). Human lymphocytes transcribe the cystic fibrosis transmembrane conductance regulator gene and exhibit CF-defective cAMP-regulated chloride current. *J. Biol. Chem.* **267,** 3242–3248.

Morris, A. P., and Frizzell, R. A. (1993a). Ca^{2+}-dependent Cl^- channels in undifferentiated human colonic cells (HT-29). I. Single-channel properties. *Am. J. Physiol.* **264,** C968–C976.

Morris, A. P., and Frizzell, R. A. (1993b). Ca^{2+}-dependent Cl^- channels in undifferentiated human colonic cells (HT-29). II. Regulation and rundown. *Am. J. Physiol.* **264,** C977–C985.

Morris, A. P., Kirk, K. L., and Frizzell, R. A. (1990). Simultaneous analysis of cell Ca^{2+} and Ca^{2+}-stimulated chloride conductance in colonic epithelial cell (HT29). *Cell Regul.* **1**(12), 953–963.

Morris, A. P., Cunningham, S. A., Benos, D. J., and Frizzell, R. A. (1992). Cellular differentiation is required for cAMP but not Ca^{2+} dependent Cl^- secretion in colonic epithelial cells expressing high levels of CFTR. *J. Biol. Chem.* **267,** 5575–5583.

Morris, A. P., Cunningham, S. A., Tousson, A., Benos, D. J., and Frizzell, R. A. (1994). Polarization-dependent apical membrane CFTR targeting underlies cAMP-stimulated Cl^- secretion in epithelial cells. *Am. J. Physiol.* **266,** C254–C268.

Musch, M. W., Nahkla, A. M., and Chang, E. B. (1990). Phorbol-ester stimulated secretion in chicken ileum: Role of arachidonic acid metabolism. *Gastroenterology* **99,** 393–400.

Nagel, G., Hwang, T. C., Natink, K. L., Nairn, A. C., and Gadsby, D. C. (1992). The protein kinase A-regulated cardiac Cl^- channel resembles the cystic fibrosis transmembrane conductance regulator. *Nature (London)* **360,** 81–84.

Parker, I., and Miledi, R. (1988). A calcium-independent chloride current activated by hyperpolarization in *Xenopus* oocytes. *Proc. R. Soc. London, Ser. B* **233,** 191–199.

Picciotto, M. R., Cohn, J. A., Bertuzzi, G., Grengard, P., and Nairn, A. C. (1992). Phosphorylation of the cystic fibrosis transmembrane conductance regulator. *J. Biol. Chem.* **267,** 12742–12752.

Pollard, C. E., Harris, A., Coleman, L., and Argent, B. E. (1991). Chloride channels on epithelial cells cultured from human fetal epididymis. *J. Membr. Biol.* **124,** 275–284.

Prince, L. S., Tousson, A., and Marchase, R. B. (1993). Cell surface labeling of CFTR in T84 cells. *Am. J. Physiol.* **264,** C491–C498.

Quinton, P. M. (1990). Cystic fibrosis: A disease in electrolyte transport. *FASEB J.* **4,** 2709–2717.

Quinton, P. M., and Reddy, M. M. (1992). Control of CFTR chloride conductance by ATP levels through non-hydrolytic binding. *Nature (London)* **360,** 79–91.

Ran, S., and Benos, D. J. (1991). Isolation and functional reconstitution of a 38-kDa chloride channel protein from bovine tracheal membranes. *J. Biol. Chem.* **266,** 4782–4788.

Ran, S., Fuller, C. M., Arrate, M. P., Latorre, R., and Benos, D. J. (1992). Functional reconstitution of a chloride channel protein from bovine trachea. *J. Biol. Chem.* **267,** 20630–20637.

Rao, M. C., and DeJonge, H. R. (1990). Ca and phospholipid-dependent protein kinases. *In* "Textbook of Secretory Diarrhea" (E. Lebenthal and M. E. Duffey, eds.), pp. 209–232. Raven Press, New York.

Rich, D. P., Anderson, M. P., Gregory, R. J., Cheng, S. H., Paul, S., Jefferson, D. M., McCann, J. D., Klinger, K. W., Smith, A. E., and Welsh, M. J. (1990). Expression of cystic fibrosis transmembrane conductance regulator corrects defective chloride channel regulation in cystic fibrosis airway epithelial cells. *Nature (London)* **347,** 358–363.

Rich, D. P., Gregory, R. J., Anderson, M. P., Manavalan, P., Smith, A. E., and Welsh, M. J. (1991). Effect of deleting the R-domain on CFTR-generated chloride channels. *Science* **253,** 205–207.

Riordan, J. R., Rommens, J. M., Kerem, B.-S., Alon, N., Rozmabel, R., Grzelczak, Z., Zielenski, J., Lok, S., Plavsic, N., Chou, J.-L., Drumm, M. L., Iannuzzi, M. C., Collins, F. S., and Tsui, L.-C. (1989). Identification of the cystic fibrosis gene: Cloning and characterization of complementary cDNA. *Science* **245,** 1066–1073.

Rodriguez-Boulan, R., and Nelson, W. J. (1989). Morphogenesis of the polarized epithelial cell phenotype. *Science* **245,** 718–725.

Schoumacher, R. A., Ram, J., Iannuzzi, M. C., Bradbury, N. A., Wallace, R. W., Tom Hon, C., Kelly, D. R., Schmid, S. M., Gelder, F. B., Rado, T. A., and Frizzell, R. A. (1990). A cystic fibrosis pancreatic adenocarcinoma cell line. *Proc. Natl. Acad. Sci. U.S.A.* **87**(10), 4012–4016.

Schultz, B. D., Venglarik, C. J., Bridges, R. J., and Frizzell, R. A. (1993). Regulation of CFTR Cl^- channels by adenine nucleotide phosphates. *FASEB J.* **7**(3), A426 (abstr.).

Sheppard, D. N., and Welsh, M. J. (1992). Effect of ATP-sensitive K^+ channel regulators on cystic fibrosis transmembrane conductance regulator chloride currents. *J. Gen. Physiol.* **100,** 573–591.

Silva, P., Stoff, J., Filed, M., Fine, L., Forrest, N., and Epstein, F. H. (1977). Mechanism of active chloride secretion by shark rectal gland: Role of Na-K-ATPase in chloride transport. *Am. J. Physiol.* **233,** F298–F306.

Simmons, N. L. (1992). Acetylcholine and kinin augmentation of Cl^- secretion stimulated by prostaglandin in a canine renal epithelial cell line. *J. Physiol. (London)* **447,** 1–15.

Singh, A. K., Afink, G. B., Venglarik, C. J., Wang, R., and Bridges, R. J. (1991). Colonic Cl channel blockade by three classes of compounds. *Am. J. Physiol.* **260,** C51–C63.

Smith, P. L., and Frizzell, R. A. (1984). Chloride secretion by canine tracheal epithelium. IV. Basolateral membrane K permeability parallels secretion rate. *J. Membr. Biol.* **77,** 187–199.

Solc, C. K., and Wine, J. J. (1991). Swelling-induced and depolarization-induced Cl^- channels in normal and cystic fibrosis epithelial cells. *Am. J. Physiol.* **261,** C658–C674.

Sorscher, E. J., Fuller, G. M., Bridges, R. J., Tousson, A., Marchase, R. B., Brinkley, B. R., Frizzell, R. A., and Benos, D. J. (1992). Identification of a membrane protein from T84 cells using antibodies made against a DIDS-binding peptide. *Am. J. Physiol.* **262,** C136–C147.

Stutts, M. J., Chinet, T. C., Mason, S. J., Fullton, J. M., Clarke, L. L., and Boucher, R. C. (1992). Regulation of Cl^- channels in normal and cystic fibrosis airway epithelial cells by extracellular ATP. *Proc. Natl. Aad. Sci. U.S.A.* **89,** 1621–1625.

Tabcharani, J. A., Jensen, T. J., Riordan, J. R., and Hanrahan, J. W. (1989). Bicarbonate permeability of the outwardly rectifying anion channel. *J. Membr. Biol.* **112,** 109–122.

Tabcharani, J. A., Low, W., Elie, D., and Hanrahan, J. W. (1990). Low-conductance chloride channel activated by cAMP in the epithelial cell line T84. *FEBS Lett.* **270,** 157–164.

Tabcharani, J. A., Chang, X.-B., Riordan, J. R., and Hanrahan, J. W. (1991). Phosphorylation-regulated Cl^- channel in CHO cells stably expressing the cystic fibrosis gene. *Nature (London)* **352,** 628–631.

Tabcharani, J. A., Chang, X.-B., Riordan, J. R., and Hanrahan, J. W. (1992). The cystic fibrosis transmembrane conductance regulator chloride channel. Iodide block and permeation. *Biophys. J.* **62,** 1–4.

Thomas, P. J., Shenbagamurthi, P., Ysern, X., and Pedersen, P. L. (1991). Cystic fibrosis transmembrane conductance regulator nucleotide binding to a synthetic peptide. *Science* **251,** 555–557.

Trapnell, B. C., Zeitlin, P. L., Chu, C. S., Yoshimura, K., Nakamura, H., Guggino, W. B., Bargon, J., Banks, T. C., Dalemans, W., Pavirani, A., and Crystal, R. G. (1991). Down-regulation of cystic fibrosis gene mRNA transcript levels and induction of the cystic fibrosis chloride secretory phenotype in epithelial cells by phorbol ester. *J. Biol. Chem.* **266,** 10319–10323.

Turner, J. T., James-Kracke, M. R., and Camden, J. M. (1990). Regulation of the neurotensin receptor and intracellular calcium mobilization in HT29 cells. *J. Pharmacol. Exp. Ther.* **253,** 1049–1056.

Valverde, M. A., Diaz, M., Sepulveda, F. V., Gill, D. R., Hyde, S. C., and Higgins, C. F. (1992). Volume-regulated chloride channels associated with the human multidrug resistance p-glycoprotein. *Nature (London)* **355,** 830–833.

Van den Berghe, N., Vaandrager, A. B., Bot, A. G. M., Parker, P. J., and DeJonge, H. R. (1992). Dual role for protein kinase C α as a regulator of ion secretion in the HT29 cl.19A human colonic cell line. *Biochem. J.* **285,** 673–679.

Venglarik, C. J., Schultz, B. D., Frizzell, R. A., and Bridges, R. J. (1994). ATP alters current fluctuation of cystic fibrosis transmembrane conductance regulator: Evidence for a three state gating mechanism. *J. Gen. Physiol.* (in press).

Wagner, J. A., Cozens, A. L., Schulman, H., Gruenert, D. C., Stryer, L., and Gardner, P. (1991). Activation of chloride channels in normal and cystic fibrosis airway epithelial cells by multifunctional calcium/calmodulin-dependent protein kinase. *Nature (London)* **349,** 793–796.

Wangemann, P., Wittner, M., DiStefano, A., Englert, H. C., Lang, H. J., Schlatter, E., and Greger, R. (1986). Cl^- channel blockers in the thick ascending limb of the loop of Henle. Structure-activity relationship. *Pfluegers Arch.* **407,** S128–S141.

Welsh, M. J., and Liedtke, C. M. (1986). Chloride and potassium channels in cystic fibrosis airway epithelia. *Nature (London)* **322,** 467–470.

Welsh, M. J., Smith, P. L., and Frizzell, R. A. (1982). Chloride secretion by canine tracheal epithelium. II. The cellular electrical potential profile. *J. Membr. Biol.* **70,** 227–238.

Welsh, M. J., Smith, P. L., and Frizzell, R. A. (1983). Chloride secretion by canine tracheal epithelium. III. Membrane resistances and electromotive forces. *J. Membr. Biol.* **71,** 209–218.

Willumsen, N. J., and Boucher, R. C. (1989). Activation of an apical Cl^- conductance by Ca^{2+} ionophores in cystic fibrosis airway epithelia. *Am. J. Physiol.* **256,** C226–C233.

Winding, B., Winding, H., and Bindslev, N. (1992). Second messengers and ion channel in acetylcholine-induced chloride secretion in hen trachea. *Comp. Biochem. Physiol. C* **103C**(1), 195–205.

Wine, J., Bradyen, D. J., Hagiwara, G., Krouse, M. E., Law, T. C., Muller, U. J., Solc, C. K., Ward, C. L., Widdicombe, J. H., and Xia, Y. (1991). Cystic fibrosis, the CFTR, and rectifying Cl^- channels. *In* "The Identification of the CF Gene: Recent Progress and New Research Strategies" (L.-C. Tsui, G. Romeo, R. Greger, and S. Giorini, eds.), pp. 253–272. Plenum, New York.

Worrell, R. T., and Frizzell, R. A. (1991). CaMKII mediates the stimulation of chloride conductance by calcium in T84 cells. *Am. J. Physiol.* **260,** C877–C882.

Worrell, R. T., Butt, A. G., Cliff, W. H., and Frizzell, R. A. (1989). A volume-sensitive chloride conductance in human colonic cell line T84. *Am. J. Physiol.* **256,** C1111–C1119.

Yajima, T., Suzuki, T., and Suzuki, Y. (1988). Synergism between calcium-mediated and cyclic-AMP-mediated activation of chloride secretion in isolated guinea pig distal colon. *Jpn. J. Physiol.* **38,** 427–443.

Yang, Y., Devor, D. C., Engelhardt, J. F., Ernst, S. A., Strong, T. V., Collins, F. S., Cohn, J. A., Frizzell, R. A., and Wilson, J. M. (1993). Molecular basis of defective anion transport in L cells expressing recombinant forms of CFTR. *Hum. Mol. Genet.* **2,** 1253–1261.

Zagotta, W. N., Hoshi, T., and Aldrich, R. W. (1990). Restoration of inactivation in mutants of Shaker potassium channels by a peptide derived from ShB. *Science* **250,** 568–571.

CHAPTER 9

$GABA_A$ Receptor-Activated Chloride Channels

David R. Burt

Department of Pharmacology, University of Maryland School of Medicine, Baltimore, Maryland 21201

I. INTRODUCTION

γ-Aminobutyric acid (GABA) is the major inhibitory neurotransmitter in the central nervous system (CNS). GABA is synthesized from glutamic acid, the major excitatory neurotransmitter, by one of two forms of glutamic acid decarboxylase (GAD) (Erlander *et al.*, 1991). About 30% of neurons in the brain, particularly small interneurons, are thought to be GABAergic (contain GAD), and most neurons will respond to GABA by reducing their firing rate. The receptors mediating this effect are designated A and B (Bormann, 1988; Sivilotti and Nistri, 1991). $GABA_B$ receptors, stimulated by baclofen, have not yet been cloned. They are G-protein linked, typically decrease calcium or increase potassium conductances, and are not covered in this review. $GABA_A$ receptors, blocked by bicuculline, are found on virtually all neurons, both pre- and postsynaptically, and include as part of their structure a chloride channel and binding sites for GABA agonists such as muscimol, for benzodiazepines, barbiturates, and certain steroids, which increase the effectiveness of GABA agonists, and for convulsant compounds such as picrotoxin and *t*-butylbicyclophosphorothionate (TBPS), which noncompetitively block the chloride channel (Olsen and Venter, 1986; Biggio *et al.*, 1992; Sieghart, 1992). Much of the interest in $GABA_A$ receptors arises from their being the major site of action of the benzodiazepines, a very important class of drugs used as anticonvulsants, sedative-hypnotics, muscle relaxants, and antianxiety agents. Not all chloride channel-linked GABA receptors are modulated by benzodiazepines, however, and not all benzodiazepine effects are GABA related. The structural basis of these observations has become clearer with the cloning of many subunits of the $GABA_A$ receptor and the partial delineation of their functional roles. This review emphasizes these recent findings.

II. STRUCTURE OF $GABA_A$ RECEPTORS

A. *Relationship to Other Receptors and Ion Channels*

$GABA_A$ receptors belong to a superfamily of ligand-gated ion channels which includes the closely related glycine receptors, also chloride channels, and more distantly related nicotinic cholinergic receptors and the $5HT_3$ subclass of serotonin (5-hydroxytryptamine) receptors (Schofield *et al.*, 1987; Barnard *et al.*, 1989; Maricq *et al.*, 1991). These latter receptors are cation (largely sodium) channels, as are most of the excitatory amino acid (glutamate and aspartate) receptors, which seem to have the same general subunit structure (four putative transmembrane domains, M1-4, starting with a long N-terminal extracellular region and with a long intracellular loop between M3 and M4—see Fig. 1) but show minimal similarity in terms of subunit amino acid sequence (Dingledine *et al.*, 1990; Sommer and Seeburg, 1992). The nicotinic and $5HT_3$ subunits show about 15–20% amino acid identity to $GABA_A$ receptor subunits (Maricq *et al.*, 1991). Cation vs. anion selectivity seems to depend on a few amino acid differences in the M2 channel domain (Galzi *et al.*, 1992). The similarity of the amino acid sequence of glycine receptor subunits to that of $GABA_A$ receptor subunits, about 30%, is almost as great as the similarity of $GABA_A$ receptor subunit families to each other (Betz, 1990). Besides the four transmembrane domains, an area of strong conservation of amino acid sequence among members of the family is a 15-residue Cys–Cys loop in the N-terminal extracellular region, thought to be important in setting up the proper conformation for ligand binding. Kosower (1988) has suggested a less-convincing similarity of $GABA_A$ receptor subunit structure to that of the anion exchange protein (band 3 protein) of erythrocytes. Interestingly, another chloride channel, similar to the voltage-dependent anion channel of mitochondria, seems to copurify with $GABA_A$ receptors from brain (Bureau *et al.*, 1992) and may be associated with it in the plasma membrane.

B. *Subunit Multiplicity*

The brain possesses a large number of $GABA_A$ receptor subunits, at least 17, in five families (α_{1-6}, β_{1-4}, γ_{1-4}, δ, ρ_{1-2}; see Table I), from which to assemble receptor complexes (Burt and Kamatchi, 1991; Cutting *et al.*, 1992; Harvey *et al.*, 1993). The families are defined by similarity of subunit amino acid sequence, about 70% among family members. As for nicotinic (Galzi *et al.*, 1991) and glycine (Langosch *et al.*, 1988) receptors, the

```
  1                                                                                    TESRMHWPGREVHEMSKKGRPQRQR   ρ1

  1  ........QPSQDELKDNTTVFTR.....ILDRLLDGYDNRLRPGLGERVTEVKTDIFVTSFGPVSDHDMEYTIDVFFRQSWKDERLKFKGPMTV.LRLNNLMASKIWT   α1
  1  ........HSSNEPSNMSYVKETV.......DRLLKGYDIRLRPDFGGPPVDVGMRIDVASIDMVSEVNMDYTLTMYFQQSWKDKRLSYSGIPLN.LTLDNRVADQLWV   β1
  1  QKSDDDYEDYASNKTWVLTPKVPEGDVTVILNNLLEGYDNKLRPDIGVKPTLIHTDMYVNSIGPVNAINMEYTIDIFFAQTWYDRRLKFNSTIKV.LRLNSNMVGKIWI   γ2
  1  ...QPHHGARAMNDIGDYVGSNLEISWLPNLDGLMEGYARNFRPGIGGPPVNVALALEVASIDHISEANMEYTMTVFLHQSWRDSRLSYNHTNET.LGLDSRFVDKLWL   δ
 26  REVHEDAHKQVSPILRRSPDITKSPLTKSEQLLRIDDHDFSMRPGFGGPAIPVGVDVQVESLDSISEVDMDFTMTLYLRHYWKDERLSFPSTNNLSMTFDGRLVKKIWV   ρ1
                                               RP  G           V S         M  T        W D RL                   W

 96  PDTFFHNGKKSVAHNMTMPNKLLRITEDGTLLYTMRLTVRAECPMHLEDFPMDAHACPLKFGSYAYTRAEVVYEWTREPARSVVVAEDGSRLNQYDLLGQTVDSGIVQS   α1
 94  PDTYFLNDKKSFVHGVTVKNRMIRLHPDGTVLYGLRITTTAACMMDLRRYPLDEQNCTLEIESYGYTTDDIEFYWNGGEGA..VTGVNKIELPQFSIVDYKMVSKKVEF   β1
109  PDTFFRNSKKADAHWITTPNRMLRIWNDGRVLYTLRLTIDAECQLQLHNFPMDEHSCPLEFSSYGYPREEIVYQWKRSSVE..VGDTRSWRLYQFSFVGLRNTTEVVKT   γ2
106  PDTFIVNAKSAWFHDVTVENKLIRLQPDGVILYSIRITSTVACDMDLAKYPMDEQECMLDLESYGYSSEDIVYYWSENQEQ..IHGLDRLQLAQFTITSYRFTTELMNF   δ
135  PDMFFVHSKRSFIHDTTTDNVMLRVQPDGKVLYSLRVTVTAMCNMDFSRFPLDTQTCSLEIESYAYTEDDLMLYWKKGNDS..LKTDERISLSQFLIQEFHTTTKLAFY   ρ1
     PD    K   H  T  N   R   DG  LY  R T    C        P D   C L    SY Y        W                  L Q

                            M1                            M2                                   M3
205  ST.GEYVVMTTHFHLKRKIGYF |VIQTYLPCIMTVILSQVSFWL| NRES |VPARTVFGVTTVLTMTTLSISA| RNSLPKVAYATA |MDWFIAVCYAFVFSALIEFATV|  α1
201  TT.GAYPRLSLSFRLKRNIGYF |ILQTYMPSTLITILSWVSFWI| NYDA |SAARVALGITTVLTMTTISTHL| RETLPKIPYVKA |IDIYLMGCFVFVFLALLEYAFV|  β1
216  TS.GDYVVMSVYFDLSRRMGYF |TIQTYIPCTLIVVLSWVSFWI| NKDA |VPARTSLGITTVLTMTTLSTIA| RKSLPKVSYVTA |MDLFVSVCFIFVFSALVEYGTL|  γ2
213  KSAGQFPRLSLHFQLRRNRGVY |IIQSYMPSVLLVAMSWVSFWI| SQAA |VPARVSLGITTVLTMTTLMVSA| RSSLPRASAIKA |LDVYFWICYVFVFAALVEYAFA|  δ
242  SSTGWYNRLYINFTLRRHIFFF |LLQTYFPATLMVMLSWVSFWI| DRRA |VPARVPLGITTVLTMSTIITGV| NASMPRVSYIKA |VDIYLWVSFVFVFLSVLEYAAV|  ρ1
      G       F L R          Q Y P        S VSFW           AR   G TTVLTM T               P      A   D       FVF      E

307  NYFT.......KRGYAWDGKSVVPEKPKKVKDPLIKKNNTYAPTATSYTPNLARGD................PGLATIAKSATIEPKEVKPETKPPEPKKT........   α1
303  NYIFFGKGPQKKGASKQDQSANEKNKLEMNKVQVDAHGNILLSTLEIRNETSGSEVLTGVSDPKSTMYSYDSASIQYRKPLSSREGYGRGLDRHGVPGKGRIRRRASQL   β1
318  HYFV............SNRKPSKDKDKKKKNPAPTIDIRPRSATIQMNNATHLQE..........RDEEYGYECLDGKDCASFFCCFEDCRTGAWRHGRIHIRIA...   γ2
316  HFNA...........DYRKKRKAKVKVTKPRAEMDVRNAIVLFSLSAAGVSQELAISRRQGRVPGNLMGSYRSVEVEAKKEGGSRPGGPGGIRSRLKP...........   δ
345  NYLT................TVQERKEQKLREKLPCTSGLPPPRTAMLDGNYSDGE................VNDLDNYMPENGEKPDRMMVQLTLASERSSPQRKSQR   ρ1
                            K
                                  M4
385  ......FNSVSKIDR |LSRIAFPLLFGIFNLVYWATY| LNREPQLKAPTPHQ   α1
412  KVKIPDLTDVNSIDK |WSRMFFPITFSLFNVVYWLYY| VH               β1
403  ...........KMDS |YARIFFPTAFCLFNLVYWVSY| LYL              γ2
405  .......IDADTIDI |YARAVFPAAFAAVNIIYWAAY| TM               δ
422  SSYVSMRIDTHAIDK |YSRIIFPAAYILFNLIYWSIF| S                ρ1
                   D     R  FP       N  YW
```

subunits are thought to be assembled five at a time to form a ring structure surrounding the central chloride channel (Nayeem *et al.*, 1994), which is lined by the second transmembrane domain of each subunit (Bertrand *et al.*, 1992; Galzi *et al.*, 1992; Xu and Akabas, 1993). As discussed in more detail elsewhere (Burt and Kamatchi, 1991), the possible number of distinct receptors which could be made by different combinations and circular permutations of these known subunits is truly astonishing, numbering in the hundreds of thousands, although the actual number of arrangements formed *in vivo* probably is fewer than 100. However, the exact subunit composition, stoichiometry, or arrangement still is not known for even 1 native GABA$_A$ receptor. Nor is it known how neuron-specific arrangements and, for a given neuron, subtype-specific arrangement(s) which are presumed to exist are actually achieved. The rules for assembling receptors are only beginning to be worked out; e.g., $\alpha_1\beta_1$ dual combinations are not formed when a third γ_2 subunit is also present for incorporation (Angelotti and Macdonald, 1993; Angelotti *et al.*, 1993; Draguhn *et al.*, 1990). Similarly, we have only a hint that subunit composition determines intracellular targeting to different locations (Perez-Velazquez and Angelides, 1993). A simpler situation may occur in fish brain, in which GABA$_A$ receptors appear to be homooligomers on protein gels (Deng *et al.*, 1991).

The listing of GABA$_A$ receptor subunits above includes ρ_1 and ρ_2 ("rho" from retina, where they are enriched) on the basis of their amino acid sequence similarity to other GABA$_A$ receptor subunits and the fact that their functional expression yields GABA-gated chloride channels. Other properties of the expressed receptors are sufficiently unique that these subunits have been suggested to belong to a new class of GABA$_C$ receptors (Shimada *et al.*, 1992; Kusama *et al.*, 1993), defined by their insensitivity to blockade by bicuculline and, secondarily, by lack of potentiation by barbiturates or benzodiazepines. Responses for expressed ρ_1-subunits are still sensitive to blockade by picrotoxin. However, the presumably related responses in *Xenopus* oocytes injected with retinal poly(A)$^+$ RNA are less sensitive to picrotoxin and TBPS than typical GABA$_A$ receptor responses (Woodward *et al.*, 1992, 1993). Additionally, some GABA agonists act

FIGURE 1 Conserved structure of GABA$_A$ receptor subunits. (Top) Aligned amino acid sequences of one representative of each family (modified from Burt and Kamatchi, 1991). The sequences are from rat, except for ρ_1, which is human. Fully conserved residues are indicated below each row. An overline connects two Cys residues thought to form a disulfide bond important in shaping the ligand binding domain. The putative transmembrane domains (M1–M4) are boxed. The eight heavy bars above the sequences show the approximate locations of introns in the coding sequences, for subunits where these are known. (Inset) Illustration of transmembrane topology of each subunit.

TABLE I

Vertebrate $GABA_A$ Receptor Subunits[a]

	Amino acids (+ signal) in rat	Sites for N-CHO	Message size in kilobases	Protein size in kilodaltons	Comments
	α: Affect benzodiazepine (BZD) pharmacology, affinity for GABA				
α_1	428 (+27)	2	4.2, 3.8	50, 51	Most abundant; gives type I BZD pharmacology; HCL = 5q31.1–q33.2
α_2	423 (+28)	3	3.6, 6.6	52, 53	Type II BZD pharmacology; HCL = 4p12p13
α_3	465 (+28)	4	4.2	56, 59 & 61	Type II BZD pharmacology; HCL = Xq28
α_4	533 (+19)	3	4.0 (bov.)	>65	Largest subunit; pharmacology like α_6
α_5	433 (+31)	4	2.8	53, 57	Gives unusual type II BZD pharmacology (has been termed α_4); HCL = 15q11q13
α_6	434 (+19)	3	1.8, 2.5	56, 57	Found only in cerebellar granule cells; selective for Ro 15-4513
	β: Needed for efficient assembly; contain conserved protein kinase A sites				
β_1	449 (+25)	3	12	52?	Less abundant than β_2 and β_3; HCL = 4p12p13
β_2	450 (+24)	3	8	55?	Probably most abundant β; HCL = 6
$\beta_{3,3a}$	448 (+25)	3	6.0, 2.5	54?	HCL = 15q11q13, near Angelman/Prader-Willi locus; 5′ end alternative splicing
β_4	459 (+25)	4	—	—	Alternative splicing by use of two splice donor sites. (Not in mammals?)
β_4'	463 by alt. spl. (chick)				
	γ: Enable benzodiazepine binding, zinc insensitivity				
γ_1	430 (+35)	3	3.8	—	Found in glia; gives atypical BZD response. HCL = 4p14–q21.1
γ_{2S}	428 (+38)	3	4.2, 2.8	43–49	Gives typical BZD response; alternative splicing (miniexon) gives new protein kinase C site; HCL = 5q31.1–q33.2
γ_{2L}	436 via alternative splicing				
γ_3	450 (+17)	2	—	—	Gives typical BZD response, with some differences from γ_2
$\gamma 4$	436 (+21) (chick)	3	5.8	—	Not in mammals?

TABLE I
Continued

	Amino acids (+ signal) in rat	Sites for N-CHO	Message size in kilobases	Protein size in kilodaltons	Comments
			δ: Functional role unknown		
δ	433 (+16)	2	2.0, 3.0	54	HCL = 1 (short arm)
		ρ: Give very atypical pharmacology ($GABA_C$?)			
ρ	458 (+15)	3	3.9 (bov.)	—	High in retina; HCl = 6q14q21
ρ_2	445 (+20)	2	—	—	HCL = 6q14q21

[a] Sites for N-CHO are number of amino-terminal extracellular consensus sequences (Asn-X-Ser/Thr) for N-linked glycosylation. Subunit protein sizes are apparent sizes on denaturing gels except α_4, for which the theoretical size without glycosylation is given. Under Comments, BZD is benzodiazepine and HCL is human chromosomal locus. Most data are for rat, but data on β_4 and γ_4 are for chicken and data on ρ_1 and ρ_2 are for human. Selected references on each subunit are as follows: α_1 (Schofield *et al.*, 1987; Buckle *et al.*, 1989; Malherbe *et al.*, 1990a; Buchstaller *et al.*, 1991b; Keir *et al.*, 1991; Wang *et al.*, 1992a; Wilcox *et al.*, 1992); α_2 (Levitan *et al.*, 1988; Buckle *et al.*, 1989; Khrestchatisky *et al.*, 1991; Wang *et al.*, 1992a); α_3 (Levitan *et al.*, 1988; Buckle *et al.*, 1989; Malherbe *et al.*, 1990c; Wang *et al.*, 1992a); α_4 (Ymer *et al.*, 1989a; Wisden *et al.*, 1991); α_5 (Khrestchatisky *et al.*, 1989; Mahlerbe *et al.*, 1990c; Pritchett and Seeburg, 1990; Knoll *et al.*, 1993; Sieghart *et al.*, 1993); α_6 (Lüddens *et al.*, 1990; Kato, 1990); β_1 (Schofield *et al.*, 1987; Buckle *et al.*, 1989; Malherbe *et al.*, 1990a; Kirkness *et al.*, 1991); β_2 (Ymer *et al.*, 1989b; Buchstaller *et al.*, 1991a; Hadingham *et al.*, 1993); β_3 (Ymer *et al.*, 1989b; Wagstaff *et al.*, 1991a, b; Kirkness and Frazer, 1993); β_4 (Bateson *et al.*, 1991; Lasham *et al.*, 1991); γ_1 (Ymer *et al.*, 1990; Wilcox *et al.*, 1992); γ_{2S} (Pritchett *et al.*, 1989b; Shivers *et al.*, 1989; Malherbe *et al.*, 1990b; Benke *et al.*, 1991a; Wilcox *et al.*, 1992); γ_{2L} (Whiting *et al.*, 1990; Kofuji *et al.*, 1991); γ_3 (Wilson-Shaw *et al.*, 1991; Knoflach *et al.*, 1991; Herb *et al.*, 1992); γ_4 (Harvey *et al.*, 1993); δ (Shivers et al., 1989; Sommer *et al.*, 1990; Zhao and Joho, 1990; Benke *et al.*, 1991b; Wang *et al.*, 1992b); ρ_1 (Cutting *et al.*, 1991,1992; Shimada *et al.*, 1992); and ρ_2 (Cutting *et al.*, 1992; Kusama *et al.*, 1993).

as antagonists. Similar responses have been described in perch retina horizontal cells (Qian and Dowling, 1993) and rat bipolar cells (Feigenspan *et al.*, 1993).

C. Genes for Subunits

A major goal in cloning genes for the different subunits is to discover the upstream control elements for subunit message transcription that permit expression in individual neurons of the appropriate combinations of subunits. An unrelated goal is to attempt to map the subunit genes on the chromosome and find if any map to the same locus as known genetic diseases. The former approach has yet to yield much illumination (see, e.g., Sommer *et al.*, 1990; Kirkness and Fraser, 1993), but the latter approach, when applied to the human β_3-subunit gene, yielded an apparent match with the Angelman/Prader-Willi region on chromosome 15 (Wagstaff *et al.*, 1991b; Saitoh *et al.*, 1992). The α_5 gene is just distal, within 100 kbp (Knoll *et al.*, 1993). Identified loci of other subunits are listed in

Table I. Some other subunit genes appear to be clustered, e.g., α_1 and γ_2 on chromosome 5 (Wilcox *et al.*, 1992); α_2, β_1, and γ_1 on chromosome 4 (Wilcox *et al.*, 1992); and ρ_1 and ρ_2 on chromosome 6 (Cutting *et al.*, 1992; also contains β_2, Hadingham *et al.*, 1993), of unknown significance. A direct role of the β_3 gene in Angelman's syndrome appears to have been eliminated by an affected patient in which both parental alleles are still present (Reis *et al.*, 1993), but in mice it maps to a cleft palate locus (Culiat *et al.*, 1993), suggesting a role in facial development.

The subunit genes which have been reported thus far have a very similar organization of 9 or 10 exons, with a highly conserved location, if not size, of introns among subunits and species (Fig. 1). The size of these genes varies from only 13 kb for the mouse δ gene (Sommer *et al.*, 1990) to more than 65 kb for the human β_1 gene (Kirkness *et al.*, 1991) and the chicken β_4 gene (Lasham *et al.*, 1991). The exon numbers or boundaries are not well conserved with nicotinic receptor subunit genes.

D. Alternative Splicing and Diversity

The potential multiplicity of $GABA_A$ receptors which could be formed from combination of 16 or more different gene products would appear to be more than sufficient without any additional means of generating diversity, but such means do exist in the form of alternative splicing of RNA transcripts and post-translational modification of the subunit proteins. Alternative splicing has been demonstrated for the γ_2-subunit (Whiting *et al.*, 1990; Kofuji *et al.*, 1991), where it arises from incorporation of a small exon of only 24 bases; for the β_4-subunit in chickens (Bateson *et al.*, 1991), where it arises from alternative splice donor site selection to add only 12 bases; for the β_2 subunit in chickens, where it arises by inclusion of a 51 base exon (Harvey *et al.*, 1994) and for the β_3-subunit (Kirkness and Fraser, 1993), where an alternative upstream exon 1a for 5′-untranslated and signal sequences is used rarely to give a protein product which barely differs from the major β_3-subunit. In the first three cases, alternative splicing occurs in the putative intracellular loop between the third and fourth transmembrane domains (M3 and M4). For the γ_2-subunit, alternative splicing creates a new site for phosphorylation by protein kinase C (Whiting *et al.*, 1990; Moss *et al.*, 1992a), and this has an important influence on whether GABA responses from expressed receptors containing the subunit are potentiated by ethanol (Wafford *et al.*, 1991; Wafford and Whiting, 1992; but see Sigel *et al.*, 1993). The γ_2 alternative splicing is regionally and developmentally regulated in several species (Whiting *et al.*, 1990; Wang and Burt, 1991; Zahniser *et al.*, 1992); the anterior pituitary gland expresses only the short form (Valerio *et al.*, 1992). The β_3 alternative splicing is also regionally and developmentally regulated (Kirkness and Fraser, 1993). Multiple protein bands with subunit-specific antibodies after N-deglycosylation have suggested possible additional alternative splicing of the α_2- and α_3-subunits (Buchstaller *et al.*, 1991b) and β_2- and β_3 (beyond that known)-subunits (Buchstaller *et al.*, 1991a; Pollard *et al.*, 1991). Extensive alternative splicing, in two regions of the extracellular domain, also exists for the *Drosophila Rdl* $GABA_A$ receptor gene associated with cyclodiene resistance (ffrench-Constant and Rocheleau, 1993).

E. Post-Translational Modifications

Post-translational modification, if cell-type specific, could help generate diversity of $GABA_A$ receptors. Apparent regional differences in effects of enzymatic carbohydrate removal on benzodiazepine binding have been reported (Sweetnam and Tallman, 1986), but this may reflect regional differences in subunit composition. More definitive is evidence for differential *O*-glycosylation of the α_5-subunit (Sieghart *et al.*, 1993). In *Xenopus* oocytes, inhibition of N-linked glycosylation by tunicamycin reduces expression of $GABA_A$ receptors (Sumikawa *et al.*, 1988), and site-directed mutagenesis of α_1-subunit glycosylation sites in 293 cells has a similar effect (Hastings *et al.*, 1992). In several other receptors, N-linked glycosylation seems to be necessary for receptors to attain an active conformation (e.g., possible guidance of subunit assembly and membrane trafficking), but is not required for maintenance of ligand binding activity (Olson and Lane, 1989).

Phosphorylation is an important means of modulating receptor function for ligand-gated ion channels (Swope *et al.*, 1992), including natural $GABA_A$ receptors (e.g., Porter *et al.*, 1990). It probably varies as much with time and circumstance as with cell type. Various $GABA_A$ receptor subunits contain potential sites for phosphorylation by protein kinases A (PKA) and C (PKC) and protein tyrosine kinases (Leidenheimer *et al.*, 1991). For some residues, e.g., Ser-409 of β_1 for PKA, Ser-410 of β_2 for PKA and PKC, Ser-327 of γ_{2S} and Ser-343 of γ_{2L} for PKC, this phosphorylation appears to influence desensitization and other aspects of receptor function (Leidenheimer *et al.*, 1992; Moss *et al.*, 1992a,b; Kellenberger *et al.*, 1992; Wafford and Whiting, 1992). Ser-343 of γ_{2L} is not necessary for the PKC-mediated enhancement of diazepam responses (Leidenheimer *et al.*, 1993).

F. Composition of Native Receptors

Evidence on the composition of native $GABA_A$ receptors is beginning to accumulate from isolation of receptors from different brain regions with subunit-specific antibodies, seeing what other subunits come along. These conclusions can be checked against data on the differential distribution among brain regions of subunits (e.g., from *in situ* hybridization of mRNAs or immunohistochemistry of proteins), seeing which ones seem to occur together, and on the functionality of artificial combinations of cloned subunits expressed in *Xenopus* oocytes or cell lines. It now appears that most receptors in the brain contain one or more α-subunits, one or more β-subunits, and usually a γ- and/or δ-subunit. The most common combination may be α_1–β_2–γ_2 (Benke *et al.*, 1991a; Fritschy *et al.*, 1992; Laurie *et al.*, 1992a; Persohn *et al.*, 1992; Wisden *et al.*, 1992).

The subunit composition of a $GABA_A$ receptor subtype would be easiest to determine for a cell line expressing only that subtype. Descriptions of cell lines expressing complete, functional $GABA_A$ receptors have been late to appear (e.g., for anamolous, partial receptors, Kasckow *et al.*, 1992; Kirkness and Fraser, 1993; Rohde and Harris, 1992) but there are now a few. These include IMR-32 human neuroblastoma cells (Anderson *et al.*, 1993), available from the American Type Culture Collection, and immortalized hypothalamic GT1-7 cells (Hales *et al.*, 1992), designed for expression of gonadotropin-releasing hormone. In both cases, the lack of a benzodiazepine response has suggested the absence of a γ_2-subunit. By use of the polymerase chain reaction (PCR), GT1-7 cells contained α_1, β_1, and β_3-subunit mRNAs, but not α_2, α_5, α_6, β_2, γ_2, or δ-subunit mRNAs (Hales *et al.*, 1992). It is unclear whether the apparent β-subunit multiplicity in the cell line implies receptor subtype heterogeneity, or whether both β_1- and β_3-subunits may be contained in the same receptor molecules.

G. Protein Chemistry

Studies of the $GABA_A$ receptor as a protein have included binding studies, photoaffinity labeling, detergent solubilization, benzodiazepine affinity purification, examination of subunit bands on denaturing gels (SDS–PAGE), and use of monoclonal and polyclonal antibodies against either native receptors or subunit-specific peptides synthesized from sequence knowledge obtained by cloning (Stephenson, 1988; Olsen and Tobin, 1990; DeLorey and Olsen, 1992). Binding studies have employed ligands directed against 3 main loci on the receptor complex: the ligand-binding site (e.g., agonists GABA, muscimol; antagonists bicuculline, gabazine), the benzodiazepine site (e.g., agonists diazepam, flunitrazepam; antagonist flumazenil), or the chloride channel (e.g., TBPS). Flunitrazepam itself is suitable as a photoaffinity label and appears to label primarily α-subunits. $GABA_A$ receptors purified from brain by steps which include detergent solubilization and use of a benzodiazepine affinity column are glycoproteins of ca. 230 kDa that retain the binding properties of membrane-associated receptors. Although they were originally thought to be composed of two α-subunits of ca. 53 kDa and two β-subunits of ca. 57 kDa (Stephenson, 1988), recent cloning of subunits has changed the picture of receptor composition from four to five subunits, as described above. More refined protein chemistry has revealed the existence of multiple α- and β-subunit isoforms, which have mostly been identified with specific cDNAs using selective antibodies (e.g., Buchstaller *et al.*, 1991a,b—see Table I).

One of the most productive lines of investigation at present, noted above, employs subunit-specific antibodies to isolate native receptors from various brain regions by immunoprecipitation or immunoaffinity chromatography and seeing what other subunits are also found in the resulting pellets or fractions. The most abundant subunit, as judged from immunoprecipitation, is α_1 (40% or more of detergent-solubilized receptor binding from whole rat brain, size 50–51 kDa, e.g., McKernan *et al.*, 1991b). There appear to be minor populations of receptors containing α_1-subunits in combination with α_2-, α_3-, and α_6-subunits (Duggan *et al.*, 1991; Endo and Olsen, 1993; Lüddens *et al.*, 1991; Pollard *et al.*, 1993; Zezula and Sieghart, 1991). Both α_2 (28%, 53 kDa)- and α_3 (24%, 59 and 61 kDa)-subunits immunoprecipitate substantial but much lower proportions of receptor binding, while other α-subunits are more minor still (McKernan *et al.*, 1991b). Two nonselective antibodies against β-subunits recognize bands at 57, 54, 53, and 52 kDa in rat brain (Endo and Olsen, 1992), but these bands are not clearly assigned to cloned sequences. One of these, against a protein kinase domain in the M3-M4 intracellular loop, immunoprecipitates 75% of solubilized [^{3}H]muscimol binding from whole rat brain. As already noted, earlier studies had seen multiple bands for both β_2- and β_3-subunits (Buchstaller *et al.*, 1991a; Pollard *et al.*, 1991). Immunoprecipitation with antibodies directed against the N-terminus of the γ_2-subunit brings down about 50% (Benke *et al.*, 1991a) or 75% (Duggan *et al.*, 1992) of solubilized, affinity-purified benzodiazepine binding in rat or bovine brain regions, at an apparent size of 43–49 kDa. Muscimol binding is precipitated in parallel, and a relatively high proportion of receptors (>20%) appear to contain α_1-, $\beta_{2/3}$-, and γ_2-subunits in common (Benke *et al.*, 1991a; Khan *et al.*, 1993). Immunoprecipitation with antisera against peptides contained in the δ-subunit brings down about 20% (just solubilized) or 30% (also affinity purified) of flumazenil binding, at an apparent size of 54 kDa (Benke *et al.*, 1991b). This had suggested coexistence of the δ-subunit and a γ-subunit in some receptors, since benzodiazepine binding requires a γ-subunit (see below). This has been confirmed by immunoaffinity chromatography for γ_2 (Mertens *et al.*, 1993). Such δ and

γ_2 coexistence is not supported by mRNA distribution data (Laurie *et al.*, 1992a; Wisden *et al.*, 1992).

III. FUNCTION OF $GABA_A$ RECEPTORS

A. Electrophysiology of Native and Recombinant Receptors

Electrophysiological studies of $GABA_A$ receptors have moved from voltage-clamp and fluctuation analysis of native receptor populations to patch-clamp studies of individual channels of known subunit composition—if not yet exact arrangement or stoichiometry. The usual response to channel opening is a net inward flux of negative chloride ions (outward current), hyperpolarizing the neuronal membrane and driving it away from firing threshold. Native $GABA_A$ receptors exhibit multiple conductance levels, complex kinetics of activation, with apparent positive cooperativity for agonist and a degree of voltage dependence, and multiple phases of desensitization (Bormann, 1988). Many of these properties vary with source of receptors. They are mimicked to varying extents by recombinant receptors of different subunit compositions (Mathers, 1991) and are shared by other ligand-gated ion channels.

Two systems have been used to date for most studies of the expression of recombinant $GABA_A$ receptors: injection of RNA transcribed *in vitro* into *Xenopus* oocytes (Sigel, 1990) and transient transfection of cDNAs into human embryonic kidney 293 cells (Pritchett *et al.*, 1988). A few studies have used stable transfection (e.g., Im *et al.*, 1992; Moss *et al.*, 1990; Petke *et al.*, 1992; Wong *et al.*, 1992). The oocyte system has also been widely used for expression of native mRNAs for $GABA_A$ receptors, often in conjunction with other methods, such as selective breeding (Wafford *et al.*, 1990), hybrid arrest (Wafford *et al.*, 1991), and careful selection of message according to source and size (Woodward *et al.*, 1992). For some glutamate receptor subunits, it has been shown that the transfection of two together into a cell line leads to the formation of homomeric and heteromeric channels simultaneously (Burnashev *et al.*, 1992), illustrating one of the difficulties of interpretation of all of these methods. Others include possible variations in lipid environment and post-translational processing and incorporation of endogenous proteins (Buller and White, 1990).

Expression of cloned $GABA_A$ receptor subunit cDNAs in oocytes (Blair *et al.*, 1988) or 293 cells (Pritchett *et al.*, 1988) indicates that single subunits poorly but detectably express many properties of $GABA_A$ receptors. (An exception is the ρ_1-subunit, which gives robust expression; Shimada *et al.*, 1992.) Various α-β pairs do so quite well, but without benzodiazepine modulation (Levitan *et al.*, 1988); α–γ pairs and β–γ pairs, in declining order, are less successful (Verdoorn *et al.*, 1990; but see Wong *et al.*, 1992, whose results may have been influenced by endogenous β_3 expression in HEK 293 cells; Kirkness and Fraser, 1993). The best approach to native receptor function is obtained with various α–β–γ combinations. In these combinations, a γ-subunit seems to be required for benzodiazepine modulation of the GABA response (Pritchett *et al.*, 1989b; Ymer *et al.*, 1990; Moss *et al.*, 1991; Puia *et al.*, 1991). As well as conferring sensitivity to modulation by benodiazepines, the γ_2-subunit seems to confer sensitivity to potentiation by lanthanum ions (Im *et al.*, 1992) and insensitivity to inhibition by zinc (Draguhn *et al.*, 1990; Smart *et al.*, 1991). Many aspects of benzodiazepine pharmacology are determined by differences between the various α-subunits (Pritchett *et al.*, 1989a; Pritchett and Seeburg, 1990; Lüddens *et al.*, 1990), which also affect GABA affinity (Levitan *et al.*, 1988). Site-directed mutagenesis has localized some α-subunit differences in functionality to single amino acid residues: Gly at α_1 200 for type I pharmacology (Pritchett and Seeburg, 1991), His at α_1 101 for benzodiazepine agonist and antagonist binding (Wieland *et al.*, 1992).

The most complete study to date of the functional properties of various subunit combinations (Sigel *et al.*, 1990) was flawed by the inadvertent use of a mutant α_1 clone (Sigel *et al.*, 1992), with a Leu substituted for a conserved Phe at residue 64. This resulted in a ca. 200-fold drop in apparent affinities for agonist GABA and antagonist bicuculline and a loss of positive cooperativity of GABA gating. Site-directed mutations in homologous sites in β_2- and γ_2-subunits resulted in only intermediate and small drops in GABA affinity, respectively, and no change in cooperativity or antagonist affinity (Sigel *et al.*, 1992). These results suggest an importance of α-subunits in binding of agonists and antagonists as well as of benzodiazepines.

Combinations giving atypical functionality include those containing α_4 and α_6, which lack classical benzodiazepine agonist binding (Lüddens *et al.*, 1990; Wisden *et al.*, 1991), and γ_1, which may yield apparent agonist activity of benzodiazepine inverse agonists (von Blankenfeld *et al.*, 1990; Puia *et al.*, 1991). γ_3-containing combinations have selectively reduced benzodiazepine agonist affinity (Herb *et al.*, 1992). As already noted, some combinations containing ρ_1-subunits are so atypical as to have been termed $GABA_C$ receptors (Shimada *et al.*, 1992).

As a channel, the $GABA_A$ receptor appears to have two sites that bind chloride and a single file transport mechanism (Bormann *et al.*, 1987). The major observed conductance state of native receptors varies with neuron type, being about 28–30 pS in spinal cord, with others at about 12, 19, and 44 pS (Bormann *et al.*, 1987). By contrast, the main conductance state in hippocampus has generally been 20 pS or less (e.g., Allen and Albuquerque, 1987). In recombinant receptors, the presence of a

γ_2-subunit seems to help confer a higher (32 pS) main single-channel conductance (Verdoorn *et al.*, 1990). In general, drugs appear to affect channel kinetics rather than channel conductances, and membrane depolarization increases probability of channel opening. Complex gating schemes for native receptors have been proposed in which a single GABA molecule is permitted to gate a short-lived open state, but entry into other, longer-duration open states requires binding of two GABA molecules (Macdonald *et al.*, 1989a; Weiss and Magleby, 1989). It is likely that many drugs preferentially bind to one or another conformational state as part of their mechanism (e.g., barbiturates, Macdonald *et al.*, 1989b).

Although most responses to $GABA_A$ receptor stimulation involve neuronal hyperpolarization and inhibition, a number of workers have observed apparent depolarizing responses mediated by $GABA_A$ receptors. Sites include hippocampal pyramidal cell dendrites (e.g., Alger and Nicoll, 1982) and hippocampal interneurons (Michelson and Wong, 1991). In the latter site, depolarization is sensitive to blockade by picrotoxin and bicuculline and appears to be due to the responsive cells possessing higher than usual levels of intracellular chloride. A similar explanation for dendritic depolarization (e.g., Avoli, 1992), involving maintenance of intracellular chloride gradients, is harder to accept. Receptor heterogeneity, with the dendritic receptor also gating another ion, such as bicarbonate, is another possibility (Lambert *et al.*, 1991). Bicarbonate's role is not well supported by recent data (Grover *et al.*, 1993), but the general idea is supported by differential immunofluorescent staining for α_1- and α_3-subunits between soma and dendrites of cerebellar Purkinje cells and hippocampal pyramidal cells (Fritschy *et al.*, 1992). There is still no evidence for differential ion selectivity of receptors made from these subunits, however.

B. Chloride Flux and Binding Measurements

Radioactive chloride flux measurements have been widely used in an attempt to replicate biochemically the chloride current measurements of electrophysiology, only with much more limited time resolution. Although the method has been applied successfully in cultured neurons and brain slices, the most usual preparation is closed membrane vesicles ("synaptoneurosomes" or "microsacs") which include postsynaptic regions (Harris and Allan, 1985; Schwartz *et al.*, 1986). Besides the usual problems with biochemical measurements (tissue heterogeneity, endogenous GABA, and other substances, *etc.*), the long half-life of ^{36}Cl (310,000 years) leads to low specific activity and high expense. Another way to measure chloride flux is by means of newer chloride-sensitive fluorescent indicators. Only limited use to date has been made of these methods for $GABA_A$ receptors (Anderson *et al.*, 1993; Engblom *et al.*, 1991; Wong *et al.*, 1992) and the similar glycine receptors (Garcia-Calvo *et al.*, 1992).

Binding measurements yield only an indirect index of function, through modulation of binding of one ligand (typically muscimol, flunitrazepam, or TBPS) in the presence of a drug of another type. Thus the "GABA shift" consists of the increased affinity of binding of benzodiazepines in the presence of GABA (Tallman *et al.*, 1978) and vice versa. These measurements can be used to predict potencies of GABA agonists or of benzodiazepines as modulators of $GABA_A$ receptors. Similarly, the displacement of [^{35}S]TBPS binding in the presence of GABA can be used to screen for modulators (e.g., Im and Blakeman, 1991).

C. Desensitization

An important phenomenon, which seems to be intrinsic to any functional combination of $GABA_A$ receptor subunits, is desensitization in response to prolonged (a few seconds) application of higher concentrations (ca. 10 μM and above) of GABA, with recovery over a few minutes. This is hard to avoid in measurements by chloride flux (Kardos and Cash, 1990). Continued application of GABA to receptors on intact cells produces a decrease in the GABA-gated current for two reasons: a decrease in the transmembrane chloride gradient and desensitization of the receptors (Huguenard and Alger, 1986). The molecular mechanisms of desensitization of $GABA_A$ receptors still are poorly understood, but desensitization is presumed to represent one or more conformational states of the receptor complex, resulting in bursts of channel opening alternating with longer silent periods (Palotta, 1991). Its speed, extent, and other properties vary with source and type of cells, age in culture, *etc.*, or, in transfected cells, with the subunit composition (e.g., slowed by γ_2-subunit; Verdoorn *et al.*, 1990; Moss *et al.*, 1992b). Among the properties which have been reported for desensitization in cultured hippocampal (Oh and Dichter, 1992) and cortical (Frosch *et al.*, 1992) neurons are independence of the charge transferred or current induced by a given concentration of GABA, but dependence on receptor activation (preexposure to low GABA concentrations does not affect response to high concentration); marked voltage dependence (slower and less extensive at depolarized potentials, almost absent at +30 mV) in whole cells, but not isolated membrane patches; and greater speed in isolated patches. Voltage dependence of desensitization is not seen in cultured retinal ganglion cells (Tauc *et al.*, 1988). Desensitization can be influenced by $GABA_A$ receptor phosphorylation state, e.g., being slowed in $\alpha_1\beta_1$ combinations by β_1-subunit phosphorylation at Ser^{409}

by cAMP-dependent protein kinase (Moss *et al.*, 1992b). In neuronal nicotinic receptors (α_7-subunit homooligomers), mutation of a conserved Leu (247) to Thr in the M2 channel domain appears to convert a desensitized state to a conducting state (Bertrand *et al.*, 1992). Change of Val (251) to Thr has a similar effect (Galzi *et al.*, 1992). Mutation of the equivalent residues have not yet been reported in $GABA_A$ receptor subunits.

D. *Internalization and Clustering*

A phenomenon which may be related to desensitization, only requiring longer periods of time, is agonist-induced downregulation, generally involving internalization followed by sequestration or proteolysis of receptors. Downregulation of $GABA_A$ receptors has been demonstrated in tissue culture (e.g., Roca *et al.*, 1990a), with a half-time of about 25 hr. Internalization of $GABA_A$ receptors has also been described (Tehrani and Barnes, 1991), based on an indirect approach using an impermeant benzodiazepine ligand and detergent permeabilization. In cultured chick cortical neurons, about 7% of benzodiazepine binding sites appeared to be intracellular, but this proportion was increased to 18% by exposure to 1 m*M* GABA for only 4 hr. Many intracellular receptors in the absence of agonist are presumably on their way out (after synthesis) rather than on their way in (internalized) (reviewed in Hopkins, 1992). These proportions of intracellular GABA receptors were somewhat lower than those (about 20%) reported earlier by another group using a similar approach without agonist (Czajkowski and Farb, 1989). This same group found the disappearance of photoaffinity-labeled receptors in cultured chick brain to occur in two phases, with half-lives of about 4 hr (42% of receptors) and 32 hr; following photoinactivation, receptor reappearance was complete in 24 hr (Borden and Farb, 1988). In rat brain, a class of $GABA_A$ receptors with low affinity for flunitrazepam is associated with clathrin-coated, presumably endocytotic vesicles (Tehrani and Barnes, 1993). In a related observation at the message level, 48-hr exposure of cultured chick neurons to 1 m*M* GABA has been shown to decrease mRNA levels for α_1- and α_2-subunits by more than 50% (Montpied *et al.*, 1991a), presumably due to effects on message transcription or stability.

In other systems, agonist-induced internalization is generally preceded by clustering of receptors on the cell surface. Interestingly, the normal state of $GABA_A$ receptors appears to be in clusters, at least in some cultured neurons, as revealed by antibodies (Ventimiglia *et al.*, 1990) or patch-clamp recording (Frosch and Dichter, 1992). The latter study demonstrated that this was true whether the neurons are innervated or not. The relationship of these clusters to synapses or synapse formation is still unknown. "Hot spots" have been seen in sections of mature brain (Fritschy *et al.*, 1992), often in association with synapses, and have also been visualized on cerebellar granule cells by freeze-fracture electron microscopy (Caruncho *et al.*, 1993). This has been an area of intense study for nicotinic cholinergic receptors at the neuromuscular junction (Froehner, 1991), and many of the mechanisms of clustering probably are similar for $GABA_A$ receptors. Cytoskeletal attachment appears to influence pharmacology of glycine receptors (Takagi *et al.*, 1992).

IV. ACTIONS OF DRUGS

A. Agonists and Antagonists

Besides GABA, direct agonists of $GABA_A$ receptors include muscimol, 3-aminopropanesulfonic acid (3-APS), piperidine-4-sulfonic acid, 4,5,6,7-tetrahydroisoxazolo[5,4-c]pyridin-3-ol (THIP), isoguvacine, and isonipecotic acid (Kerr and Ong, 1992). The subject of extensive structure-activity studies, they can be of potency equal to or greater than that of GABA for various responses, and many have been tried as binding ligands (Olsen and Venter, 1986). As noted above, binding of two agonist molecules appears to be needed for efficient gating of native receptors (but not all recombinant ones; Mathers, 1991), and it still is not clear which subunits preferentially bind agonists. Site-directed mutagenesis has suggested importance of both β (Amin and Weiss, 1993) and α (Sigel *et al.*, 1992) subunits.

The classical competitive antagonist of the GABA recognition site is the alkaloid bicuculline. Newer antagonists include gabazine (SR-95531, 2-[3-carboxypropyl]-3-amino-6-*p*-methoxyphenylpyridazinium bromide), pitrazepine, securinine, and the steroid RU5135 (Kerr and Ong, 1992; Rognan *et al.*, 1992). D-Tubocurarine, a classical nicotinic antagonist, also can act at the same site. Except for RU5135 (which acts at glycine receptors also), these drugs are not very potent.

Another class of antagonist is noncompetitive "chloride channel blockers," prominently including the alkaloid picrotoxin (active component picrotoxinin) and synthetic cage convulsants such as TBPS as well as, surprisingly, penicillin G. These exhibit use-dependent blockade, stabilizing closed forms of the Cl^- channel, and decreasing probability of channel opening (e.g., Newland and Cull-Candy, 1992; Woodward *et al.*, 1992). Many other drugs also appear to act at the picrotoxin site, including the convulsant pentylenetetrazol (De Deyn and Macdonald, 1989) and γ-butyrolactones (Holland *et al.*, 1991), which include "inverse agonists" (relative to picrotoxin, i.e., augment GABA function) as well as "agonists" (like picrotoxin) and "antagonists" (block Cl^- channel blockade by picrotoxin). This site may also be a target for certain endogenous steroids.

B. Benzodiazepines and Related Compounds

As already noted, the most important class of drugs acting on $GABA_A$ receptors is the benzodiazepines. An example is diazepam (Valium). They increase frequency of channel opening as well as cause a modest increase in channel mean open time (Study and Barker, 1981). The complex developments in this area introduced a new term into pharmacology, "inverse agonist," to describe drugs which have an effect on $GABA_A$ receptors opposite from that of conventional benzodiazepines (agonists), which increase the probability of channel opening. An example of such drugs, which are better termed negative allosteric effectors, is the β-carboline DMCM (methyl-6,7-dimethoxyl-4-ethyl-β-carboline-3-carboxylate). They are anxiogenic and proconvulsant. There are also conventional antagonists, such as flumazenil (Ro 15-1788, ethyl-8-fluoro-5,6-dihydro-5-methyl-6-oxo-4H-imidaza [1,5-a] [1,4]benzodiazepine-3-carboxylate), which prevent the actions of both positive and negative allosteric effectors. Partial agonists, such as clonazepam, bretazenil, divaplon, and abecarnil, with high affinity but low efficacy, have generated the most recent interest as optimal therapeutic agents for anxiety and panic reactions with low incidence of side effects such as sedation, ataxia, or tolerance (Haefely *et al.*, 1990; Scaf, 1991; Puia *et al.*, 1992). The benzodiazepine binding site accomodates a variety of drugs of diverse structure, making the subject, which is full of arbitrary drug company designations and obscure abbreviations, too complex for more than a brief consideration here (see Gardner *et al.*, 1993, for a more complete review).

Benzodiazepine receptors have been divided into two classes, type I and type II, based on their affinity for the triazolopyridazine CL 218,872 (I, high affinity), and the type I pharmacology is associated with the presence of the α_1-subunit (Pritchett *et al.*, 1989a; Zezula and Sieghart, 1991). Another drug, the imidazopyridine zolpidem, further subdivides the type II receptors, giving intermediate affinity for combinations containing α_2- and α_3-subunits and very low affinity for those containing α_5-subunits (Pritchett and Seeburg, 1990; Puia *et al.*, 1991). Similarly, the partial inverse agonist (and behavioral alcohol antagonist) Ro 15-4513 (sapmazenil) binds with high affinity only to subunit combinations containing α_4 (Wisden *et al.*, 1991)- and α_6 (Lüddens *et al.*, 1990)-subunits, from which it cannot be competed by typical benzodiazepine agonists. The requirement for γ-subunits, generally γ_2, for any of these effects has already been described. Imidazoquinoxalines (Petke *et al.*, 1992) are another class of drugs, not already mentioned, which interact with benzodiazepine receptors. They include both agonists (U-79098) and antagonists (U-78875).

The actions of benzodiazepines are of such high affinity and specificity, paralleling observations for opiates, that the discovery of endogenous opioids soon led to the search for endogenous benzodiazepines. Initial interest centered on purines such as inosine, but more recent attention has turned to polypeptides such as diazepam binding inhibitor (DBI), β-carbolines such as ethyl-β-carboline-3-carboxylate (β-CCE), originally extracted from human urine (and formed by cyclization of tryptophan), actual benzodiazepines (perhaps from diet), and "endozepines" with presumably related but undetermined structures (Rothstein *et al.*, 1992). DBI, containing 86 amino acids, is a flumazenil-sensitive, negative allosteric modulator at $GABA_A$ receptors (Costa and Guidotti, 1991), as are several smaller peptides derived from it [e.g., octadecaneuropeptide (ODN), DBI:33–50 and triakontatetraneuropeptide (TTN), DBI:17–50]. The β-carbolines include positive as well as negative allosteric effectors; e.g., abecarnil is anxiolytic. The functional significance of any of these compounds remains to be established.

The above discussion has ignored another whole class of benzodiazepine receptors, the peripheral benzodiazepine receptors, which are associated with mitochondria (Gavish *et al.*, 1992; McEnery *et al.*, 1992; Parola *et al.*, 1993) and have a distinct pharmacology, including affinity for DBI. Effects include modulation of steroid biosynthesis (Cavallaro *et al.*, 1992; Krueger and Papadopoulos, 1992), making an interesting connection with the effects of steroids on $GABA_A$ receptors (see below). Some benzodiazepine effects *in vivo* could be related to interactions with still other receptors, e.g., for adenosine (Phillis and O'Regan, 1988) or thyrotropin-releasing hormone (Sharif and Burt, 1984).

C. Barbiturates

Barbiturates are important general anesthetics, sedatives, and hypnotic drugs, although much less so than before the discovery of benzodiazepines and other newer, safer drugs. One of them, phenobarbital, still is an important anticonvulsant. There are not thought to be any endogenous barbiturates, a blessing, although endogenous steroids described in the next section may amount to the same thing. Barbiturates stabilize $GABA_A$ receptors in the open and conducting configuration (Study and Barker, 1981), at the single-channel level prolonging burst duration by favoring longer-lived open states (Macdonald *et al.*, 1989b), and also are capable of opening these channels by themselves at higher concentrations. The latter activity is more characteristic of pentobarbital, used as a general anesthetic, than of phenobarbital, useful in epilepsy, and depends on different structural determinants than GABA agonist activation (Amin and Weiss, 1993). Barbiturates seem to be active on virtually any subunit combination, with the exception of ρ_1 homooligomers (Shimada *et al.*, 1992), as already mentioned.

D. Steroids

Certain endogenous steroids such as progesterone, deoxycorticosterone, and some of their metabolites produce barbiturate-like effects on $GABA_A$ receptors of the central nervous system at concentrations similar to those found in the brain (reviewed in Majewska, 1992; Paul and Purdy, 1992). This mechanism is distinct from the usual steroid actions in the nucleus (McEwen, 1991). So-called neurosteroids, which include pregnenolone, dehydroepiandrosterone, and their sulfates and fatty acid esters, seem to be synthesized *de novo* in the brain, primarily in glia (oligodendrocytes). Some steroids, like some barbiturates, have been used as anesthetics (e.g., alphaxalone). *In vivo,* these steroids may contribute to known influences of stress or reproductive status on behavior, e.g., seizure threshold (Wilson, 1992). The most potent steroid activator of $GABA_A$ receptors (n*M* potency) is 5α-pregnane-3α-ol-20-one (tetrahydroprogesterone, allopregnanolone), which increases in response to stress (Purdy *et al.,* 1991). Steroids resemble barbiturates in acting from outside the cell, in being insensitive to flumazenil, in prolonging GABA-induced chloride conductance (burst duration) of $GABA_A$ receptors, and in opening channels alone in higher (μM) concentration. However, they appear to bind to a slightly different site of the $GABA_A$ receptor chloride channel complex in order to do so, as deduced from additivity of barbiturate and steroid effects on binding and chloride flux (e.g., Turner *et al.,* 1989) and the fact that they also increase channel opening frequencies (Twyman and Mcdonald, 1992). Some steroid derivatives, including pregnenolone sulfate, dehydroepiandrosterone sulfate, cortisol, and the synthetic RU5135, inhibit $GABA_A$ receptor function noncompetitively (μM potency for first three; RU5135 is low nM), through actions at more than one class of site. Different combinations of clones subunits appear to differ in their response to steroids (e.g., Zaman *et al.,* 1992b), correlating with reported regional differences in steroid potency (Lan *et al.,* 1991). The picture has been complicated by the recent discovery of other potential sites of neurosteroid action, including glycine receptors (Wu *et al.,* 1990; Prince and Simmonds, 1992), calcium channels (Ffrench-Mullen and Spence, 1991), and the *N*-

methyl-D-aspartic acid (NMDA) subclass of excitatory amino acid receptors (Wu *et al.*, 1991).

E. Other General Anesthetics

A number of general anesthetic agents are thought to act, at least in part, by enhancing GABAergic inhibitory neurotransmission (Tanelian *et al.*, 1993). Besides barbiturates and steroids, discussed above, these include the injectable anesthetics etomidate and althesin (Saffan) and several inhalational anesthetics (Nakahiro *et al.*, 1989; Jones *et al.*, 1992). Recently added to the list is propofol (2,6-diisopropylphenol), a new short-acting intravenous general anesthetic which resembles barbiturates and steroids in its ability to prolong burst duration, to open channels by itself at higher concentrations (bicuculline sensitive), and to produce these effects without block by the benzodiazepine antagonist flumazenil (Hales and Lambert, 1991). It was inactive intracellularly and produced a modest enhancement of glycine-induced currents. Most of these observations have been paralleled in binding and chloride flux measurements and in some *in vivo* observations of inhibition of neuronal firing (Peduto *et al.*, 1991). Findings include lack of effect on flunitrazepam binding, enhancement of GABA binding and of TBPS binding in the presence of GABA (inhibits without GABA), and flumazenil-insensitive enhancement of chloride flux into membrane vesicles in the presence of GABA (no effect without). Data suggest that γ-subunits are involved in some of these responses as well as responses to benzodiazepines (L.-H. Lin *et al.*, 1993a; Uchida *et al.*, 1993). Other potential targets for general anesthetics include excitatory amino acid receptors (L.-H. Lin *et al.*, 1993b).

F. Alcohol

Alcohol is a very important drug in terms of its social use and abuse. There is a large and rather controversial literature on the effects on ethanol on $GABA_A$ receptors (reviewed, with other potential targets, in Deitrich *et al.*, 1989; Harris and Allan, 1989; Gonzales and Hoffman, 1991). Besides $GABA_A$ receptors, prime contenders for sites of ethanol's actions include NMDA receptors (e.g., Lima-Landman and Albuquerque, 1989) and $5HT_3$ receptors (e.g., Lovinger and White, 1991). The potentiating effect of ethanol on $GABA_A$ receptors seems to depend on many factors, including brain region (e.g., cerebellum is more sensitive than hippocampus; Proctor *et al.*, 1992), adrenergic modulation (A. M.-Y. Lin *et al.*, 1993), genetic background (e.g., selection for long sleep, LS, vs short sleep, SS, mice; Allan and Harris, 1986; Wafford *et al.*, 1990), subunit composition (e.g., need for γ_{2L}-subunit, Wafford *et al.*, 1991; but see Sigel *et al.*, 1993), and post-translational modification (Wafford and Whiting, 1992). Subunit composition differences would appear to explain regional and genetic differences, but nature may not be so cooperative. Thus the hippocampus contains γ_{2L}-subunits, and there is no difference in γ_{2L}-subunit mRNA levels in different brain regions between LS and SS mice (Zahniser *et al.*, 1992, by quantitative PCR; we have verified this in whole brains by RNase protection, J. B. Wang and D. R. Burt, unpublished results). A possible explanation is in terms of differential $GABA_A$ receptor post-translational processing between LS and SS mice, particularly phosphorylation of the serine residue in the alternatively spliced γ_{2L}-subunit. In a different model of alcohol sensitivity, the alcohol tolerant (AT) and alcohol nontolerant (ANT) rats, the major explanation of another behavioral difference, sensitivity to diazepam-induced ataxia in ANT rats, appears to be in terms of an Arg to Gln change in the α_6-subunit extracellular domain at residue 100 (Korpi *et al.*, 1993). This Gln residue is equivalent to 101 in the α_1-subunit, where a His residue similarly provides diazepam sensitivity (Wieland *et al.*, 1992). The change correlates with the increased efficacy of benzodiazepine agonists in competing for cerebellar binding of [^{3}H]Ro 15-4513 binding in ANT rats (Uusi-Oukari and Korpi, 1991). In both models, there could also be contributions involving other subunits not yet examined, e.g., α_4, or other potential targets for ethanol such as nicotinic, NMDA, kainate, or $5HT_3$ receptors.

G. Calcium, Zinc, Lysine

There are a few miscellaneous substances in the body which affect $GABA_A$ receptors and may be of physiological relevance. The most obviously relevant is calcium, an important intracellular mediator of signal transduction which affects phosphorylation status. Firing-evoked calcium entry reduces $GABA_A$ currents in hippocampal pyramidal cells (Pitler and Alger, 1992), perhaps by inducing dephosphorylation (Chen *et al.*, 1990; but see Shirasaki *et al.*, 1992, for a possible direct effect of ATP). Zinc is an ion with a less-defined role which also inhibits $GABA_A$ receptor function in some cells, mainly by reducing the frequency of channel opening (Legendre and Westbrook, 1991; Smart, 1992). As already noted, the γ_2-subunit seems to confer zinc insensitivity. The amino acid lysine has GABA-enhancing or -mimicking effects *in vivo* and *in vitro* (Chang *et al.*, 1991).

H. Regulation by Chronic Drugs

On chronic administration, several of the drugs which affect $GABA_A$ receptors acutely seem to produce regulatory changes in the properties of the receptors distinct from the agonist-induced immediate desensitization and slower downregulation already described. Rather, there appears to be a functional "uncoupling" of the drug from the receptor. The most likely explanations for these effects, which are not unlike those observed for other receptors (Hadcock and Malbon, 1991), are in terms of regulatory changes in receptor subunit composition, phosphorylation, etc., and such changes are now becoming established. The two classes of drugs which have drawn the most attention, both of obvious clinical relevance, are chronic benzodiazepines and chronic alcohol. *In vivo* and *in vitro*, both benzodiazepines (e.g., Roca *et al.*, 1990b) and alcohol (e.g., Mhatre and Ticku, 1989; Morrow *et al.*, 1990) produce changes in $GABA_A$ receptor-related binding and function which accompany tolerance and dependence, and these changes have been shown to be accompanied further by changes (up and down) in subunit mRNA levels (for benzodiazepines, Heninger *et al.*, 1990; Kang and Miller, 1991; O'Donovan *et al.*, 1992; for alcohol, Buck *et al.*, 1991; Montpied *et al.*, 1991b). Benzodiazepine inverse agonists seem to produce changes opposite those of agonists (Primus and Gallager, 1992), and, in rats chronically treated with alcohol, results include a cerebellar decrease in α_1 message accompanied by an increase in α_6 message (Mhatre and Ticku, 1992a; Morrow *et al.*, 1992). These subunit mRNA changes, which suggest regulatory subunit switching, are being followed up at the protein level (Mhatre *et al.*, 1993). There is no simple correlation among the behavioral, biochemical, and molecular biological measurements in terms of timing, so that it is likely that several mechanisms are occurring at once for both alcohol and benzodiazepines. These and drugs with similar actions listed above (e.g., steroids, barbiturates, other anesthetics) produce various degrees of cross-tolerance and dependence (e.g., Buck and Harris, 1990; Crabbe *et al.*, 1991; Allan *et al.*, 1992), suggesting some mechanisms in common.

V. DISTRIBUTION OF GABA RECEPTORS

A. Central Nervous System

$GABA_A$ receptors are essentially ubiquitous in the brain and spinal cord. A major enterprise at present is to map the distribution of individual subunits, through their mRNAs (*in situ* hybridization; e.g., Laurie *et al.*, 1992a; Persohn *et al.*, 1992; Wisden *et al.*, 1992) or subunit-specific antibodies (immunohistochemistry, e.g., Benke *et al.*, 1991a,b; Zimprich *et al.*, 1991; Fritschy *et al.*, 1992). The goal is not only to help establish the subunit composition of native receptors in different neurons or brain regions, by seeing which subunits go together, as discussed above, but also to see which combinations of subunits occur in neurons or regions of known function as a possible guide to developing more selective drugs for those combinations and functions.

Notable findings in this arena include the almost exclusive localization (in adults; Poulter *et al.*, 1992) of α_6-subunits to synapses of cerebellar granule cells (Kato, 1990; Lüddens *et al.*, 1990; Baude *et al.*, 1992; α_1 and α_5 also occur there, Bovolin *et al.*, 1992a,b) and predominant localization of ρ_1-subunits to the retina (Cutting *et al.*, 1991). In each case, as already noted, pharmacological peculiarities of $GABA_A$ receptors in the host neurons accompany the subunit's presence. Similarly, there is a frequent association of δ with α_4- or α_6-subunit mRNAs in neurons (thalamic nuclei or cerebellar granule cells, respectively) not responding to benzodiazepine agonists. Many past studies of $GABA_A$ receptor heterogeneity have contrasted hippocampus and cerebellum, and newer studies of subunit distribution find clear differences between these regions, including a preponderance of α_5 and β_1 in the hippocampus and of α_1 and $\beta_{2/3}$ in the cerebellum. Space considerations preclude a more detailed description of the distribution of the subunits. The interested reader is referred to Laurie *et al.*, (1992a), Persohn *et al.* (1992), and Wisden *et al.* (1992) for the most complete studies of subunit mRNAs to date and to Fritschy *et al.* (1992) for the most complete study of antibody staining.

B. Peripheral Nervous System

Many cells of the peripheral nervous system also respond to GABA through $GABA_A$ receptors (Erdö and Wolff, 1990; Ong and Kerr, 1990). Prominent loci include dorsal root ganglia, where receptor presence probably reflects GABAergic synapses on primary afferent terminals in the dorsal horn of the spinal cord, and the enteric nervous system, where GABA neurons affect gut motility (mostly $GABA_B$). Other sites where GABA appears to play a role in peripheral innervation include the gallbladder, urinary bladder, and lung. In sympathetic ganglia (e.g., Newland and Cull-Candy, 1992), GABA is taken up from and released from glia, but is also present in preganglionic neural elements.

C. Glial Cells

An important locus of nonneuronal $GABA_A$ receptors is glial cells of the CNS and periphery (reviewed in von Blankenfeld and Kettenmann, 1992). Most glial cells, e.g., cultured astrocytes and oligodendrocytes, have a relatively high, pump-maintained, internal chloride concentration of about 40 mM, so that net $GABA_A$ receptor currents are inward (depolarizing, corresponding to chloride efflux). The properties of these responses are generally similar to those of the typical hyperpolarizing responses in neurons (e.g., blocked by bicuculline and picrotoxin and potentiated by benzodiazepines, barbiturates, and steroids), but there are exceptions. Astrocytic $GABA_A$ receptors differ from most in that DMCM slightly enhances GABA-induced currents (Bormann and Kettenmann, 1988), a pharmacology mimicked by various γ_1-subunit-containing combinations (von Blankenfeld *et al.*, 1990; Puia *et al.*, 1991) and associated with the presence of γ_1-subunits in astrocytes (Bovolin *et al.*, 1992b), along with α_1, α_2, β_1, and β_3 (and some γ_{2S} and γ_3). Oligodendrocytic $GABA_A$ receptors differ from most by showing lack of positive cooperativity for GABA (von Blankenfeld *et al.*, 1991), also found for many double-subunit combinations (Sigel *et al.*, 1990). The functional role of glial $GABA_A$ receptors remains unknown, but suggestions have included buffering extracellular chloride, similar to the presumed glial role in buffering extracellular potassium (Ritchie, 1992).

D. GABA Receptors Outside the Nervous System

$GABA_A$ receptors are found on many types of cells outside the central and peripheral nervous systems (Erdö and Wolff, 1990; Ong and Kerr, 1990). Especially prominent are selected components of the endocrine system, including glucagon- and somatostatin-secreting cells of pancreatic islets (Michalik and Erecinska, 1992), catecholamine-secreting chromaffin cells of the adrenal medulla, anterior pituitary cells of several types (Valerio *et al.*, 1992), especially lactotrophs, and melanotrophs of the pituitary intermediate lobe (e.g., Louiset *et al.*, 1992). Pancreatic GABA probably comes from GAD in insulin-secreting β cells. This GAD appears to be the initial target of autoantibodies in insulin-dependent (juvenile or type I) diabetes (Baekkeskov *et al.*, 1990). In most of these sites, $GABA_A$ receptors serve to reduce hormone secretion. The adrenal medulla contains $GABA_A$ receptor α_1- and β_{1-3}-subunits but not α_2- or α_3-subunits, as determined from cDNA library screening and Northern blots (Ymer *et al.*, 1989b). A PCR-based study of subunits present in the rat anterior pituitary gland (Valerio *et al.*, 1992) found evidence for messages for α_1-, β_1-, β_3-, and γ_{2S}-subunits, but not α_5-, β_2-, γ_{2L}-, and δ-subunits. The direct biochemical evidence fo $GABA_A$ receptors in pancreatic islets is thus far just from antibodies against brain β-subunits (Rorsman *et al.*, 1989), with good functional data as well. Another site for GABA and $GABA_A$ receptors, apparently without neuronal involvement, is the female reproductive system, particulary ovary, fallopian tube, and uterus. Regulation by reproductive steroids of $GABA_A$ receptors in this site (Majewska and Vaupel, 1991) makes sense, but such regulation even seems to occur in the brain (Finn and Gee, 1993). Other sites with some evidence for $GABA_A$ receptors include pancreatic exocrine cells, renal tubular epithelium, hepatocytes, gastric mucosa, cerebral arteries, oocytes, sperm, platelets, thymus, and spleen.

E. Developmental Changes

There is some evidence that GABA functions as a trophic factor during development (e.g., Madtes and Redburn, 1983; Wolff *et al.*, 1993). Indeed, many embryonic neurons appear to go through a stage of transient expression of GABA immunoreactivity (Schaffner *et al.*, 1993). Although there is no clear division into embryonic and adult subunits, the early appearance of some $GABA_A$ receptor subunits in embryos may be related to this function. Moreover, GABA is frequently depolarizing early in nervous system development, with little desensitization, and this provides a major source of excitation to developing neurons which may be important in appropriate circuit formation (Cherubini *et al.*, 1991). A developmental increase in benzodiazepine sensitivity of the hippocampus has been reported (Rovira and Ben-Ari, 1991), presumably reflecting changes in γ_2 or other subunits (e.g., Killisch *et al.*, 1991). A similar developmental decrease in sensitivity to zinc inhibition of GABA responses in cultured superior cervical ganglion neurons (Smart, 1992) may again involve the γ_2-subunit. Possible developmental exchanges among functionally dissimilar subunits would resemble apparent switching among subunits during drug tolerance and dependence, as discussed above, and may reflect changes in innervation patterns and receptor roles, while exchanges among functionally similar subunits could change receptor regulatory properties. Among α-subunit proteins in several regions of rat brain, α_1- and α_5-subunits are very low at birth and increase dramatically with age up to about 3 months, whereas α_2 and α_3 are present in high levels at birth and continue to be high, with a tendency to decline in older rats (McKernan *et al.*, 1991a). These results generally agree with changes in α-subunit mRNA levels in rats (e.g., Garrett *et al.*, 1990; Gambarana *et al.*, 1991) and mice (Wang *et al.*, 1992a), except that α_5 mRNAs tend to start high at birth and drop with development (MacLennan *et al.*, 1991; Bovolin *et al.*, 1992a; Laurie *et al.*, 1992b; Poulter *et al.*, 1992). For rat β-subunit mRNAs, β_1 is high at birth and declines to adult levels by 5–7 days of age (Garrett *et al.*, 1990), while β_2 and β_3 increase following birth to a maximum at about 3 weeks (Gambarana *et al.*, 1991). The β_3 mRNA appears relatively early in embryonic develoment (Laurie *et al.*, 1992b; Poulter *et al.*, 1993). In mouse cerebellum, β_1 mRNA appears later than β_2 or β_3 (Zdilar *et al.*, 1992). Total γ_2 mRNAs show a modest general increase with age (e.g., Gambarana *et al.*, 1991, in rats). In both mice (Wang and Burt, 1991) and rats (Bovolin *et al.*, 1992a), the short form (γ_{2S}) is high at birth and changes little with development, while the long form (γ_{2L}) shows a dramatic developmental increase. In rats, γ_1 and γ_3 mRNAs drop with development (Laurie *et al.*, 1992b). In mice, the γ_1 mRNA shows little change with postnatal development (Wang *et al.*, 1994). The δ mRNA shows a developmental increase similar to that of α_1 in mice (Wang *et al.*, 1992b) and rats (Laurie *et al.*, 1992b). In the prenatal rat spinal cord, a detailed study of 13 receptor subunit mRNAs by *in situ* hybridization and qualitative PCR (Ma *et al.*, 1993) has revealed three major patterns of expression: (1) The $\alpha_4\beta_1\gamma_1$ combination is expressed early (embryonic day 13, about as soon as GABA appears) and transiently, and appears to be associated with proliferating zones and neurogenesis. (2) The $\alpha_5\beta_2\gamma_3$ combination is also transiently expressed, only in the mantle zone. (3) The α_2, α_3, β_3, and γ_2 subunits are widely coexpressed in gray matter, with dramatic early increases and a persistence into adulthood. These four subunits are most clearly associated with spinal synapses. Developmental changes of $GABA_A$ receptors in the brain continue into old age, with major decreases in α_1 message in cortex but not cerebellum, where α_6 message actually increases (Mhatre and Ticku, 1992b).

F. Invertebrate GABA Receptors

GABA is a major inhibitory neurotransmitter in invertebrates as well as in vertebrates, but differs in playing a role in neuromuscular transmission as well as in neuron–neuron communication. Most described invertebrate GABA receptors resemble vertebrate $GABA_A$ receptors in being chloride channels, some of which are modulated by benzodiazepines and barbiturates, but there are important pharmacological differences as well (Sattelle *et al.*, 1991). Although many GABA responses are blocked by picrotoxin, most are insensitive to bicuculline. Some responses are excitatory and are carried by sodium. Invertebrate GABA receptors appear to be important targets for certain insecticides, including lindane and cyclodienes such as dieldrin, chlordane, and endrin (Eldefrawi and Eldefrawi, 1987).

The ability of insects to develop resistance to cyclodienes has been used indirectly to clone a putative $GABA_A$ receptor subunit from the fruit fly *Drosophila melanogaster* (ffrench-Constant *et al.*, 1991). Resistant strains were crossed to sensitive ones to map the single locus (*Rdl*) conferring resistance, and chromosome walking in conjunction with X-ray-induced chromosome rearrangement was used to identify a cosmid clone which conferred sensitivity to resistant flies receiving it through P-element-mediated germ-line transformation. This genomic clone was used to screen a cDNA library, yielding the final cDNA. When translated, this gave a $GABA_A$ receptor subunit-like protein (27–29% identity to mammalian subunits), whose differences from mammalian $GABA_A$ receptor subunits include 80 extra amino acids, mostly glycines, in the M3–M4 intracellular loop. In the absence of an intron in the middle of M2, the structure of the gene (>25 kbp) resembles mammalian genes for glycine receptors more than those for $GABA_A$ receptors (ffrench-Constant and Rocheleau, 1992), and there is also a slightly higher (31–32%) amino acid identity to mammalian glycine receptor subunits. The gene shows extensive alternative splicing (ffrench-Constant and Rocheleau, 1993). The mutation which prevents channel block by dieldrin or picrotoxinin is an Ala to Ser or Gly transition in M2 (ffrench-Constant *et al.*, 1993a,b).

Another invertebrate $GABA_A$ receptor subunit has been cloned by homology screening of a genomic library from the pond snail *Lymnaea stagnalis* with a bovine β_1-subunit clone (Harvey *et al.*, 1991). The corresponding cDNA was isolated by a PCR variant (rapid amplification of cDNA ends, RACE), after unsuccessful attempts to screen cDNA libraries. Not suprisingly, the resulting subunit is very β like (ca. 50% amino acid identity). Indeed, in combination with bovine α_1-subunits, it can replace β-subunits to form heteromeric functional channels in *Xenopus* oocytes. By itself it forms homomeric channels with a finite probability of spontaneous opening in the absence of GABA, sensitive to blockade by bicuculline and cyclodienes, and directly activated by peripheral benzodiazepine receptor ligands (Aaman *et al.*, 1992a). These results emphasize the very strong conservation of $GABA_A$ receptor subunits through evolution.

VI. GABA RECEPTORS AND DISEASE

A. Epilepsy

Many considerations argue for an involvement of GABA in at least some forms of inherited and acquired epilepsy (Meldrum, 1989; Tasker and Dudek, 1991). Clinically, some of the most effective drugs in controlling seizures, benzodiazepines and barbiturates, augment GABA-mediated inhibition through actions at $GABA_A$ receptors. Experimentally, seizures can be induced by a variety of drugs which interfere with GABA's actions, whether at the level of its biosynthesis, release, or receptors. These effects are mediated at discrete sites (Gale, 1989). More generalized changes in parameters related to GABA, including $GABA_A$ receptors, have been reported in several animal models of inherited epilepsy. For instance, binding and chloride flux changes accompany susceptibility to audiogenic seizures in DBA/2 mice (e.g., Yu *et al.*, 1986; Schwartz *et al.*, 1989), and there is a greater reduction in α_1-subunit mRNA following chronic ethanol in withdrawal seizure prone (wsp) mice (genetically selected for susceptibility to handling-induced seizures following ethanol withdrawal) than in withdrawal seizure resistant (wsr) mice (Buck *et al.*, 1991). A similar connection has been made for $GABA_B$ receptors and absence seizures in *lethargic* mice (Hosford *et al.*, 1992). Many genes contributing to epilepsy are being mapped in mice (e.g., Rise *et al.*, 1991) and humans (e.g., Durner *et al.*, 1991), and some such genes are likely to be related to $GABA_A$ receptors.

B. Alcoholism

As noted above, $GABA_A$ receptors appear to be an important site for the acute effects of alcohol, and there are regulatory changes in receptor function and subunit message levels that accompany alcohol tolerance and dependence, suggesting a role for $GABA_A$ receptors in these phenomena. There are changes in other potential alcohol targets, e.g., NMDA receptors and voltage-dependent calcium channels (Buck and Harris, 1991), as well. Although obviously related, tolerance and dependence are distinct from alcoholism, or the abuse of alcohol, and there is less evidence linking $GABA_A$ receptors to alcoholism (Samson and Harris, 1992). The reward aspects of alcohol consumption and abuse are more likely to involve another neurotransmitter system, such as dopamine (Koob, 1992).

C. Hepatic Encephalopathy

Liver disease, particularly with shunting of intestinal blood so as to bypass the liver, is frequently associated with a severe depression of brain function termed hepatic encephalopathy. Although the disease is commonly related to disturbances of nitrogen metabolism (excess ammonia production), there is considerable evidence for excess stimulation of $GABA_A$ receptors, perhaps by elevated levels of endogenous benzodiazepine agonists (Basile *et al.*, 1991). The evidence includes amelioration of symptoms in humans by flumanzenil, a benzodiazepine receptor antagonist. Available data do not exclude a more direct contribution from changes in $GABA_A$ receptors themselves.

D. Other Diseases

Drugs acting at $GABA_A$ receptors influence a variety of behaviors, presumably reflecting the wide distribution of these receptors in the brain. Both factors suggest that disorders of $GABA_A$ receptors could contribute to a varitety of brain disorders, including anxiety, depression, schizophrenia, and disorders of movement, eating, sleeping, agression, memory, etc. Genetic mapping may be expected to connect many other diseases to $GABA_A$ receptor subunits.

VII. COMPARISON WITH GLYCINE RECEPTORS

As already noted, glycine is another inhibitory neurotransmitter in the central nervous system, particularly important in the spinal cord and brain stem. It is also an important modulator of NMDA receptors. Glycine receptors are perhaps best known as the major site of action of the antagonist strychnine, a convulsant. Like $GABA_A$ receptors, they are chloride channels, with a similar set of four conductance states (Hamill *et al.*, 1983). (Also like $GABA_A$ receptors, they are potently blocked by the synthetic steroid RU5135.) Several subunits have now been cloned by the group of Betz in Germany (Betz, 1990, 1991) and shown to resemble closely $GABA_A$ receptors in their primary amino acid sequences. These include at least four α-subunits, which bind glycine and strychnine (weakly for an α_2 variant, α_2[E167G], sometimes present in neonatal rat spinal cord; α_3 and α_4 are only minor constituents at all ages), and at least one β-subunit, whose mRNA appears to be much more widely distributed in brain than mRNAs for known α-subunits. Other similarites of glycine receptors to $GABA_A$ receptors include desensitization by agonist and alternative splicing of α_1 and α_2 mRNAs. In the case of the α_1 (adult)-subunit, the latter adds eight amino acids including a possible phosphorylation site (by alternative splice acceptor site selection) in almost the same place as eight amino acids are added to the $GABA_A$ receptor γ_2-subunit (and four to $GABA_A$ β_4), just after M3. The glycine receptor α_2-subunit (neonatal) alternative splicing involves alternate exon usage in the extracellular domain, to produce little apparent difference in the final proteins. As is the case for $GABA_A$ receptors, phosphorylation by protein kinase A has been reported to modulate glycine receptor function (Song and Huang, 1990), in this case increasing the probability of channel opening. Reduced expression of the adult isoform of the glycine receptor, detected by monoclonal antibody 2b against an epitope on the N-terminus of the α_1 subunit, has been linked to the mouse mutant *spastic* (Becker *et al.*, 1992), which exhibits symptoms resembling strychnine poisoning. No accompanying change in α_1-subunit message levels was detected by Northern blots. Some interactions between $GABA_A$ and glycine receptors probably occur at the cellular level. Many neurons contain both types of receptor (some even contain both transmitters, Todd and Sullivan, 1990), and crossed desensitization between the two receptors has been reported (Barker and McBurney, 1979; Hamill *et al.*, 1983).

VIII. FUTURE PROSPECTS

After all of the $GABA_A$ receptor subunit genes are mapped and characterized, the use of embryonic stem cells and homologous recombination to facilitate receptor subunit gene deletion or modification one at a time in transgenic mice (Zimmer, 1992) should facilitate understanding of the roles of the many $GABA_A$ receptor subunits *in vivo* and perhaps provide mouse models of human disease. Similarly, in cultured neurons, use of *Herpes simplex*-based vectors (Geller *et al.*, 1991) to add or subtract subunits, or just antisense oligonucleotides in the medium for subtraction (Listerud *et al.*, 1991), could provide a useful correlate *in vitro*. Further progress may be expected in the development of cell lines which express defined combinations of subunits, either through immortalization of a given cell type (Cepko, 1989), as for GT1 cells (Hales *et al.*, 1992), or permanent transfection of subunits into an existing cell line (Moss *et al.*, 1990). These should prove invaluable in the empirical development of more selective drugs. Similarly, increased understanding of endogenous compounds acting on $GABA_A$ receptors, including various peptides, endozapines, and steroids, could lead to new classes of drugs. Drug design will be aided by site-directed mutagenesis to improve knowledge of key residues in drug interactions.

The major immediate problem remains just to sort out the subunit arrangements of native $GABA_A$ receptors, ideally starting with cell lines or tissues expressing a single subtype. From these data we should be able to learn the rules for assembling subunits. Then will come the goal of determining how the rules are set through transcriptional controls and subunit–subunit interactions during assembly. There will be a lot more site-directed mutagenesis in all of this. A long-term goal, of course, is complete three-dimensional structure determination.

Acknowledgments

I thank Drs. Jay Yang and Ganesan Kamatchi for helpful comments. This work was supported in part by USPHS Grant AA07559 and by a Special Research Initiative award from the UMAB Designated Research Initiative Fund.

References

Alger, B. E., and Nicoll, R. A. (1982). Pharmacological evidence for two kinds of GABA receptor on rat hippocampal pyramidal cells studied in vitro. *J. Physiol.* (*London*) **328,** 125–141.

Allan, A. M., and Harris, R. A. (1986). Gamma-aminobutyric acid and alcohol actions: Neurochemical studies of long sleep and short sleep mice. *Life Sci.* **39,** 2005–2015.

Allan, A. M., Zhang, X., and Baier, L. D. (1992). Barbiturate tolerance: Effects on GABA-operated chloride channel function. *Brain Res.* **588,** 255–260.

Allen, C. N., and Albuquerque, E. X. (1987). Conductance properties of GABA-activated chloride currents recorded from cultured hippocampal neurons. *Brain Res.* **410,** 159–163.

Amin, J., and Weiss, D. S. (1993). $GABA_A$ receptor needs two homologous domains of the β-subunit for activation by GABA but not by pentobarbital. *Nature* (*London*) **366,** 565–569.

Anderson, S. M. P., De Souza, R. J., and Cross, A. J. (1993). The human neuroblastoma cell line, IMR-323, possesses a $GABA_A$ receptor lacking the benzodiazepine modulatory site. *Neuropharmacology* **32,** 455–460.

Angelotti, T. P., and Macdonald, R. L. (1993). Assembly of $GABA_A$ receptor subunits: $\alpha_a\beta_1$ and $\alpha_1\beta_1\gamma_{2S}$ subunits produce unique ion channels with dissimilar single-channel properties. *J. Neurosci.* **13,** 1429–1440.

Angelotti, T. P., Uhler, M. D., and Macdonald, R. L. (1993). Assembly of $GABA_A$ receptor subunits: Analysis of transient single-cell expression using a fluorescent substrate/marker gene technique. *J. Neurosci.* **13,** 1418–1428.

Avoli, M. (1992). Synaptic activation of $GABA_A$ receptors causes a depolarizing potential under physiological conditions in rat hippocampal cells. *Eur. J. Neurosci.* **4,** 16–26.

Baekkeskov, S., Aanstoot, H.-J., Christgau, S., Reetz, A., Solimena, M., Cascalho, M., Folli, F., Richter-Olesen, H., DeCamilli, P., and DeCamilli, P. (1990). Identification of the 64K autoantigen in insulin-dependent diabetes as the GABA-synthesizing enzyme glutamic acid decarboxylase. *Nature* (*London*) **347,** 151–156.

Barker, J. L., and McBurney, R. N. (1979). GABA and glycine may share the same conductance channel on cultured mammalian neurones. *Nature* (*London*) **277,** 234–236.

Barnard, E. A., Burt, D. R., Darlison, M. G., Fujita, N., Levitan, E. S., Schofield, P. R., Seeburg, P. H., Squire, M. D., and Stephenson, F. A. (1989). Molecular biology of the

$GABA_A$ receptor. *In* "Allosteric Modulation of Amino Acid Receptors: Therapeutic Implications" (E. A. Barnard and E. Costa, eds.), pp. 19–30. Raven Press, New York.

Basile, A. S., Jones, E. A., and Skolnick, P. (1991). The pathogenesis and treatment of hepatic encephalopathy: Evidence for the involvement of benzodiazepine receptor ligands. *Pharmacol. Rev.* **43,** 27–71.

Bateson, A. N., Lasham, A., and Darlison, M. G. (1991). γ-Aminobutyric $acid_A$ receptor heterogeneity is increased by alternative splicing of a novel β-subunit gene transcript. *J. Neurochem.* **56,** 1437–1440.

Baude, A., Sequier, J.-M., McKernan, R. M., Olivier, K. R., and Somogyi, P. (1992). Differential subcellular distribution of the $\alpha 6$ versus the $\alpha 1$ and $\beta 2/3$ subunits of the $GABA_A$/benzodiazepine receptor complex in granule cells of the cerebellar cortex. *Neuroscience* **51,** 739–748.

Becker, C.-M., Schmieden, V., Tarroni, P., Strasser, U., and Betz, H. (1992). Isoform-selective deficit of glycine receptors in the mouse mutant *spastic*. *Neuron* **8,** 283–289.

Benke, D., Mertens, S., Trzeciak, A., Gillessen, D., and Mohler, H. (1991a). $GABA_A$ receptors display association of $\gamma 2$-subunit with $\alpha 1$ and $\beta 2/3$-subunits. *J. Biol. Chem.* **266,** 4478–4483.

Benke, D., Mertens, S., Trzeciak, A., Gillessen, D., and Mohler, H. (1991b). Identification and immunohistochemical mapping of $GABA_A$ receptor subtypes containing the δ-subunit in rat brain. *FEBS Lett.* **283,** 145–149.

Bertrand, D., Devillers-Thiéry, A., Revah, F., Galzi, J.-L., Hussy, N., Mulle, C., Bertrand, S., Ballivet, M., and Changeux, J.-P. (1992). Unconventional pharmacology of a neuronal nicotinic receptor mutated in the channel domain. *Proc. Natl. Acad. Sci. U.S.A.* **89,** 1261–1265.

Betz, H. (1990). Ligand-gated ion channels in the brain: The amino acid receptor superfamily. *Neuron* **5,** 383–392.

Betz, H. (1991). Glycine receptors: Heterogeneous and widespread in the mammalian brain. *Trends Neurosci.* **14,** 458–461.

Biggio, G., Concas, A., and Costa, E., eds. (1992). "GABAergic Synaptic Transmission: Molecular Pharmacological, and Clinical Aspects," Adv. Biochem. Psychopharmacol., Vol. 47. Raven Press, New York.

Blair, L. A. C., Levitan, E. S., Marshall, J., Dionne, V. E., and Barnard, E. A. (1988). Single subunits of the $GABA_A$ receptor form ion channels with properties of the native receptor. *Science* **242,** 577–579.

Borden, L. A., and Farb, D. H. (1988). Mechanism of γ-aminobutyric acid/benzodiazepine receptor turnover in neuronal cells: Evidence for non-lysosomal degradation. *Mol. Pharmacol.* **34,** 354–362.

Bormann, J. (1988). Electrophysiology of $GABA_A$ and $GABA_B$ receptor subtypes. *Trends Neurosci.* **11,** 112–116.

Bormann, J., and Kettenmann, H. (1988). Patch-clamp study of γ-aminobutyric acid receptor Cl^- channels in cultured astrocytes. *Proc. Natl. Acad. Sci. U.S.A.* **85,** 9336–9340.

Bormann, J., Hamill, O. P., and Sakmann, B. (1987). Mechanism of anion permeation through channels gated by glycine and γ-aminobutyric acid in mouse cultured spinal neurones. *J. Physiol. (London).* **385,** 243–286.

Bovolin, P., Santi, M.-R., Memo, M., Costa, E., and Grayson, D. R. (1992a). Distinct developmental patterns of expression of rat α_1, α_5, γ_{2S}, and γ_{2L} γ-aminobutyric $acid_A$ receptor subunit mRNAs in vivo and in vitro. *J. Neurochem.* **59,** 62–72.

Bovolin, P., Santi, M. R., Puia, G., Costa, E., and Grayson, D. (1992b). Expression patterns of γ-aminobutyric acid type A receptor subunit mRNAs in primary cultures of granule neurons and astrocytes from neonatal rat cerebella. *Proc. Natl. Acad. Sci. U.S.A.* **89,** 9344–9348.

Buchstaller, A., Adamiker, D., Fuchs, K., and Sieghart W. (1991a). N-Deglycosylation and immunological identification indicates the existence of β-subunit isoforms of the rat $GABA_A$ receptor. *FEBS Lett.* **287,** 27–30.

Buchstaller, A., Fuchs, K., and Sieghart, W. (1991b). Identification of $\alpha 1$ subunit, $\alpha 2$ subunit and $\alpha 3$ subunit isoforms of the $GABA_A$ benzodiazepine receptor in the rat brain. *Neurosci. Lett.* **120,** 237–241.

Buck, K. J., and Harris, R. A. (1990). Benzodiazepine agonist and inverse agonist actions on $GABA_A$ receptor-operated chloride channels. II. Chronic effects of ethanol. *J. Pharmacol. Exp. Ther.* **253,** 713–719.

Buck, K. J., and Harris, R. A. (1991). Neuroadaptive responses to chronic ethanol. *Alcohol: Clin. Exp. Res.* **15,** 460–470.

Buck, K. J., Hahner, L., Sikela, J., and Harris, R. A. (1991). Chronic ethanol treatment alters brain levels of γ-aminobutyric $acid_A$ receptor subunit mRNAs: Relationship to genetic differences in ethanol withdrawal seizure severity. *J. Neurochem.* **57,** 1452–1455.

Buckle, V. J., Fujita, N., Ryder-Cook, A. S., Derry, J. M. J., Barnard, P. J., Lebo, R. V., Schofield, P. R., Seeburg, P. H., Bateson, A. N., Darlison, M. G., and Barnard, E. A. (1989). Chromosomal localization of $GABA_A$ receptor subunit genes: Relationship to human genetic disease. *Neuron* **3,** 647–654.

Buller, A. L., and White, M. M. (1990). Functional acetylcholine receptors expressed in *Xenopus* oocytes after injection of *Torpedo* β, γ, and δ subunit RNAs are a consequence of endogenous oocyte gene expression. *Mol. Pharmacol.* **37,** 423–428.

Bureau, M., Khrestchatisky, M., Heeren, M. A., Zambrowicz, E. B., Kim, H., Grisar, T. M., Colmbini, M., Tobin, A. J., and Olsen, R. W. (1992). Isolation and cloning of a voltage-dependent anion channel-like M_r 36,000 polypeptide from mammalian brain. *J. Biol. Chem.* **267,** 8679–8684.

Burnashev, N., Monyer, H., Seeburg, P. H., and Sakmann, B. (1992). Divalent ion permeability of AMPA receptor channels is dominated by the edited form of a single subunit. *Neuron* **8,** 189–198.

Burt, D. R., and Kamatchi, G. L. (1991). $GABA_A$ receptor subtypes: From pharmacology to molecular biology. *FASEB J.* **5,** 2916–2923.

Caruncho, H. J., Puia, G., Slobodyansky, E., da Silva, P. P., and Costa, E. (1993). Freeze-fracture immunocytochemical study of the expression of native and recombinant $GABA_A$ receptors. *Brain Res.* **603,** 234–242.

Cavallaro, S., Korneyev, A., Guidotti, A., and Costa, E. (1992). Diazepam-binding inhibitor (DBI)-processing products, acting at the mitochondrial DBI receptor, mediate adrenocorticotropic hormone-induced steroidogenesis in rat adrenal gland. *Proc. Natl. Acad. Sci. U.S.A.* **89,** 10598–10602.

Cepko, C. L. (1989). Immortalization of neural cells via retrovirus-mediated oncogene transduction. *Annu. Rev. Neurosci.* **12,** 47–65.

Chang, Y. F., Gao, X. M., and Chen, J. S. (1991). Correlation between enhancement of [^{3}H]flunitrazepam binding and suppression of pentylenetetrazol-induced seizures by L-lysine. *Eur. J. Pharmacol.* **193,** 239–247.

Chen, Q. X., Stelzer, A., Kay, A. R., and Wong, R. K. S. (1990). $GABA_A$ recptor function is regulated by phosphorylation in acutely dissociated guinea-pig hippocampal neurones. *J. Physiol. (London)* **420,** 207–221.

Cherubini, E., Gaiarsa, J. L., and Ben-Ari, Y. (1991). GABA: An excitatory transmitter in early postnatal life. *Trends Neurosci.* **14,** 515–519.

Costa, E., and Guidotti, A. (1991). Diazepam binding inhibitor (DBI): A peptide with multiple biological actions. *Life Sci.* **49,** 325–344.

Crabbe, J. C., Merrill, C., and Belknap, J. K. (1991). Acute dependence on depressant drugs is determined by common genes in mice. *J. Pharmacol. Exp. Ther.* **257,** 663–667.

Culiat, C. T., Stubbs, L., Nicholls, R. D., Montgomery, C. S., Russell, L. B., Johnson, D. K., and Rinchik, E. M. (1993). Concordance between isolated cleft palate in mice and alterations within a region including the gene encoding the β_3 subunit of the type A γ-aminobutyric acid receptor. *Proc. Natl. Acad. Sci. U.S.A.* **90,** 5105–5109.

Cutting, G. R., Lu, L., O'Hara, B. F., Kasch, L. M., Montrose-Rafizadeh, C., Donovan, D. M., Shimada, S., Antonarakis, S. E., Guggino, W. B., Uhl, G. R., and Kazazian, H. H., Jr. (1991). Cloning of the γ-aminobutyric acid (GABA) ρ_1 cDNA: A GABA receptor subunit highly expressed in the retina. *Proc. Natl. Acad. Sci. U.S.A.* **88,** 2673–2677.

Cutting, G. R., Curristin, S., Zoghbi, H., O'Hara, B., Seldin, M. F., and Uhl, G. R. (1992). Identification of a putative γ-aminobutyric acid (GABA) receptor subunit rho_2 cDNA and colocalization of the gene encoding rho_2 (GABRR2) and rho_1 (GABRR1) to human chromosome 6q14–q21 and mouse chromosome 4. *Genomics* **12,** 801–806.

Czajkowski, C., and Farb, D. H. (1989). Identification of an intracellular pool of γ-aminobutyric $acid_A$/benzodiazepine receptors en route to the cell surface of brain neurons in culture. *Mol. Pharmacol.* **35,** 183–188.

De Deyn, P. P., and Macdonald, R. L. (1989). Effect of antiepileptic drugs on GABA responses and on reduction of GABA responses by PTZ and DMCM on mouse neurons in cell culture. *Epilepsia* **30,** 17–25.

Deitrich, R. A., Dunwiddie, T. V., Harris, R. A., and Erwin, V. (1989). Mechanism of action of ethanol: initial central nervous system actions. *Pharmacol. Rev.* **41,** 491–537.

DeLorey, T. M., and Olsen, R. W. (1992). γ-Aminobutyric $acid_A$ receptor structure and function. *J. Biol. Chem.* **267,** 16747–16750.

Deng, L., Nielsen, M., and Olsen, R. W. (1991). Pharmacological and biochemical properties of the γ-aminobutyric acid-benzodiazepine receptor from codfish brain. *J. Neurochem.* **56,** 968–977.

Dingledine, R., Myers, S. J., and Nicholas, R. A. (1990). Molecular biology of mammalian amino acid receptors. *FASEB J.* **4,** 2636–2645.

Draguhn, A., Verdorn, T. A., Ewert, M., Seeburg, P. H., and Sakmann, B. (1990). Functional and molecular distinction between recombinant rat $GABA_A$ receptor subtypes by Zn^{2+}. *Neuron* **5,** 781–788.

Duggan, M. J., Pollard, S., and Stephenson, F. A. (1991). Immunoaffinity purification of $GABA_A$ receptor α-subunit iso-oligomers. Demonstration of receptor populations containing $\alpha1\alpha2$, $\alpha1\alpha3$, and $\alpha2\alpha3$ subunit pairs. *J. Biol. Chem.* **266,** 24778–24784.

Duggan, M. J., Pollard, S., and Stephenson, F. A. (1992). Quantitative immunoprecipitation studies with anti-γ-aminobutyric $acid_A$ receptor $\gamma2$ 1–15 Cys antibodies. *J. Neurochem.* **58,** 72–77.

Durner, M., Sander, T., Greenberg, D. A., Johnson, K., Beck-Mannagetta, G., and Janz, D. (1991). Localization of idiopathic generalized epilepsy on chromosome 6p in families of juvenile myoclonic epilepsy patients. *Neurology* **41,** 1651–1655.

Eldefrawi, A. T., and Eldefrawi, M. E. (1987). Receptors for γ-aminobutyric acid and voltage-dependent chloride channels as targets for drugs and toxicants. *FASEB J.* **1,** 262–271.

Endo, S., and Olsen, R. W. (1992). Preparation of antibodies to β subunits of γ-aminobutyric $acid_A$ receptors. *J. Neurochem.* **59,** 1444–1451.

Endo, S., and Olsen, R. W. (1993). Antibodies specific for α-subunit subtypes of $GABA_A$ receptors reveal brain regional heterogeneity. *J. Neurochem.* **60,** 1388–1398.

Engblom, A. C., Holopainen, I., and Akermann, K. E. (1991). Ethanol-induced Cl^- flux in rat cerebellar granule cells as measured by a fluorescent probe. *Brain Res.* **568,** 55–60.

Erdö, S. L., and Wolff, J. R. (1990). γ-Aminobutyric acid outside the mammalian brain. *J. Neurochem.* **54,** 363–372.

Erlander, M. G., Tillakaratne, N. J. K., Feldblum, S., Patel, N., and Tobin, A. J. (1991). Two genes encode distinct glutamate decarboxylases. *Neuron* **7,** 91–100.
Feigenspan, A., Wassle, H., and Bormann, J. (1993). Pharmacology of GABA receptor Cl^- channels in rat retinal bipolar cells. *Nature (London)* **361,** 159–162.
ffrench-Constant, R. H., and Rocheleau, T. (1992). *Drosophila* cyclodiene resistance gene shows conserved genomic organization with vertebrate γ-aminobutyric $acid_A$ receptors. *J. Neurochem.* **59,** 1562–1565.
ffrench-Constant, R. H., and Rocheleau, T. (1993). *Drosophila* γ-aminobutyric acid receptor gene *Rdl* shows extensive alternative splicing. *J. Neurochem.* **60,** 2323–2326.
ffrench-Constant, R. H., Mortlock, D. P., Shaffer, C. D., MacIntyre, R. J., and Roush, R. T. (1991). Molecular cloning and transformation of cyclodiene resistance in *Drosophila:* An invertebrate γ-aminobutyric acid subtype A receptor locus. *Proc. Natl. Acad. Sci. U.S.A.* **88,** 7209–7213.
ffrench-Constant, R. H., Steichen, J. C., Rocheleau, T. A., Aronstein, K., and Roush, R. T. (1993a). A single-amino acid substitution in a γ-aminobutyric acid subtype A receptor locus is associated with cyclodiene insecticide resistance in *Drosophila* populations. *Proc. Natl. Acad. Sci. U.S.A.* **90,** 1957–1961.
ffrench-Constant, R. H., Rocheleau, T. A., Steichen, J. C., and Chalmers, A. E. (1993b). A point mutation in a *Drosophila* GABA receptor confers insecticide resistance. *Nature (London)* **363,** 449–451.
Ffrench-Mullen, J. H. M., and Spence, K. (1991). Neurosteroids block Ca^{2+} channel current in freshly isolated hippocampal CA1 neurons. *Eur. J. Pharmacol.* **202,** 269–272.
Finn, D. A., and Gee, K. W. (1993). The influence of estrus cycle on neurosteroid potency at the γ-aminobutyric $acid_A$ receptor complex. *J. Pharmacol. Exp. Ther.* **265,** 1374–1379.
Fritschy, J.-M., Benke, D., Mertens, A., Oertel, W. H., Bachi, T., and Möhler, H. (1992). Five subtypes of type A γ-aminobutyric acid receptors identified in neurons by double and triple immunofluorescence staining with subunit-specific antibodies. *Proc. Natl. Acad. Sci. U.S.A.* **89,** 6726–6730.
Froehner, S. C. (1991). The submembrane machinery for nicotinic acetylcholine receptor clustering. *J. Cell Biol.* **114,** 1–7.
Frosch, M. P., and Dichter, M. A. (1992). Non-uniform distribution of GABA activated chloride channels in cultured cortical neurons. *Neurosci. Lett.* **138,** 59–62.
Frosch, M. P., Lipton, S. A., and Dichter, M. A. (1992). Desensitization of GABA-activated currents and channels in cultured cortical neurons. *J. Neurosci.* **12,** 3042–3053.
Gale, K. (1989). GABA in epilepsy: The pharmacologic basis. *Epilepsia* **30,** Suppl. 3, S1–S11.
Galzi, J.-L., Revah, F., Bessis, A., and Changeux, J.-P. (1991). Functional architecture of the nicotinic acetylcholine receptor: From electric organ to brain. *Annu. Rev. Pharmacol.* **31,** 37–72.
Galzi, J.-L., Devillers-Thiéry, A., Hussy, N., Bertrand, S., Changeux, J.-P., and Bertrand, D. (1992). Mutations in the channel domain of a neuronal nicotinic receptor convert ion selectivity from cationic to anionic. *Nature (London)* **359,** 500–505.
Gambarana, C., Beattie, C. E., Rodriguez, Z. R., and Siegel, R. E. (1991). Region-specific expression of messenger RNAs encoding $GABA_A$ receptor subunits in the developing rat brain. *Neuroscience* **45,** 423–432.
Garcia-Calvo, M., Valdivieso, F., Mayor, F., Jr., and Vasquez, J. (1992). Sensitive procedures for measuring chloride fluxes mediated by the purified glycine receptor incorporated into phospholipid vesicles. *Neurosci. Lett.* **136,** 102–104.
Gardner, C. R., Tully, W. R., and Hedgecock, C. J. R. (1993). The rapidly expanding range of neuronal benzodiazepine receptor ligands. *Prog. Neurobiol.* **40,** 1–61.
Garrett, K. M., Saito, N., Duman, R. S., Abel, M. S., Ashton, R. A., Fujimori, S., Beer,

R., Tallman, J. F., Vitek, M. P., and Blume, A. J. (1990). Differential expression of γ-aminobutyric $acid_A$ receptor subunits. *Mol. Pharmacol.* **37,** 652–657.

Gavish, M., Katz, Y., Barami, S., and Weizman, R. (1992). Biochemical, physiological, and pathological aspects of the peripheral benzodiazepine receptor. *J. Neurochem.* **58,** 1589–1601.

Geller, A. I., During, M. J., and Neve, R. L. (1991). Molecular analysis of neuronal physiology by gene transfer into neurons with herpes simplex virus vectors. *Trends Neurosci.* **14,** 428–432.

Gonzales, R. A., and Hoffman, P. L. (1991). Receptor-gated ion channels may be selective CNS targets for ethanol. *Trends Pharmacol. Sci.* **12,** 1–3.

Grover, L. M., Lambert, N. A., Schwartzkroin, P. A., and Teyler, T. J. (1993). Role of HCO_3^- ions in depolarizing $GABA_A$ receptor-mediated responses in pyramidal cells of rat hippocampus. *J. Neurophysiol.* **69,** 1541–1555.

Hadcock, J. R., and Malbon, C. C. (1991). Regulation of receptor expression by agonists: Transcriptional and post-transcriptional controls. *Trends Neurosci.* **14,** 242–247.

Hadingham, K. L., Wingrove, P. B., Wafford, K. A., Bain, C., Kemp, J. A., Palmer, K. J., Wilson, A. W., Wilcox, A. S., Sikela, J. M., Ragan, C. I., and Whiting, P. J. (1993). Role of the β subunit in determining the pharmacology of human γ-aminobutyric acid type A receptors. *Mol. Pharmacol.* **44,** 1211–1218.

Haefely, W., Martin, J. R., and Schoch, P. (1990). Novel anxiolytics that act as partial agonists at benzodiazepine receptors. *Trends Pharmacol. Sci.* **11,** 452–456.

Hales, T. G., and Lambert, J. J. (1991). The actions of propofol on inhibitory amino acid receptors of bovine adrenomedullary chromaffin cells and rodent central neurones. *Br. J. Pharmacol.* **104,** 619–628.

Hales, T. G., Kim, H., Longoni, B., Olsen, R. W., and Tobin, A. J. (1992). Immortalized hypothalamic GT1-7 neurons express functional γ-aminobutyric acid type A receptors. *Mol. Pharmacol.* **42,** 197–202.

Hamill, O. P., Bormann, J., and Sakmann, B. (1983). Activation of multiple conductance state channels in spinal neurones by glycine and GABA. *Nature (London)* **305,** 805–808.

Harris, R. A., and Allan, A. M. (1985). Functional coupling of γ-aminobutryic acid receptors to chloride channels in brain membranes. *Science* **228,** 1108–1110.

Harris, R. A., and Allan, A. M. (1989). Alcohol intoxication: Ion channels and genetics. *FASEB J.***3,** 1689–1695.

Harvey, R. J., Vreugdenhil, E., Zaman, S. H., Bhandal, N. S., Usherwood, P. N. R., Barnard, E. A., and Darlison, M. G. (1991). Sequence of a functional invertebrate $GABA_A$ receptor subunit which can form a chimeric receptor with a vertebrate α subunit. *EMBO J.* **10,** 3239–3245.

Harvey, R. J., Kim, H.-Y., and Darlison, M. D. (1993). Molecular cloning reveals the existence of a fourth γ subunit of the vertebrate brain $GABA_A$ receptor. *FEBS Lett.* **331,** 211–216.

Harvey, R. J., Chinchetru, M. A., and Darlison, M. D. (1994). Alternative splicing of a 51-nucleotide exon that encodes a putative protein kinase C phosphorylation site generates two forms of the chicken γ-aminobutyric $acid_A$ receptor β2 subunit. *J. Neurochem.* **62,** 10–16.

Hastings, G. A., Buller, A. L., Kirkness, E. F., and Fraser, C. M. (1992). Mutagenesis of N-linked glycosylation sites of the $GABA_A$/benzodiazepine receptor α1 subunit decreases the level of R015-1788 binding in transfected 293 cells. *Soc. Neurosci. Abstr.* **18,** 262.

Heninger, C., Saito, N., Tallman, J. F., Garrett, K. M., Vitek, M. P., Duman, R. S., and Gallager, D. W. (1990). Effects of continuous diazepam administration on $GABA_A$ receptor subunit mRNA in rat brain. *J. Mol. Neurosci.* **2,** 101–107.

Herb, A., Wisden, W., Lüddens, H., Puia, G., Vicini, S., and Seeburg, P. H. (1992). The third γ subunit of the γ-aminobutyric acid type A receptor family. *Proc. Natl. Acad. Sci. U.S.A.* **89,** 1433–1437.

Holland, K. D., Yoon, K.-W., Ferrendelli, J. A., Covey, D. F., and Rothman, S. M. (1991). γ-Butyrolactone antagonism of the picrotoxin receptor: Comparison of a pure antagonist and a mixed antagonist/inverse agonist. *Mol. Pharmacol.* **39,** 79–84.

Hopkins, C. R. (1992). Selective membrane protein trafficking: Vectorial flow and filter. *Trends Biochem. Sci.* **17,** 27–32.

Hosford, D. A., Clark, S., Cao, Z., Wilson, W. A., Jr., Lin, F., Morrisett, R. A., and Huin, A. (1992). The role of $GABA_B$ receptor activation in absence seizures of lethargic (*lh/lh*) mice. *Science* **257,** 398–401.

Huguenard, J. R., and Alger, B. E. (1986). Whole-cell voltage-clamp study of the fading of GABA-activated currents in acutely dissociated hippocampal neurons. *J. Neurophysiol.* **56,** 1–18.

Im, M. S., Hamilton, B. J., Carter, D. B., and Im, W. B. (1992). Selective potentiation of GABA-mediated Cl^- current by lanthanum ion in subtypes of cloned $GABA_A$ receptors. *Neurosci. Lett.* **144,** 165–168.

Im, W. B., and Blakeman, D. P. (1991). Correlation between γ-aminobutyric $acid_A$ receptor ligand-induced changes in [^{35}S]t-butylbicyclophosphorothionate binding and $^{36}Cl^-$ uptake in rat cerebrocortical membranes. *Mol. Pharmacol.* **39,** 394–398.

Jones, M. V., Brooks, P. A., and Harrison, N. L. (1992). Enhancement of γ-aminobutyric acid-activated Cl^- currents in cultured rat hippocampal neurones by three volatile anesthetics. *J. Physiol.* (*London*) **449,** 279–293.

Kang, I., and Miller, L. G. (1991). Decreased $GABA_A$ receptor subunit mRNA concentrations following chronic lorazepam administration. *Br. J. Pharmacol.* **103,** 1285–1287.

Kardos, J., and Cash, D. J. (1990). Transmembrane $^{36}Cl^-$ flux measurements and desensitization of the γ-aminobutyric $acid_A$ receptor. *J. Neurochem.* **55,** 1095–1099.

Kasckow, J. W., Tillakaratne, N. J. K., Kim, H., Strecker, G. J., Tobin, A. J., and Olsen, R. W. (1992). Expression of $GABA_A$ receptor polypeptides in clonal rat cell lines. *Brain Res.* **581,** 143–147.

Kato, K. (1990). Novel $GABA_A$ receptor α subunit is expressed only in cerebellar granule cells. *J. Mol. Biol.* **214,** 619–624.

Keir, W. J., Kozak, C. A., Chakraborti, A., Deitrich, R. A., and Sikela, J. M. (1991). The cDNA sequence and chromosomal location of the murine $GABA_A$ $\alpha 1$ receptor gene. *Genomics* **9,** 390–395.

Kellenberger, S., Malherbe, P., and Sigel, E. (1992). Function of the $\alpha 1\beta 2\gamma 2S$ γ-aminobutyric acid type A receptor is modulated by protein kinase C via multiple phosphorylation sites. *J. Biol. Chem.* **267,** 25660–25663.

Kerr, D. I. B., and Ong, J. (1992). GABA agonists and antagonists. *Med. Res. Rev.* **12,** 593–636.

Khan, Z. U., Fernando, L. P., Escriba, P., Busquets, X., Mallet, J., Miralles, C. P., Filla, M., and De Blas, A. L. (1993). Antibodies to the human γ_2 subunit of the γ-aminobutyric $acid_A$/benzodiazepine receptor. *J. Neurochem.* **60,** 961–971.

Khrestchatisky, M., MacLennan, A. J., Chiang, M.-Y., Xu, W., Jackson, M. B., Brecha, N., Sternini, C., Olsen, R. W., and Tobin, A. J. (1989). A novel α subunit in rat brain $GABA_A$ receptors. *Neuron* **3,** 745–753.

Khrestchatisky, M., MacLennan, A. J., Tillakaratne, N. J. K., Chiang, M.-Y., and Tobin, A. J. (1991). Sequence and regional distribution of the mRNA encoding the α_2 polypeptide of rat γ-aminobutyric $acid_A$ receptors. *J. Neurochem.* **56,** 1717–1722.

Killisch, I., Dotti, C. G., Laurie, D. J., Luddens, H., and Seeburg, P. H. (1991). Expression pattern of $GABA_A$ receptor subtypes in developing hippocampal neurons. *Neuron* **7,** 927–936.

Kirkness, E. F., and Fraser, C. M. (1993). A strong promoter element is located between alternative exons of a gene encoding the human γ-aminobutyric acid-type A receptor β3 subunit (GABRB3). *J. Biol. Chem.* **268,** 4420–4428.

Kirkness, E. F., Kusiak, J. W., Fleming, J. T., Menninger, J., Gocayne, J. D., Ward, D. C., and Venter, J. C. (1991). Isolation, characterization, and localization of human genomic DNA encoding the β1 subunit of the $GABA_A$ receptor (GABRB1). *Genomics* **10,** 985–995.

Knoflach, F., Rhyner, T., Villa, M., Kellenberger, S., Drescher, U., Malherbe, P., Sigel, E., and Möhler, H. (1991). The γ3-subunit of the $GABA_A$-receptor confers sensitivity to benzodiazepine receptor ligands. *FEBS Lett.* **293,** 191–194.

Knoll, J. H. M., Sinnett, D., Wagstaff, J., Glatt, K., Wilcox, A. S., Whiting, P. M., Wingrove, P., Sikela, J. M., and Lalande, M. (1993). FISH ordering of reference markers and of the gene for the α5 subunit of the γ-aminobutyric acid receptor (GABRA5) within the Angelman and Prader-Willi syndrome chromosomal regions. *Hum. Mol. Genet.* **2,** 183–189.

Kofuji, P., Wang, J. B., Moss, S. J., Huganir, R. L., and Burt, D. R. (1991). Generation of two forms of the γ-aminobutyric $acid_A$ receptor γ_2-subunit in mice by alternative splicing. *J. Neurochem.* **56,** 713–715.

Koob, G. F. (1992). Drugs of abuse: Anatomy, pharmacology and function of reward pathways. *Trends Pharmacol. Sci.* **13,** 177–184.

Korpi, E. R., Kleingoor, C., Kettenmann, H., and Seeburg, P. H. (1993). Benzodiazepine-induced motor impairment linked to point mutation in cerebellar $GABA_A$ receptor. *Nature (London)* **361,** 356–359.

Kosower, E. M. (1988). A partial structure for the γ-aminobutyric acid ($GABA_A$) receptor is derived from the model for the nicotinic acetylcholine receptor. The anion-exchange protein of cell membranes is related to the $GABA_A$ receptor. *FEBS Lett.* **231,** 5–10.

Krueger, K. E., and Papadopoulos, V. (1992). Mitochondrial benzodiazepine receptors and the regulation of steroid biosynthesis. *Annu. Rev. Pharmacol.* **32,** 211–237.

Kusama, T., Wang, T.-L., Guggino, W. B., Cutting, G. R., and Uhl, G. R. (1993). GABA ρ_2 receptor pharmacological profile: GABA recognition site similarities to ρ_1. *Eur. J. Pharmacol. Mol. Pharmacol.* **245,** 83–84.

Lambert, N. A., Borroni, A. M., Grover, L. M., and Teyler, T. J. (1991). Hyperpolarizing and depolarizing $GABA_A$ receptor-mediated dendritic inhibition in area CA1 of the rat hippocampus. *J. Neurophysiol.* **66,** 1538–1548.

Lan, N. C., Gee, K. W., Bolger, M. B., and Chen, J. S. (1991). Differential responses of expressed recombinant human $GABA_A$ receptors to neurosteroids. *J. Neurochem.* **57,** 1818–1821.

Langosch, D., Thomas, L., and Betz, H. (1988). Conserved quaternary structure of ligand-gated ion channels: The postsynaptic glycine receptor is a pentamer. *Proc. Natl. Acad. Sci. U.S.A.* **85,** 7394–7398.

Lasham, A., Vreugdenhil, E., Bateson, A. N., Barnard, E. A., and Darlison, M. G. (1991). Conserved organization of γ-aminobutyric $acid_A$ receptor genes: Cloning and analysis of the chicken β_4-subunit gene. *J. Neurochem.* **57,** 352–355.

Laurie, D. J., Seeburg, P. H., and Wisden, W. (1992a). The distribution of 13-$GABA_A$ receptor subunit messenger RNAs in the rat brain. 2. Olfactory bulb and cerebellum. *J. Neurosci.* **12,** 1063–1076.

Laurie, D. J., Wisden, W., and Seeburg, P. H. (1992b). The distribution of 13 $GABA_A$

receptor subunit messenger RNAs in the rat brain. 3. Embryonic and postnatal development. *J. Neurosci.* **12,** 4151–4172.

Legendre, P., and Westbrook, G. L. (1991). Noncompetitive inhibition of γ-aminobutyric $acid_A$ channels by zinc. *Mol. Pharmacol.* **39,** 267–274.

Leidenheimer, N. J., Browning, M. D., and Harris, R. A. (1991). $GABA_A$ receptor phosphorylation: Multiple sites, actions and artifacts. *Trends Pharmacol. Sci.* **12,** 84–87.

Leidenheimer, N. J., McQuilkin, S. J., Hahner, L. D., Whiting, P., and Harris, R. A. (1992). Activation of protein kinase C selectively inhibits the γ-aminobutyric $acid_A$ receptor: Role of desensitization. *Mol. Pharmacol.* **41,** 1116–1123.

Leidenheimer, N. J., Whiting, P. J., and Harris, R. A. (1993). Activation of calcium-phospholipid-dependent protein kinase enhances benzodiazepine and barbiturate potentiation of the $GABA_A$ receptor. *J. Neurochem.* **60,** 1972–1975.

Levitan, E. S., Schofield, P. R., Burt, D. R., Rhee, L. M., Wisden, W., Köhler, M., Fujita, N., Rodriguez, H. F., Stephenson, A., Darlison, M. G., Barnard, E. A., and Seeburg, P. H. (1988). Structural and functional basis for $GABA_A$ receptor heterogeneity. *Nature (London)* **335,** 76–79.

Lima-Landman, M. T. R., and Albuquerque, E. X. (1989). Ethanol potentiates and blocks NMDA-activated single-channel currents in rat hippocampal pyramidal cells. *FEBS Lett.* **247,** 61–67.

Lin, A. M.-Y., Freund, R. K., and Palmer, M. R. (1993). Sensitization of γ-aminobutyric acid-induced depressions of cerebellar Purkinje neurons to the potentiative effects of ethanol by *beta*-adrenergic mechanisms in rat brain. *J. Pharmacol. Exp. Ther.* **265,** 426–432.

Lin, L.-H., Whiting, P., and Harris, R. A. (1993a). Molecular determinants of general anesthetic action: Role of $GABA_A$ receptor structure. *J. Neurochem.* **60,** 1548–1553.

Lin, L.-H., Chen, L. L., and Harris, R. A. (1993b). Enflurane inhibits NMDA, AMPA, and kainate-induced currents in Xenopus oocytes expressing mouse and human brain mRNA. *FASEB J.* **7,** 479–485.

Listerud, M., Brussaard, A. B., Devay, P., Colman, D. R., and Role, L. W. (1991). Functional contribution of neuronal AChR subunits revealed by antisense oligonucleotides. *Science* **254,** 1518–1521.

Louiset, E., Valentijn, J. A., Vaudry, H., and Cazin, L. (1992). Central-type benzodiazepines modulate $GABA_A$ receptor chloride channels in cultured pituitary melanotrophs. *Mol. Brain Res.* **12,** 1–6.

Lovinger, D. M., and White, G. (1991). Ethanol potentiation of 5-$hydroxytryptamine_3$ receptor-mediated ion current in neuroblastoma cells and isolated adult mammalian neurons. *Mol. Pharmacol.* **40,** 263–270.

Lüddens, H., Pritchett, D. B., Köhler, M., Killisch, I., Keinänen, K., Monyer, H., Sprengel, R., and Seeburg, P. H. (1990). Cerebellar $GABA_A$ receptor selective for a behavioural alcohol antagonist. *Nature (London)* **346,** 648–651.

Lüddens, H., Killisch, I., and Seeburg, P. H. (1991). More than one alpha variant may exist in a $GABA_A$/benzodiazepine receptor complex. *J. Recept. Res.* **11,** 535–551.

Ma, W., Saunder, P. A., Somogyi, R., Poulter, M. O., and Barker, J. L. (1993). Ontogeny of $GABA_A$ receptor subunit mRNAs in rat spinal cord and dorsal root ganglia. *J. Comp. Neurol.* **338,** 337–359.

Macdonald, R. L., Rogers, C. J., and Twyman, R. E. (1989a). Kinetic properties of the $GABA_A$ receptor main conductance state of mouse spinal neurones in culture. *J. Physiol. (London)* **410,** 479–499.

Macdonald, R. L., Rogers, C. J., and Twyman, R. E. (1989b). Barbiturate regulation of the kinetic properties of the $GABA_A$ receptor channel of mouse spinal neurones in culture. *J. Physiol. (London)* **417,** 483–500.

MacLennan, A. J., Brecha, N., Khrestchatisky, M., Sternini, C., Tillakaratne, N. J. K., Chang, M.-Y., Anderson, K., Lai, M., and Tobin, A. J. (1991). Independent cellular and ontogenetic expression of mRNAs encoding 3 α polypeptides of the rat $GABA_A$ receptor. *Neuroscience* **43,** 369–380.

Madtes, P., Jr., and Redburn, D. (1983). GABA as a trophic factor during development. *Life Sci.* **33,** 979–984.

Majewska, M. D. (1992). Neurosteroids: Endogenous bimodal modulators of the $GABA_A$ receptor. Mechanism of action and physiological significance. *Prog. Neurobiol.* **38,** 379–395.

Majewska, M. D., and Vaupel, B. (1991). Steroid regulation of uterine motility via $GABA_A$ receptor in the rabbit: A novel mechanism? *J. Endocrinol.* **131,** 427–434.

Malherbe, P., Draguhn, A., Multhaup, G., Beyreuther, K., and Möhler, H. (1990a). $GABA_A$-receptor expressed from rat brain α- and β-subunit cDNAs displays potentiation by benzodiazepine receptor ligands. *Mol. Brain Res.* **8,** 199–208.

Malherbe, P., Sigel, E., Baur, R., Persohn, E., Richards, J. G., and Möhler, H. (1990b). Functional characteristics and sites of gene expression of the $\alpha_1\beta_1\gamma_{2-}$-isoform of the rat $GABA_A$-receptor. *J. Neurosci.* **10,** 2330–2337.

Malherbe, P., Sigel, E., Baur, R., Persohn, E., Richards, J. G., and Möhler, H. (1990c). Functional expression and sites of gene transcription of a novel α-subunit of the $GABA_A$ receptor in rat brain. *FEBS Lett.* **260,** 261–265.

Maricq, A. V., Peterson, A. S., Brake, A. J., Myers, R. M., and Julius, D. (1991). Primary structure and functional expression of the $5HT_3$ receptor, a serotonin-gated ion channel. *Science* **254,** 432–437.

Mathers, D. A. (1991). Activation and inactivation of the $GABA_A$ receptor: Insights from comparison of native and recombinant subunit assemblies. *Can. J. Physiol. Pharmacol.* **69,** 1057–1063.

McEnery, M. W., Snowman, A. M., Trifiletti, R. R., and Snyder, S. H. (1992). Isolation of the mitochondrial benzodiazepine receptor: Association with the voltage-dependent anion channel and the adenine nucleotide carrier. *Proc. Natl. Acad. Sci. U.S.A.* **89,** 3170–3174.

McEwen, B. S. (1991). Non-genomic and genomic effects of steroids on neural activity. *Trends Pharmacol. Sci.* **12,** 141–146.

McKernan, R. M., Cox, P., Gillard, N. P., and Whiting, P. (1991a). Differential expression of $GABA_A$ receptor α-subunits in rat brain during development. *FEBS Lett.* **286,** 44–46.

McKernan, R. M., Quirk, K., Prince, R., Cox, P. A., Gillard, N. P., Ragan, C. I., and Whiting, P. (1991b). $GABA_A$ receptor subtypes immunopurified from rat brain with α subunit-specific antibodies have unique pharmacological properties. *Neuron* **7,** 667–676.

Meldrum, B. S. (1989). GABAergic mechanisms in the pathogenesis and treatment of epilepsy. *Br. J. Clin. Pharmacol.* **27,** 3S-11S.

Mertens, S., Benke, D., and Mohler, H. (1993). $GABA_A$ receptor populations with novel subunit combinations and drug binding profiles identified in brain by α5- and δ-subunit-specific immunoprecipitation. *J. Biol. Chem.* **268,** 5965–5973.

Mhatre, M. C., and Ticku, M. K. (1989). Chronic ethanol treatment selectively increases the binding of inverse agonists for benzodiazepine binding sites in cultured spinal cord neurons. *J. Pharmacol. Exp. Ther.* **251,** 164–168.

Mhatre, M. C., and Ticku, M. K. (1992a). Chronic ethanol administration alters γ-aminobutyric $acid_A$ receptor gene expression. *Mol. Pharmacol.* **42,** 415–422.

Mhatre, M. C., and Ticku, M. K. (1992b). Aging related alterations in $GABA_A$ receptor subunit mRNA levels in Fisher rats. *Mol. Brain Res.* **14,** 71–78.

Mhatre, M. C., Pena, G., Sieghart, W., and Ticku, M. K. (1993). Antibodies specific for

$GABA_A$ receptor α subunits reveal that chronic alcohol treatment down-regulates α-subunit expression in rat brain regions. *J. Neurochem.* **61,** 1620–1625.
Michalik, M., and Erecinska, M. (1992). GABA in pancreatic islets: Metabolism and function. *Biochem. Pharmacol.* **44,** 1–9.
Michelson, H. B., and Wong, R. K. S. (1991). Excitatory synaptic responses mediated by $GABA_A$ receptors in the hippocampus. *Science* **253,** 1420–1423.
Montpied, P., Ginns, E. I., Martin, B. M., Roca, D., Farb, D. H., and Paul, S. M. (1991a). γ-Aminobutyric acid (GABA) induces a receptor-mediated reduction in $GABA_A$ receptor α subunit messenger RNAs in embryonic chick neurons in culture. *J. Biol. Chem.* **266,** 6011–6014.
Montpied, P., Morrow, A. L., Karanian, J. W., Ginns, E. I., Martin, B. M., and Paul, S. M. (1991b). Prolonged ethanol inhalation decreases γ-aminobutyric $acid_A$ receptor α subunit mRNAs in the rat cerebral cortex. *Mol. Pharmacol.* **39,** 157–163.
Morrow, A. L., Montpied, P., Lingford-Hughes, A., and Paul, S. M. (1990). Chronic ethanol and pentobarbital administration in the rat: Effects on $GABA_A$ receptor function and expression in brain. *Alcohol* **7,** 237–244.
Morrow, A. L., Herbert, J. S., and Montpied, P. (1992). Differential effects of chronic ethanol administration on $GABA_A$ receptor α_1 and α_6 subunit mRNA levels in rat cerebellum. *Mol. Cell. Neurosci.* **3,** 251–258.
Moss, S. J., Smart, T. G., Porter, N. M., Nayeem, N., Devine, J., Stephenson, F. A., Macdonald, R. L., and Barnard, E. A. (1990). Cloned GABA receptors are maintained in a stable cell line: Allosteric and channel properties. *Eur. J. Pharmacol.* **189,** 77–88.
Moss, S. J., Ravindran, A., Mei, L., Wang, J. B., Kofuji, P., Huganir, R. L., and Burt, D. R. (1991). Characterization of recombinant $GABA_A$ receptors produced in transfected cells from murine α_1, β_1, and γ_2 subunit cDNAs. *Neurosci. Lett.* **123,** 265–268.
Moss, S. J., Doherty, C. A., and Huganir, R. L. (1992a). Identification of the cAMP-dependent protein kinase and protein kinase C phosphorylation sites within the major intracellular domains of the β_1, γ_{2S}, and γ_{2L} subunits of the γ-aminobutyric acid type A receptor. *J. Biol. Chem.* **267,** 14470–14476.
Moss, S. J., Smart, T. G., Blackstone, C. D., and Huganir, R. L. (1992b). Functional modulation of $GABA_A$ receptors by cAMP-dependent protein phosphorylation. *Science* **257,** 661–665.
Nakahiro, M., Yeh, J. Z., Brunner, E., and Narahashi, T. (1989). General anesthetics modulate GABA receptor channel complex in rat dorsal root ganglion neurons. *FASEB J.* **3,** 1850–1854.
Nayeem, N., Green, T. P., Martin, I. L., and Barnard, E. A. (1994). Quaternary structure of the native $GABA_A$ receptor determined by electron microscopic image analysis. *J. Neurochem.* **62,** 815–818.
Newland, C. F., and Cull-Candy, S. G. (1992). On the mechanism of action of picrotoxin on GABA receptor channels in dissociated sympathetic neurons of the rat. *J. Physiol. (London)* **447,** 191–213.
O'Donovan, M. C., Buckland, P. R., Spurlock, G., and McGuffin, P. (1992). Bi-directional changes in the levels of messenger RNAs encoding γ-aminobutyric $acid_A$ receptor α subunits after flurazepam treatment. *Eur. J. Pharmacol. Mol. Pharmacol.* **226,** 335–341.
Oh, D. J., and Dichter, M. A. (1992). Desensitization of GABA-induced currents in cultured rat hippocampal neurons. *Neuroscience* **49,** 571–576.
Olsen, R. W., and Tobin, A. J. (1990). Molecular biology of $GABA_A$ receptors. *FASEB J.* **4,** 1469–1480.
Olsen, R. W., and Venter, J. C., eds. (1986). "Benzodiazepine/GABA Receptors and Chloride Channels: Structural and Functional Properties," Recept. Biochem. Methodol., Vol. 5. Alan R. Liss, New York.

Olson, T. S., and Lane, M. D. (1989). A common mechanism for posttranslational activation of plasma membrane receptors. *FASEB J.* **3,** 1618–1624.

Ong, J., and Kerr, D. I. B. (1990). GABA-receptors in peripheral tissues. *Life Sci.* **46,** 1489–1501.

Palotta, B. S. (1991). Single ion channel's view of classical receptor theory. *FASEB J.* **5,** 2035–2043.

Parola, A. L., Yamamura, H. I., and Laird, H. E., II (1993). Peripheral-type benzodiazepine receptors. *Life Sci.* **52,** 1329–1342.

Paul, S. M., and Purdy, R. H. (1992). Neuroactive steroids. *FASEB J.* **6,** 2311–2322.

Peduto, V. A., Concas, A., Santoro, G., Biggio, G., and Gessa, G. L. (1991). Biochemical and electrophysiological evidence that propofol enhances GABAergic transmission in the rat brain. *Anesthesiology* **75,** 1000–1009.

Perez-Velazquez, J. L., and Angelides, K. J. (1993). Assembly of $GABA_A$ receptor subunits determines sorting and localization in polarized cells. *Nature (London)* **361,** 457–460.

Persohn, E., Malherbe, P., and Richards, J. G. (1992). Comparative molecular neuroanatomy of cloned $GABA_A$ receptor subunits in the rat CNS. *J. Comp. Neurol.* **326,** 193–216.

Petke, J. D., Im, H. K., Im, W. B., Blakeman, D. P., Pregenzer, J. F., Jacobsen, E. J., Hamilton, B. J., and Carter, D. B. (1992). Characterization of functional interactions of imidazoquinoxaline derivatives with benzodiazepine-γ-aminobutyric $acid_A$ receptors. *Mol. Pharmacol.* **42,** 294–301.

Phillis, J. W., and O'Regan, M. H. (1988). Benzodiazepine interaction with adenosine systems explains some anomalies in GABA hypothesis. *Trends Pharmacol. Sci.* **9,** 153–154.

Pitler, T. A., and Alger, B. E. (1992). Postsynaptic spike firing reduces synaptic $GABA_A$ responses in hippocampal pyramidal cells. *J. Neurosci.* **12,** 4122–4132.

Pollard, S., Duggan, M. J., and Stephenson, F. A. (1991). Promiscuity of $GABA_A$-receptor $\beta 3$ subunits as demonstrated by their presence in $\alpha 1$, $\alpha 2$ and $\alpha 3$ subunit-containing receptor populations. *FEBS Lett.* **295,** 81–83.

Pollard, S., Duggan, M. J., and Stephenson, F. A. (1993). Further evidence for the existence of α subunit heterogeneity within discrete γ-aminobutyric $acid_A$ receptor subpopulations. *J. Biol. Chem.* **268,** 3753–3757.

Porter, N. M., Twynan, R. E., Uhler, M. D., and Macdonald, R. L. (1990) Cyclic AMP-dependent protein kinase decreases $GABA_A$ receptor current in mouse spinal neurons. *Neuron* **5,** 789–796.

Poulter, M. O., Barker, J. L., O'Carroll, A.-M., Lolait, S. J., and Mahan, L. C. (1992). Differential and transient expression of $GABA_A$ receptor α-subunit mRNAs in the developing rat CNS. *J. Neurosci.* **12,** 2888–2900.

Poulter, M. O., Barker, J. L., O'Carroll, A.-M., Lolait, S. J., and Mahan, L. C. (1993). Co-existent expression of $GABA_A$ receptor β_2, β_3 and γ_2 subunit messenger RNAs during embryogenesis and early postnatal development of the rat central nervous system. *Neuroscience* **53,** 1019–1033.

Primus, R. J., and Gallager, D. W. (1992). $GABA_A$ receptor subunit mRNA levels are differentially influenced by chronic FG 7142 and diazepam exposure. *Eur. J. Pharmacol. Mol. Pharmacol.* **226,** 21–28.

Prince, R. J., and Simmonds, M. A. (1992). Steroid modulation of the strychnine-sensitive glycine receptor. *Neuropharmacology* **31,** 201–205.

Pritchett, D. B., and Seeburg, P. H. (1990). γ-Aminobutyric $acid_A$ receptor α_5-subunit creates novel type II benzodiazepine receptor pharmacology. *J. Neurochem.* **54,** 1802–1804.

Pritchett, D. B., and Seeburg, P. H. (1991). γ-Aminobutyric acid type A receptor point mutation increases the affinity of compounds for the benzodiazepine site. *Proc. Natl. Acad. Sci. U.S.A.* **88,** 1421–1425.

Pritchett, D. B., Sontheimer, H., Gorman, C. M., Kettenmann, H., Seeburg, P. H., and Schofield, P. R. (1988). Transient expression shows ligand gating and allosteric potentiation of $GABA_A$ receptor subunits. *Science* **242,** 1306–1308.

Pritchett, D. B., Lüddens, H., and Seeburg, P. H. (1989a). Type I and type II $GABA_A$-benzodiazepine receptors produced in transfected cells. *Science* **245,** 1389–1392.

Pritchett, D. B., Sontheimer, H., Shivers, B. D., Ymer, S., Kettenmann, H., Schofield, P. R., and Seeburg, P. H. (1989b). Importance of a novel $GABA_A$ receptor subunit for benzodiazepine pharmacology. *Nature (London)* **338,** 582–585.

Proctor, W. R., Allan, A. M., and Dunwiddie, T. V. (1992). Brain region-dependent sensitivity of $GABA_A$ receptor-mediated responses to modulation by ethanol. *Alcohol: Clin. Exp. Res.* **16,** 480–489.

Puia, G., Vicini, S., Seeburg, P. H., and Costa, E. (1991). Influence of recombinant γ-aminobutyric acid-$_A$ receptor subunit composition on the action of allosteric modulators of γ-aminobutyric acid-gated Cl^- currents. *Mol. Pharmacol.* **39,** 691–696.

Puia, G., Ducic, I., Vicini, S., and Costa, E. (1992). Molecular mechanisms of the partial allosteric modulatory effects of bretazenil at γ-aminobutyric acid type A receptor. *Proc. Natl. Acad. Sci. U.S.A.* **89,** 3620–3624.

Purdy, R. H., Morrow, A. L., Moore, P. H., Jr., and Paul, S. M. (1991). Stress-induced elevations of γ-aminobutyric acid type A receptor-active steroids in rat brain. *Proc. Natl. Acad. Sci. U.S.A.* **88,** 4553–4557.

Qian, H., and Dowling, J. E. (1993). Novel GABA responses from rod-driven retinal horizontal cells. *Nature (London)* **361,** 162–164.

Reis, A., Kunze, J., Ladanyi, L., Enders, H., Klein-Vogler, U., and Niemann, G. (1993). Exclusion of the $GABA_A$ receptor $\beta 3$ subunit gene as the Angelman's syndrome gene. *Lancet* **341,** 122–123.

Rise, M. L., Frankel, W. N., Coffin, J. M., and Seyfried, T. N. (1991). Genes for epilepsy mapped in the mouse. *Science* **253,** 669–673.

Ritchie, J. M. (1992). Voltage-gated ion channels in Schwann cells and glia. *Trends Neurosci.* **15,** 345–350.

Roca, D. J., Rozenberg, I., Farrant, M., and Farb, D. H. (1990a). Chronic agonist exposure induces down-regulation and allosteric uncoupling of the γ-aminobutyric acid/benzodiazepine receptor complex. *Mol. Pharmacol.* **37,** 37–43.

Roca, D. J., Friedman, L., Schiller, G. D., Rozenberg, I., Gibbs, T. T., and Farb, D. H. (1990b). γ-Aminobutyric $acid_A$ receptor regulation in culture: Altered allosteric interactions following prolonged exposure to benzodiazepines, barbiturates, and methylxanthines. *Mol. Pharmacol.* **37,** 710–719.

Rognan, D., Boulanger, T., Hoffmann, R., Vercauteren, D. P., Andre, J.-M., Durant, F., and Wermuth, C.-G. (1992). Structure and molecular modeling of $GABA_A$ receptor antagonists. *J. Med. Chem.* **35,** 1969–1977.

Rohde, B. H., and Harris, R. A. (1992). GABA and flunitrazepam binding to neuroblastoma cell membranes: Effects of growth conditions, ethanol and pentobarbital. *Brain Res.* **253,** 133–141.

Rorsman, P., Berggren, P.-O., Bokvist, K., Ericson, H., Möhler, H., Östenson, C.-G., and Smith, P. A. (1989). Glucose-inhibition of glucagon secretion involves activation of $GABA_A$-receptor chloride channels. *Nature (London)* **341,** 233–236.

Rothstein, J. D., Garland, W., Puia, G., Guidotti, A., Weber, R. J., and Costa, E. (1992). Purification and characterization of naturally occurring benzodiazepine receptor ligands in rat and human brain. *J. Neurochem.* **58,** 2102–2115.

Rovira, C., and Ben-Ari, Y. (1991). Benzodiazepines do not potentiate GABA responses in neonatal hippocampal neurons. *Neurosci. Lett.* **130,** 157–161.

Saitoh, S., Kubota, T., Ohta, T., Jinno, Y., Niikawa, N., Sugimoto, T., Wagstaff, T., and LaLande, M. (1992). Familial Angelman syndrome caused by imprinted submicroscopic deletion encompassing $GABA_A$ receptor β_3-subunit gene. *Lancet* **339,** 366–367.

Samson, H. H., and Harris, R. A. (1992). Neurobiology of alcohol abuse. *Trends Pharmacol. Sci.* **13,** 206–211.

Sattelle, D. B., Lummis, S. C. R., Wong, J. F. H., and Rauh, J. J. (1991). Pharmacology of insect GABA receptors. *Neurochem. Res.* **16,** 363–374.

Scaf, A. H. J. (1991). Role of partial agonism in selectivity of benzodiazepines. *Trends Pharmacol. Sci.* **12,** 182–185.

Schaffner, A. E., Behar, T., Nadi, S., and Barker, J. L. (1993). Quantitative analysis of transient GABA expression in embryonic and early postnatal rat spinal cord neurons. *Dev. Brain Res.* **72,** 265–276.

Schofield, P. R., Darlison, M. G., Fujita, N., Burt, D. R., Stephenson, F. A., Rodriguez, H., Rhee, L. M., Ramachandran, J., Reale, V., Glencorse, T. A., Seeburg, P. H., and Barnard, E. A. (1987). Sequence and functional expression of the $GABA_A$ receptor shows a ligand-gated receptor super-family. *Nature (London)* **328,** 221–227.

Schwartz, R. D., Sudzak, P. D., and Paul, S. M. (1986). *Gamma*-aminobutyric acid and barbiturate-mediated $^{36}Cl^-$ uptake in rat brain synaptoneurosomes. *Mol. Pharmacol.* **30,** 419–426.

Schwartz, R. D., Seale, T. W., Skolnick, P., and Paul, S. M. (1989). Differential seizure sensitivities to picrotoxinin in two inbred strains of mice (DBA/2J and BALB/c ByJ): Parallel changes in GABA receptor-mediated chloride flux and receptor binding. *Brain Res.* **481,** 169–174.

Sharif, N. A., and Burt, D. R. (1984). Modulation of binding affinity of receptors for thyrotropin-releasing hormone by benzodiazepines: Brain regional differences. *J. Neurochem.* **43,** 742–746.

Shimada, S., Cutting, G., and Uhl, G. R. (1992). γ-Aminobutyric acid A or C receptor? γ-Aminobutyric acid ρ_1 receptor RNA induces bicuculline-, barbiturate-, and benzodiazepine-insensitive γ-aminobutyric acid responses in *Xenopus* oocytes. *Mol. Pharmacol.* **41,** 683–687.

Shirasaki, T., Aibara, K., and Akaike, N. (1992). Direct modulation of $GABA_A$ receptor by intracellular ATP in dissociated nucleus tractus solitarii neurones of rat. *J. Physiol. (London)* **449,** 551–572.

Shivers, B. D., Killisch, I., Sprengel, R., Sontheimer, H., Köhler, M., Schofield, P. R., and Seeburg, P. H. (1989). Two novel $GABA_A$ receptor subunits exist in distinct neuronal subpopulations. *Neuron* **3,** 327–337.

Sieghart, W. (1992). $GABA_A$ receptors: Ligand-gated Cl^- ion channels modulated by multiple drug-binding sites. *Trends Pharmacol. Sci.* **13,** 446–450.

Sieghart, W., Item, C., Buchstaller, A., Fuchs, K., Höger, H., and Adamiker, D. (1993). Evidence for the existence of differential *O*-glycosylated α_5-subunits of the γ-aminobutyric $acid_A$ receptor in the rat brain. *J. Neurochem.* **60,** 93–98.

Sigel, E. (1990). Use of *Xenopus* oocytes for the functional expression of plasma membrane proteins. *J. Membr. Biol.* **117,** 201–221.

Sigel, E., Baur, R., Trube, G., Möhler, H., and Malherbe, P. (1990). The effect of subunit composition of rat brain $GABA_A$ receptors on channel function. *Neuron* **5,** 703–711.

Sigel, E., Baur, R., Kellenberger, S., and Malherbe, P. (1992). Point mutations affecting antagonist affinity and agonist dependent gating of $GABA_A$ receptor channels. *EMBO J.* **11,** 2017–2023.

Sigel, E., Baur, R., and Malherbe, P. (1993). Recombinant $GABA_A$ receptor function and ethanol. *FEBS Lett.* **324,** 140–142.

Sivilotti, L., and Nistri, A. (1991). GABA receptor mechanisms in the central nervous system. *Prog. Neurobiol.* **36,** 35–92.

Smart, T. G. (1992). A novel modulatory binding site for zinc on the $GABA_A$ receptor complex in cultured rat neurones. *J. Physiol.* (*London*) **447,** 587–625.

Smart, T. G., Moss, S. J., Xie, X., and Huganir, R. L. (1991). $GABA_A$ receptors are differentially sensitive to zinc: Dependence on subunit composition. *Br. J. Pharmacol.* **103,** 1837–1839.

Sommer, B., and Seeburg, P. H. (1992). Glutamate receptor channels: Novel properties and new clones. *Trends Pharmacol. Sci.* **13,** 291–296.

Sommer, B., Poustka, A., Spurr, N. K., and Seeburg, P. H. (1990). The murine $GABA_A$ receptor δ-subunit gene: Structure and assignment to human chromosome 1. *DNA Cell Biol.* **9,** 561–568.

Song, Y., and Huang, L.-Y. M. (1990). Modulation of glycine receptor chloride channels by cAMP-dependent protein kinase in spinal trigeminal neurons. *Nature* (*London*) **384,** 242–245.

Stephenson, F. A. (1988). Understanding the $GABA_A$ receptor: A chemically gated ion channel. *Biochem. J.* **249,** 21–32.

Study, R. E., and Barker, J. L. (1981). Diazepam and (−)-pentobarbital: Fluctuation analysis reveals different mechanisms for potentiation of γ-aminobutyric acid responses in cultured central neurons. *Proc. Natl. Acad. Sci. U.S.A.* **78,** 7180–7184.

Sumikawa, K., Parker, I., and Miledi, R. (1988). Effects of tunicamycin on the expression of functional brain neurotransmitter receptors and voltage-operated channels in *Xenopus* oocytes. *Mol. Brain Res.* **4,** 191–199.

Sweetnam, P. M., and Tallman, J. F. (1986). Regional differences in brain benzodiazepine receptor carbohydrates. *Mol. Pharmacol.* **29,** 299–306.

Swope, S. L., Moss, S. J., Blackstone, C. D., and Huganir, R. L. (1992). Phosphorylation of ligand-gated ion channels: A possible mode of synaptic plasticity. *FASEB J.* **6,** 2514–2523.

Takagi, T., Pribilla, I., Kirsch, J., and Betz, H. (1992). Coexpression of the receptor-associated protein gephyrin changes ligand binding affinities of α_2 glycine receptors. *FEBS Lett.* **303,** 178–180.

Tallman, J. F., Thomas, J. W., and Gallagher, D. W. (1978). GABAergic modulation of benzodiazepine binding site sensitivity. *Nature* (*London*) **274,** 383–385.

Tanelian, D. L., Kosek, P., Mody, I., and MacIver, M. B. (1993). The role of the $GABA_A$ receptor/chloride channel complex in anesthesia. *Anesthesiology* **78,** 757–776.

Tasker, J. G., and Dudek, F. E. (1991). Electrophysiology of GABA-mediated synaptic transmission and possible roles in epilepsy. *Neurochem. Res.* **16,** 251–262.

Tauc, D., Frosch, M., and Lipton, S. (1988). Characterization of GABA and glycine-induced currents of solitary rodent retinal ganglion cells in culture. *Neuroscience* **27,** 193–203.

Tehrani, M. H. J., and Barnes, E. M., Jr. (1991). Agonist-dependent internalization of γ-aminobutyric $acid_A$/benzodiazepine receptors in chick cortical neurons. *J. Neurochem.* **57,** 1307–1312.

Tehrani, M. H. J., and Barnes, E. M., Jr. (1993). Identification of $GABA_A$/benzodiazepine receptors on clathrin-coated vesicles from rat brain. *J. Neurochem.* **60,** 1755–1761.

Todd, A. J., and Sullivan, A. C. (1990). Light microscopic study of the coexistence of GABA-like and glycine-like immunoreactivities in the spinal cord of the rat. *J. Comp. Neurol.* **296,** 496–505.

Turner, D. M., Ransom, R. W., Yang, J. S.-J., and Olsen, R. W. (1989). Steroid anesthetics and naturally occurring analogs modulate the γ-aminobutyric acid receptor complex at a site distinct from barbiturates. *J. Pharmacol. Exp. Ther.* **248,** 960–966.

Twyman, R. E., and Macdonald, R. L. (1992). Neurosteroid regulation of $GABA_A$ receptor single-channel kinetic properties of mouse spinal neurons in culture. *J. Physiol. (London)* **456,** 215–245.

Uchida, I., Kamatchi, G., Burt, D. R., and Yang, J. (1993). Receptor subunit dependent etomidate potentiation of $GABA_A$ current. *Biophys. J.* **64,** A326.

Uusi-Oukari, M., and Korpi, E. R. (1991). Specific alterations in the cerebellar $GABA_A$ receptors of an alcohol-sensitive ANT rat line. *Alcohol: Clin. Exp. Res.* **15,** 241–248.

Valerio, A., Tinti, C., Spano, P. F., and Memo, M. (1992). Rat pituitary cells selectively express mRNA encoding the short isoform of the $\gamma 2$ $GABA_A$ receptor subunit. *Mol. Brain Res.* **13,** 145–150.

Ventimiglia, R., Grierson, J. P., Gallombardo, P., Sweetnam, P. M., Tallman, J. F., and Geller, H. M. (1990). Cultured rat neurons and astrocytes express immunologically related epitopes of the $GABA_A$/benzodiazepine receptor. *Neurosci. Lett.* **115,** 131–136.

Verdoorn, T. A., Draguhn, A., Ymer, S., Seeburg, P. H., and Sakmann, B. (1990). Functional properties of recombinant rat $GABA_A$ receptors depend upon subunit composition. *Neuron* **4,** 919–928.

von Blankenfeld, G., and Kettenmann, H. (1992). Glutamate and GABA receptors in vertebrate glial cells. *Mol. Neurobiol.* **5,** 31–43.

von Blankenfeld, G., Ymer, S., Pritchett, D. B., Sontheimer, H., Ewert, M., Seeburg, P. H., and Kettenmann, H. (1990). Differential benzodiazepine pharmacology of mammalian recombinant $GABA_A$ receptors. *Neurosci. Lett.* **115,** 269–273.

von Blankenfeld, G., Trotter, J., and Kettenmann, H. (1991). Expression and developmental regulation of a $GABA_A$ receptor in cultured murine cells of the oligodendrocyte lineage. *Eur. J. Neurosci.* **3,** 310–316.

Wafford, K. A., and Whiting, P. J. (1992). Ethanol potentiation of $GABA_A$ receptors requires phosphorylation of the alternatively spliced variant of the $\gamma 2$ subunit. *FEBS Lett.* **313,** 113–117.

Wafford, K. A., Burnett, D. M., Dunwiddie, T. V., and Harris, R. A. (1990). Genetic differences in the ethanol sensitivity of $GABA_A$ receptors expressed in *Xenopus* oocytes. *Science* **249,** 291–293.

Wafford, K. A., Burnett, D. M., Leidenheimer, N. J., Burt, D. R., Wang, J. B., Kofuji, P., Dunwiddie, T. V., Harris, R. A., and Sikela, J. M. (1991). Ethanol sensitivity of the $GABA_A$ receptor expressed in Xenopus oocytes requires eight amino acids contained in the γ_{2L} subunit. *Neuron* **7,** 27–33.

Wagstaff, J., Chaillet, J. R., and Lalande, M. (1991a). The $GABA_A$ receptor β_3 subunit gene: Characterization of a human cDNA from chromosome 15q11q13 and mapping to a region of conserved synteny on mouse chromosome 7. *Genomics* **11,** 1071–1078.

Wagstaff, J., Knoll, J. H. M., Fleming, J., Kirkness, E. F., Martin-Gallardo, A., Greenberg, F., Graham, J. M., Jr., Menninger, J., Ward, D., Venter, J. C., and Lalande, M. (1991b). Localization of the gene encoding the $GABA_A$ receptor $\beta 3$ subunit to the Angelman/Prader-Willi region of chromosome 15. *Am. J. Hum. Genet.* **49,** 330–337.

Wang, J. B., and Burt, D. R. (1991). Differential expression of two forms of $GABA_A$ receptor γ_2-subunit in mice. *Brain Res. Bull.* **27,** 731–735.

Wang, J. B., Kofuji, P., Fernando, J. C. R., Moss, S. J., Huganir, R. L., and Burt, D. R. (1992a). The α_1, α_2, and α_3 subunits of $GABA_A$ receptors: Comparison in seizure prone and resistant mice and during development. *J. Mol. Neurosci.* **3,** 177–184.

Wang, J. B., Kofuji, P., and Burt, D. R. (1992b). Strain comparisons and developmental profile of the delta subunit of the murine $GABA_A$ receptor. *Brain Res. Bull.* **29,** 119–123.

Wang, J. B., Kofuji, P., Moss, S. L., ,and Burt, D. R. (1994). In preparation.

Weiss, D. S., and Magleby, K. L. (1989). Gating scheme for single GABA-activated Cl^-

channels determined from stability plots, dwell-time distributions, and adjacent-interval durations. *J. Neurosci.* **9,** 1314–1324.

Whiting, P., McKernan, R. M., and Iversen, L. L. (1990). Another mechanism for creating diversity in γ-aminobutyrate type A receptors: RNA splicing directs expression of two forms of $\gamma 2$ subunit, one of which contains a protein kinase C phosphorylation site. *Proc. Natl. Acad. Sci. U.S.A.* **87,** 9966–9970.

Wieland, H. A., Lüddens, H., and Seeburg, P. H. (1992). A single histidine in $GABA_A$ receptors is essential for benzodiazepine agonist binding. *J. Biol. Chem.* **267,** 1426–1429.

Wilcox, A. S., Warrington, J. A., Gardiner, K., Berger, R., Whiting, P., Altherr, M. R., Wasmuth, J. J., Patterson, D., and Sikela, J. M. (1992). Human chromosomal localization of genes encoding the $\gamma 1$ and $\gamma 2$ subunits of the γ-aminobutyric acid receptor indicates that members of this gene family are often clustered in the genome. *Proc. Natl. Acad. Sci. U.S.A.* **89,** 5857–5861.

Wilson, M. A. (1992). Influences of gender, gonadectomy, and estrous cycle on GABA/BZ receptors and benzodiazepine responses in rats. *Brain Res. Bull.* **29,** 165–172.

Wilson-Shaw, D., Robinson, M., Gambarana, C., Siegel, R. E., and Sikela, J. M. (1991). A novel γ subunit of the $GABA_A$ receptor identified using the polymerase chain reaction. *FEBS Lett.* **284,** 211–215.

Wisden, W., Herb, A., Wieland, H., Keinänen, K., Lüddens, H., and Seeburg, P. H. (1991). Cloning, pharmacological characteristics and expression pattern of the rat $GABA_A$ receptor $\alpha 4$ subunit. *FEBS Lett.* **289,** 227–230.

Wisden, W., Laurie, D. J., Monyer, H., and Seeburg, P. H. (1992). The distribution of 13 $GABA_A$ receptor subunit mRNAs in the rat brain. I. Telencephalon, diencephalon, mesencephalon. *J. Neurosci.* **12,** 1040–1062.

Wolff, J. R., Joo, F., and Kasa, P. (1993). Modulation by GABA of neuroplasticity in the central and peripheral nervous system. *Neurochem. Res.* **18,** 453–461.

Wong, G., Sei, Y., and Skolnick, P. (1992). Stable expression of type I γ-aminobutyric $acid_A$/benzodiazepine receptors in a transfected cell line. *Mol. Pharmacol.* **42,** 996–1003.

Woodward, R. M., Polenzani, L., and Miledi, R. (1992). Characterization of bicuculline/baclofen-insensitive γ-aminobutyric acid receptors expressed in *Xenopus* oocytes I. Effects of Cl^- channel inhibitors. *Mol. Pharmacol.* **42,** 165–173.

Woodward, R. M., Polenzani, L., and Miledi, R. (1993). Characterization of bicuculline/baclofen-insensitive (ρ-like) γ-aminobutyric acid receptors expressed in *Xenopus* oocytes. II. Pharmacology of γ-aminobutyric $acid_A$ and γ-aminobutyric $acid_B$ agonists and antagonists. *Mol. Pharmacol.* **43,** 609–625.

Wu, F.-S., Gibbs, T. T., and Farb, D. H. (1990). Inverse modulation of γ-aminobutyric acid- and glycine-induced currents by progesterone. *Mol. Pharmacol.* **37,** 597–602.

Wu, F.-S., Gibbs, T. T., and Farb, D. H. (1991). Pregnenolone sulfate: A positive allosteric modulator at the *N*-methyl-D-aspartate receptor. *Mol. Pharmacol.* **40,** 333–336.

Xu, M., and Akabas, M. H. (1993). Amino acids lining the channel of the γ-aminobutyric acid type A receptor identified by cysteine substitution. *J. Biol. Chem.* **268,** 21505–21508.

Ymer, S., Draguhn, A., Köhler, M., Schofield, P. R., and Seeburg, P. H. (1989a). Sequence and expression of a novel $GABA_A$ receptor α subunit. *FEBS Lett.* **258,** 119–122.

Ymer, S., Schofield, P. R., Draguhn, A., Werner, P., Köhler, M., and Seeburg, P. (1989b). $GABA_A$ receptor β subunit heterogeneity: Functional expression of cloned cDNAs. *EMBO J.* **8,** 1665–1670.

Ymer, S., Draguhn, A., Wisden, W., Werner, P., Keinänen, K., Schofield, P. R., Sprengel, R., Pritchett, D. B., and Seeburg, P. H. (1990). Structural and functional characterization of the γ_1 subunit of the $GABA_A$/benzodiazepine receptors. *EMBO J.* **9,** 3261–3267.

Yu, O., Ito, M., Chiu, T. H., and Rosenberg, H. C. (1986). GABA-gated chloride ion influx and receptor binding studies in C57BL6J and DBA2J mice. *Brain Res.* **399,** 374–378.

Zahniser, N. R., Buck, K. J., Curella, P., McQuilkin, S. J., Wilson-Shaw, D., Miller, C. L., Klein, R. L., Heidenreich, K. A., Keir, W. J., Sikela, J. M., and Harris, R. A. (1992). $GABA_A$ receptor function and regional analysis of subunit mRNAs in long-sleep and short-sleep mouse brain. *Mol. Brain Res.* **14,** 196–206.

Zaman, S. H., Harvey, R. J., Barnard, E. A., and Darlison, M. G. (1992a). Unusual effects of benzodiazepines and cyclodiene insecticides on an expressed invertebrate $GABA_A$ receptor. *FEBS Lett.* **307,** 351–354.

Zaman, S. H., Shingai, R., Harvey, R. J., Darlison, M. G., and Barnard, E. A. (1992b). Effects of subunit types of the recombinant $GABA_A$ receptor on the response to a neurosteroid. *Eur. J. Pharmacol. Mol. Pharmacol.* **225,** 321–330.

Zdilar, D., Luntz-Leybman, V., Frostholm, A., and Rotter, A. (1992). Differential expression of $GABA_A$/benzodiazepine receptor β_1, β_2, and β_3 subunit mRNAs in the developing mouse cerebellum. *J. Comp. Neurol.* **326,** 580–594.

Zezula, J., and Sieghart, W. (1991). Isolation of type I and type II $GABA_A$/benzodiazepine receptors by immunoaffinity chromatography. *FEBS Lett.* **284,** 15–18.

Zhao, Z.-Y., and Joho, R. H. (1990). Isolation of distantly related members in a multigene family using the polymerase chain reaction technique. *Biochem. Biophys. Res. Commun.* **167,** 174–182.

Zimmer, A. (1992). Manipulating the genome by homologous recombination in embryonic stem cells. *Annu. Rev. Neurosci.* **15,** 115–137.

Zimprich, F., Zezula, J., Sieghart, W., and Lassman, H. (1991). Immunohistochemical localization of the α_1-, α_2-, and α_3- subunit of the $GABA_A$ receptor in the rat brain. *Neurosci. Lett.* **127,** 125–128.

CHAPTER 10

Chloride Channels along the Nephron

Erik M. Schwiebert, Anibal G. Lopes,* and William B. Guggino
Department of Physiology, Johns Hopkins University School of Medicine, Baltimore, Maryland 21205; and *Instituto de Biofisica Carlos Chagas Filho, Universidade do Rio de Janeiro, Rio de Janeiro, RJ, Brazil

I. INTRODUCTION

Chloride (Cl^-) transport along the nephron is involved in several key physiological processes such as tubuloglomerular feedback (TGF), macula densa-mediated renin release, urinary acidification, cell volume control, and acidification of intracellular compartments. Cl^- is also the most abundant inorganic anion in body fluids. Under normal conditions, the human kidney filters about 18 mol of Cl^- per day across the glomeruli, the majority of which is reabsorbed along the nephron by mechanisms specific to different proximal and distal nephron segments (Berry and Rector, 1991; Koeppen and Stanton, 1992). Figure 1 depicts the relative percentage of Cl^- reabsorption in different segments along the nephron. The majority

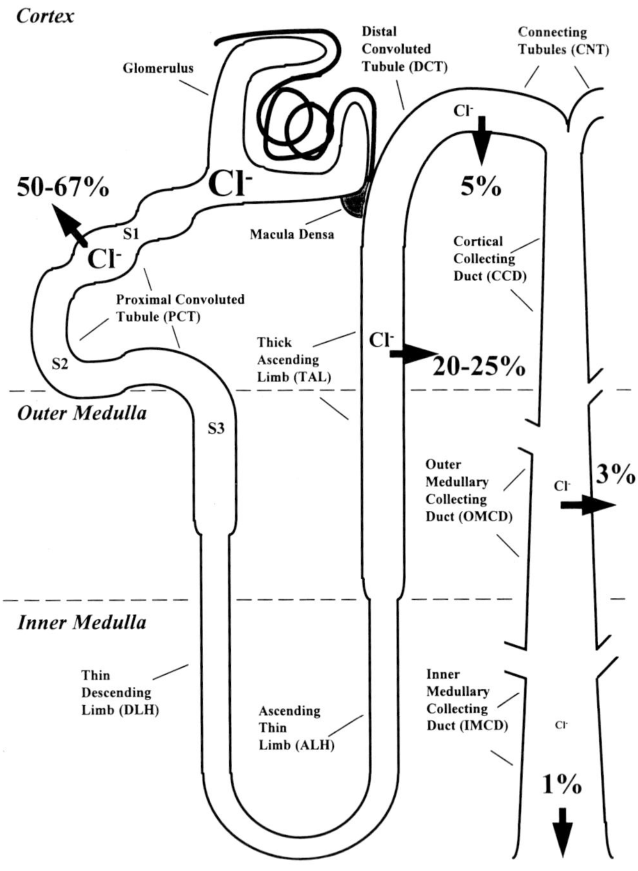

FIGURE 1 Illustration of the handling of Cl^- ions by segments along the nephron. This figure illustrates Cl^- reabsorption; however, putative pathways and physiological conditions where Cl^- secretion may occur in various nephron segments are discussed where appropriate and are depicted in cell models included in Figs. 2 and 3.

of Cl^- (approximately 50–67% of the filtered load) is reabsorbed by the proximal tubule in parallel with sodium (Na^+) (Berry and Rector, 1991; Weinstein, 1992; Koeppen and Stanton, 1992). In the thick ascending limb, Na^+ and Cl^- are reabsorbed in concert; although, less salt (approximately 20–25% of the filtered load) is reabsorbed in this segment (Berry and Rector, 1991; Reeves and Andreoli, 1992a; Koeppen and Stanton, 1992). Only 5% of the filtered load of Cl^- is reabsorbed in the distal tubules, while approximately 3% is reabsorbed in the collecting duct (Berry and Rector, 1991; Koeppen and Stanton, 1992). It is the distal nephron that is responsible for "fine" regulation of NaCl balance (Berry and Rector, 1991; Koeppen and Stanton, 1992), while it is the proximal nephron that is responsible for "bulk" reabsorption of NaCl. Less than 1% of the filtered load is excreted in the urine under physiological conditions (Berry and Rector, 1991; Koeppen and Stanton, 1992).

The objective of this review is to provide a synopsis of the mechanisms of Cl^- transport in specialized renal cell types such as mesangial cells and macula densa (MD) cells, nephron segments such as proximal tubule (PT), thick ascending limb (TAL), cortical collecting duct (CCD), and, in a few examples, cell culture models of renal ion transport such as the Madin–Darby canine kidney (MDCK) and A6 cells. It is our intent to focus each part of this review on specific cell types or segments of the nephron. Our discussion of the cellular mechanisms of Cl^- transport in each segment or cell type consists of: (1) original studies which identified Cl^- transport in a cell type or nephron segment; (2) studies which identified the specific mechanisms by which Cl^- transport was achieved (Cl^- channel or Cl^- transporter); and (3) patch-clamp studies which characterized the properties of Cl^- channels in a nephron segment or renal cell type. When possible and where defined, the functional role of Cl^- channels as well as the regulation of channels are discussed.

II. CHLORIDE TRANSPORT IN SPECIALIZED CELLS ASSOCIATED WITH THE GLOMERULUS

A. Chloride Transport in Macula Densa Cells of the Juxtaglomerular Apparatus

A group of specialized cells in the late TAL of Henle's loop attach firmly to the extraglomerular mesangium at the angle formed by the afferent and efferent arterioles of the glomerulus (see also Fig. 1). This group of cells forms the MD, and the association of these cells with the glomerulus of the same nephron is called the juxtaglomerular apparatus (JGA) (Schnermann and Briggs, 1992). The JGA is a complex structure that contains a complete feedback regulatory system comprising sensory, integrative, and effector elements (Schnermann and Briggs, 1992). Changes in luminal fluid and electrolyte composition, specifically fluid delivery and NaCl concentration, at the MD lead to changes in both the filtration rate of the associated glomerulus and the rate of renin secretion from granular cells in the MD.

Due to the development of an *in vitro* microperfusion system consisting of isolated, cortical TALs containing MD cells and an attached glomerulus, this preparation allowed for the direct study of the mechanism by which these cells sense Cl^- concentration in the tubular fluid (Skott and Briggs, 1987). Using this preparation, where effects from renal nerves or local hemodynamics were eliminated, Skott and Briggs (1987) measured renin secretion of the entire juxtaglomerular apparatus and observed a prompt stimulation of renin release rate by the MD cells in response to a decrease in tubular NaCl concentration in the cortical TAL. Therefore, studies were undertaken to examine the mechanisms of NaCl transport in MD cells, the regulation of renin release from renin-secreting cells, and the link between NaCl transport and renin secretion in this region of the nephron. Several original studies depleted or removed NaCl from the tubular fluid and stimulated renin release from the JGA (Abboud *et al.*, 1979; Kirchner *et al.*, 1978; Rostand *et al.*, 1985).

To determine the cellular mechanisms underlying Cl^- sensing in the JGA, MD cells were impaled with microelectrodes. Schlatter *et al.* (1989) demonstrated that the basolateral membrane potential (V_{bl}) was hyperpolarized by three maneuvers: (1) reduction of luminal NaCl concentration; (2) application of furosemide, an inhibitor of $Na^+,K^+,2Cl^-$ co-transporter, to the luminal membrane; and (3) application of NPPB, a Cl^- channel blocker, to the basolateral membrane. Depolarization of V_{bl} was achieved by reduction of basolateral Cl^- concentration. These experiments suggested that the electrical characteristics of MD cells were similar to those observed in the TAL. Additional data supporting this hypothesis were provided by Salomonsson *et al.* (1991), who measured intracellular Cl^- activity ($[Cl^-]_i$) in MD cells with the Cl^--sensitive dye SPQ. This group observed a sharp decrease of $[Cl^-]_i$ when furosemide was added to the tubule fluid, suggesting that the entry step across the apical membrane was via a Na^+, $K^+,2Cl^-$ co-transporter. On the other hand, NPPB added to the basolateral membrane increased $[Cl^-]_i$ three-fold, providing evidence that the exit step across the basolateral membrane of MD cells was mediated by a Cl^- channel. In MD cells, Bell *et al.* (1989) and Lapointe *et al.* (1991) used intracellular microelectrode techniques to discover a Cl^- conductance in the basolateral membrane of these cells. The Cl^-

conductance was much larger (53 ± 19%) than conductances for K^+ (23 ± 8%) and Na^+ (17 ± 10%).

Taken together, these results are compatible with a model for MD cells which is similar to that proposed for cells of the TAL (Greger, 1985). The model includes an electroneutral, furosemide-sensitive, Na^+,K^+,$2Cl^-$ co-transporter in the luminal membrane and a Cl^- channel in the basolateral membrane (see Fig. 2). Until now, patch-clamp studies of the basolateral membrane of the macula densa or MD cells in culture have not been performed to characterize the type of Cl^- channel present. Development of a cell culture of renin-secreting JGA cells would facilitate these studies.

Because the Na^+,K^+,$2Cl^-$ co-transporter has a low K_m for Na^+, luminal Cl^- concentration has been proposed as the critical "sensory" criterion for the macula densa in tubuloglomerular feedback (TGF) (Schlatter *et al.*, 1989; Persson *et al.*, 1991). In order to demonstrate that this criterion existed, a series of studies by Lorenz *et al.* (1990,1991) used techniques of *in vitro* microperfusion of cortical TAL that had an attached MD and glomerulus and found that renin secretion responded to luminal Cl^- concentration variations but was much less sensitive to changes of luminal Na^+ concentration. Moreover, it was shown that luminal application of bumetanide, an inhibitor of Na^+,K^+,$2Cl^-$ co-transport, stimulated renin secretion. These results suggest that the initial signal for the control of renin secretion was a decrease in the rate of luminal Na^+,K^+,$2Cl^-$ co-transport in MD cells and that Cl^- transport was the rate-limiting step in sensing whether to release renin or to decrease renin production.

To define further the signaling mechanisms underlying the link between Cl^- transport and renin secretion in the juxtaglomerular, renin-secreting cells from isolated mouse glomerulus, Kurtz (Kurtz and Penner, 1989) measured both intracellular Ca^{2+} using Ca^{2+}-sensitive dye, Fura-2, and Cl^- currents. The results demonstrated that agonists which trigger an increase in intracellular Ca^{2+} either via activation of phospholipase C and production of inositol 1,4,5-trisphosphate (IP_3) (Kurtz and Penner, 1989) stimulate Cl^- currents. This result suggested that Cl^- channels in these cells are activated by Ca^{2+}. The rise in intracellular Ca^{2+} concentration caused a marked Cl^- efflux which may inhibit renin release by some unknown mechanism. A similar mechanism involving a rise in intracellular Ca^{2+} could underlie the known "osmometric" behavior of JGA cells, in which cell shrinkage is paralleled by inhibition of renin secretion (Frederiksen *et al.*, 1975; Skott, 1988).

Finally, numerous studies have described a role for adenosine, a local metabolite and autacoid, in the tubuloglomerular feedback system (Arend *et al.*, 1984; Churchill and Churchill, 1985; Churchill *et al.*, 1979; Cook and Churchill, 1984; Itoh *et al.*, 1985; Osswald *et al.*, 1975; Tagawa and

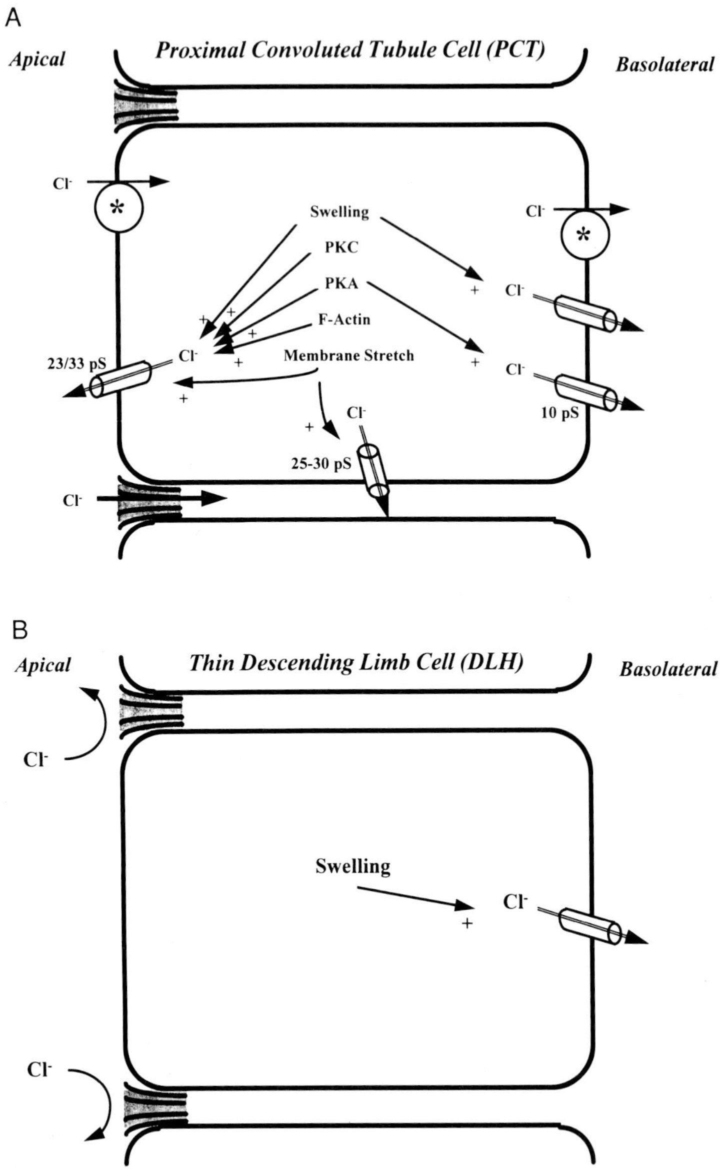

FIGURE 2 Cell models illustrating the mechanisms of Cl^- transport in cell types within the proximal nephron of the kidney. Out of convenience, cell models from the proximal nephron (glomerulus through thick ascending limb) are shown in this figure, and cell models from the distal nephron (distal convoluted tubule through collecting duct) are shown in Fig. 3. The JGA is used as the benchmark which separates proximal and distal nephrons.

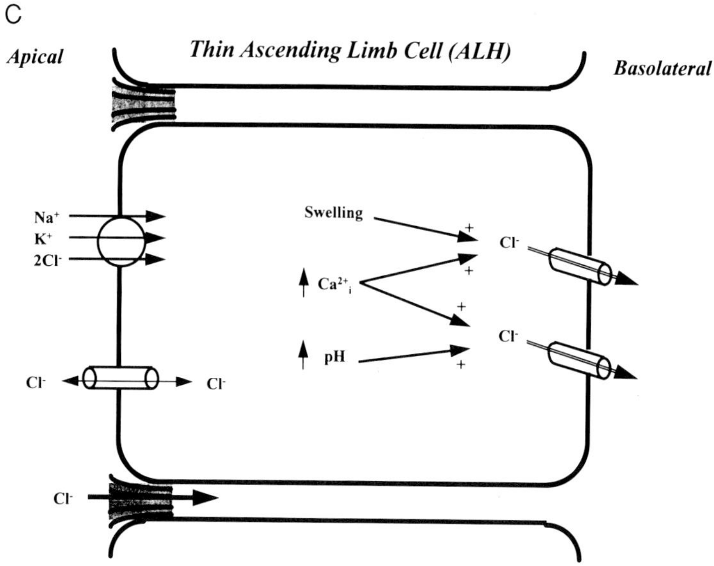

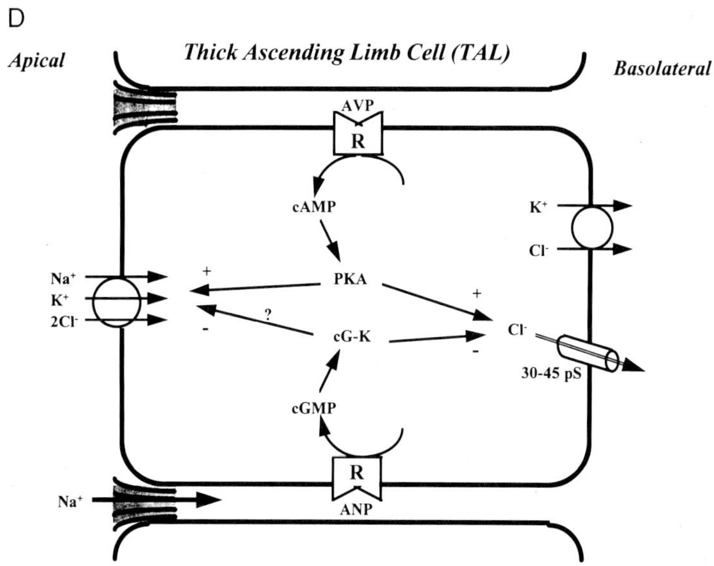

FIGURE 2 *Continued* For simplicity and clarity, only Cl^- transport mechanisms are shown. The asterisks in (A) represent multiple Cl^- transporters/exchangers expressed in each membrane (see text for details).

Vander, 1970; Weihprecht *et al.*, 1990). N^6-cyclohexyladenosine (CHA), an adenosine analogue and A_1 adenosine receptor agonist, inhibits the ability of low NaCl concentrations in the TAL to stimulate renin release. The selective A_1 adenosine receptor antagonist, 8-cyclopentyl-1,3-dipropylxanthine (CPX) was shown to reverse the inhibitory effect of CHA (Weihprecht *et al.*, 1990). Moreover, the inhibitory effect of high tubular fluid NaCl at the MD was blocked by CPX. Finally, CPX did not significantly affect renin secretion when luminal NaCl was low, suggesting that adenosine formation and/or release is dependent upon luminal NaCl concentration (Weihprecht *et al.*, 1990). Further studies are necessary to elucidate the cellular mechanisms by which luminal NaCl concentration modulates renin release and adenosine production and by which angiotensin II causes renin release from granular cells of the MD.

B. Chloride Transport in Mesangial Cells of the Glomerulus

To study the integration of the signal from MD cells to the glomerulus, Persson measured Cl^- activity in the juxtaglomerular interstitial spaces of *Amphiuma* during perfusion of the early distal tubule of the same nephron (Persson *et al.*, 1988). It was observed that $[Cl^-]$ was higher in the interstitial space than in tubular fluid or plasma and increased when perfusion rate in the adjacent tubule was increased, reaching values higher than five-fold of isotonic values. This buildup of hypertonicity could be inhibited by bumetamide. Simultaneous measurements of single-nephron blood flow showed typical feedback inhibition parallel to the hypertonicity. Of interest, MD cells are in contact with extracellular mesangial cells, both of which form a microtissue not supplied with capillaries (Goormaghtigh, 1937,1939). Thus, this morphology is permissive for the generation of the hypertonicity. A close association of MD and extracellular mesangial cells to form a restricted space is important, because MD cells have a low Na^+/K^+-ATPase activity (Kashgarian *et al.*, 1985; Schnermann and Marver, 1986) and are thought to have a low capacity for transport of NaCl. The ability of MD cells to define the $[Cl^-]$ in this space may represent the integrative step of this system. Both mesangial cells and renin-secreting cells are in intimate contact with the JGA environment and respond to $[Cl^-]$.

Mesangial cells reside in the intercapillary space of the glomerulus and have contractile fibers similar to those found in vascular smooth muscle cells. There is substantial experimental evidence supporting the idea that mesangial cells participate, through contraction, in the regulation of intraglomerular hemodynamics and, therefore, glomerular filtration rate (GFR)

(Blantz *et al.*, 1976; Schlondorff, 1987). The physiological role of mesangial cells in glomerular hemodynamics was examined directly by inducing mesangial cell lysis with an antithymocyte antibody serum (Blantz *et al.*, 1993). These results suggest an important role for these cells in the regulation of the glomerular ultrafiltration coefficient, capillary hydrostatic pressure, and efferent arteriolar resistance (Blantz *et al.*, 1993).

It is well established that mesangial cell contraction, evoked by angiotensin II (AII) and vasopressin (AVP), depends on the production of IP_3, a transient increase in Ca^{2+}_i, and prostaglandin E_2 (PGE_2) synthesis (Bonventre *et al.*, 1986; Pfeilschifter, 1986; Pfeilschifter and Bauer, 1986; Hassid *et al.*, 1986). Different experimental approaches have shown an increase in the Cl^- conductance across the plasma membrane of mesangial cells in response to AII and AVP. The observation that a decrease in external chloride concentration $[Cl^-]_o$ supressed the production of IP_3 and the resultant increase in the Ca^{2+}_i induced by AII and AVP led to the hypothesis that $[Cl^-]_o$ may be the signal that, by modulating mesangial cell contraction, triggers tubuloglomerular feedback (Okuda *et al.*, 1986,1989; Kremer *et al.*, 1989).

Using the patch-clamp technique, Ling *et al.* (1992b) found in rat glomerular mesangial cells (GMC) grown in the presence of insulin that 12% of cell-attached patches contained a 4-pS Cl^--selective channel with a current–voltage (*I–V*) relationship that exhibited slight outward rectification. Mean NP_o at resting potential was <0.1. Acute application of 100 n*M* AII or 0.25 μM thapsigargin (to release Ca^{2+}_i) to the extracellular bath stimulated Cl^- channel activity (Ling *et al.*, 1992b). In excised inside-out patches, the rise of Ca^{2+} from 10^{-8} to 10^{-4} *M* on the cytoplasmic side increased the open probability ~100-fold (Ling *et al.*, 1992b). When cells were grown in the absence of insulin, these channels were observed in only 4% of the cell-attached patches. Also, they were unresponsive to AII and to thapsigargin and were less sensitive to activation by cytoplasmic Ca^{2+}. Since a nonselective cation channel with a similar response to AII and Ca^{2+}_i was observed under the same conditions, the authors postulate that insulin-dependent Cl^- efflux and cation influx should promote depolarization of GMC cells and lead to voltage-dependent Ca^{2+} channel activation, Ca^{2+} influx, and GMC contraction (Ling *et al.*, 1992b). These observations are in accordance with the observation that GMC contraction in response to vasoactive peptides also requires insulin (Ling *et al.*, 1992b).

Taken together, these results show that Cl^- channels as well as cation channels and calcium channels act in concert to depolarize the mesangial cell and allow for mesangial cell contraction. Studies are beginning to elucidate the complex regulation of these channels by insulin, angiotensin II (a mediator of tubuloglomerular feedback), and vasopressin. Since insulin binds to a receptor which displays intrinsic as well as associated tyrosine kinase activity and because increases in phosphoinositide turnover and intracellular calcium are important for the mesangial cell contractile response, substantial crosstalk between Ca^{2+}-dependent protein kinases and tyrosine kinases may exist within the mesangial cell during the contractile response. In the coming years, the mesangial cell should serve as an excellent model for the study of signal transduction in the kidney. Moreover, the integration of mesangial cell and macula densa granular cell functions in the tubuloglomerular feedback response provides a unique opportunity for the study of linked intracellular and intercellular signaling pathways with Cl^- channels and Cl^-transporters serving as functional end points for these pathways.

III. CHLORIDE TRANSPORT IN THE PROXIMAL NEPHRON: PROXIMAL CONVOLUTED TUBULE THROUGH THICK ASCENDING LIMB

A. Chloride Transport in the Proximal Tubule

More than 50% of the filtered load of Cl^- is reabsorbed as NaCl in the proximal tubule, specifically in the late proximal convoluted tubule (PCT) (Chantrelle *et al.*, 1985; Berry and Rector, 1991; Weinstein, 1992). Many studies have examined the relative importance of passive and active mechanisms underlying Cl^- reabsorption along the PT. Chantrelle examined the active and passive components of NaCl absorption in doubly perfused proximal convoluted tubules of rat kidney. Under conditions where the lumen was perfused with high Cl^- and the peritubular capillaries were perfused with an ultrafiltrate-like solution, approximately 50% of NaCl and water absorption was passive, driven by anion gradients, and 50% of the above was active. When the PCT lumen and peritubular capillaries were perfused with similar high Cl^- solutions, anion gradients were absent and all NaCl absorption was active (Chantrelle *et al.*, 1985). With the knowledge obtained in these initial studies, laboratories have focused on the characterization of transcellular mechanisms of Cl^- transport across the proximal tubule.

Numerous reviews have described electroneutral transporters or parallel exchange mechanisms in both the apical and the basolateral membranes of proximal tubule cells (Berry and Rector, 1991; Weinstein, 1992). Due to the limited length of this review and the overwhelming evidence for several mechanisms of Cl^- transport to the proximal tubule, the evidence for Cl^- transport and exchange mechanisms is summarized. In the apical membrane, evidence exists for electroneutral NaCl transport in amphibian proximal tubule (Spring and Kimura, 1978; Guggino *et al.*, 1982), whereas studies have also provided evidence for parallel Na^+/H^+ and Cl^-/OH^- exchange mechanisms (Warnock and Yee, 1981; Burnham *et al.*, 1982; Shiuan and Weinstein, 1984; Chen *et al.*, 1988; Karniski and Aronson, 1985,1987). In the basolateral membrane, studies suggest that several transporters mediate Cl^- transport including KCl co-transport (Sasaki *et al.*, 1988; Ishibashi *et al.*, 1990; Eveloff and Warnock, 1987), Na^+-dependent Cl^-/HCO_3^- exchange (Ishibashi *et al.*, 1990,1993; Alpern and Chambers, 1987; Grassl *et al.*, 1987; Sasaki and Yoshiyama, 1988; Kondo and Fromter, 1990; Guggino *et al.*, 1983), and Na^+-independent Cl^-/HCO_3^- exchange (Ishibashi *et al.*, 1990; Alpern and Chambers, 1987; Grassl *et al.*, 1987; Chen *et al.*, 1988; Sasaki and Yoshiyama, 1988; Kondo and Fromter, 1990; Kurtz, 1989; Grassl and Aronson, 1986).

Several studies have focused on the contribution of Cl^- conductive pathways to Cl^- transport in the proximal tubule. Studies using either conventional or ion-sensitive microelectrodes showed a minor or negligible Cl^- conductance in the luminal and basolateral membranes of PCT and PST in rabbit and rat, suggesting that Cl^- channels played a small role in proximal tubule Cl^- transport (Berry and Rector, 1991; Burckhardt *et al.*, 1984; Cardinal *et al.*, 1984; Cassola *et al.*, 1983; Ishibashi *et al.*, 1990; Kuwahara *et al.*, 1988; Sasaki *et al.*, 1988; Shindo and Spring, 1981; Chen *et al.*, 1988). However, several studies summarized below have found Cl^- conductances and the channels which underlie these conductances in the basolateral membrane as well as the luminal membrane of proximal tubule cells. Welling and O'Neil (1990a) used simultaneous measurements of tubule volume via videooptical imaging techniques and intracellular microelectrodes recording to monitor the effect of rapid changes in peritubular concentration of Cl^- on the basolateral membrane voltage, V_{bl}, in nonperfused rabbit proximal straight tubules (S2). Under isosmotic conditions, these authors observed an instantaneous and sustained depolarization of V_{bl} of +5.3 mV upon reduction of peritubular Cl^- from 120 to 4 m*M* which was not affected by bumetamide. This result was consistent with other studies above that demonstrated a small Cl^- conductance with an apparent transference number for chloride (T_{Cl}) calculated to be 0.06.

However, previous studies by Welling had illustrated the importance of Cl^- and K^+ to regulatory volume decrease in rabbit proximal straight tubule. Depletion of Cl^- from the tubules or inhibition of K^+ channels with barium attenuated regulatory volume decrease (RVD) induced by hypotonic cell swelling (Welling and Linshaw, 1988). Using similar methods of tubule volume measurement and intracellular microelectrode recording, hyposmotic challenge increased Cl^- and K^+ conductances in the basolateral membrane by approximately 100% during the peak of cell swelling (Welling and O'Neil, 1990b). These results are compatible with the presence of a swelling-activated Cl^- channel in the basolateral membrane of proximal tubule cells similar to those found in other tissues. Schild *et al.* (1991) and Macri *et al.* (1990,1993) also showed evidence for a swelling-sensitive Cl^- conductance in the basolateral membrane of rabbit proximal convoluted tubule.

These studies provide an example of the minor contribution or the complete lack of conductances (active channels) measured under basal or steady-state conditions in the membranes of tubules or cultured cells. From such observations, the logical conclusion is that Cl^- conductive pathways are unimportant in transcellular ion and fluid transport. However, certain stimuli such as cell swelling or agonist activation may induce the activation of channels that were closed due to tonic inhibition or cause the insertion of channels held in vesicle pools beneath the cell membrane. Examples of Cl^- conductances which may be induced in the apical and basolateral cell membranes of the proximal tubule are discussed in the following sections.

With the observation that a basolateral Cl^- conductance could be induced in the proximal tubule, the patch-clamp technique was utilized to characterize the channels that underlie this condutance. Using whole-cell patch-clamp recordings in isolated cells of *Ambystoma* proximal tubule, a macroscopic Cl^- conductance was identified that was enhanced reversibly by forskolin, an agonist for adenylyl cyclase, or by a membrane-permeant analogue of cyclic AMP, dibutyryl-cAMP (dB-cAMP). The current could be inhibited by diphenylamine carboxylate (DPC), a Cl-channel blocker. In single-channel recordings performed in symmetrical Cl^- solutions, a 10-pS channel was found with a linear I–V relationship in patches excised from the basolateral membrane (Segal and Boulpaep, 1990,1992). In cell-attached patches, dB-cAMP and forskolin activated as many as 10 channels in a single patch. In excised, inside-out patches, the channel was stimulated by ATP applied to the cytoplasmic face of the patch; ATP-elicited channel activity was enhanced by addition of the catalytic subunit of protein kinase A. The open probability of the channel was also increased with depolarization and the anion selectivity sequence was determined to be $I^- > Br^- > Cl^- >> F^- >$ aspartate$^-$ and $Cl^- >> Na^+$ (Segal and Boulpaep, 1992). It is interesting that this channel shares some properties with the CFTR (cystic fibrosis transmembrane conductance regulator) Cl^- channel (Riordan *et al.*, 1989) characterized in great detail in other, nonrenal epithelial cells (Cliff *et al.*, 1992; Egan *et al.*, 1992; Gray *et al.*, 1989; Tabcharani *et al.*, 1990) (see also Chapters 7 and 8 by Frizzell and Morris and Welsh and co-workers, this volume). However, this Cl^-

channel in proximal tubule is activated by depolarization and has a different permselectivity sequence than CFTR. Nevertheless, it may represent an amphibian version of CFTR, a new isoform or splice variant of CFTR, or a channel closely related to CFTR.

Gogelein and Greger (1986) described a voltage-sensitive channel in cell-attached and excised patches of the lateral membrane of rabbit proximal tubules perfused *in vitro*. This channel has a conductance of 30 pS, does not discriminate between Na^+ and K^+, and is twice as permeable to cations than to Cl^-. In cell-attached patches, the channel is nearly inactive at 0 mV, but is activated by depolarization. Despite a greater permeability to cations, experiments with Cl^- channel inhibitors showed that DPC blocks channel activity reversibly, whereas SITS inhibits channel activity in a partially reversible manner. Amiloride, an inhibitor of Na^+ channels, has no effect on this channel. Of interest, Hunter (1990) described a swelling- and stretch-activated 25-pS channel that was permeable to cations as well as anions in the basolateral membrane of single proximal tubule cells isolated from *Rana temporaria*. This channel is also sensitive to voltage with open probability increasing with depolarization. Taken together, these data are compatible with the presence of a Cl^- channel in the basolateral membrane of proximal tubule cells. Additional studies are necessary in mammalian proximal tubules to characterize the types of Cl^- channels present and their sensitivity to various stimuli.

A few studies have suggested that a Cl^- conductive pathway can also be induced in the apical membrane of proximal tubule cells. In a preparation of brush-border membrane vesicles, the appearance of a Cl^- conductance was demonstrated after stimulation by cAMP (Lipkowitz and Abramson, 1987). Patch-clamp recordings have also demonstrated the presence of Cl^- channels in the apical membrane of proximal tubule cells. Suzuki *et al.* (1991) found a Cl^- channel in the apical membranes of rabbit proximal tubule cells in primary culture that was activated by parathyroid hormone (PTH) in cell-attached patches and by prolonged depolarization in inside-out patches. This channel expresses the following characteristics: a moderately outwardly rectifying I–V relationship, a conductance of 23 pS at negative potentials and 33 pS at positive potentials, and a permeability ratio of chloride versus sodium of 10:1 ($P_{Cl^-}:P_{Na^+} = 10:1$). Since the channel is stimulated in a dose–response manner by PTH, by the catalytic subunit of protein kinase A (CSU-PKA), by purified protein kinase C (PKC), and by phorbol ester, a known PKC activator, the authors conclude that PTH activates the channel by signaling pathways which involve PKA and PKC (Suzuki *et al.*, 1991).

Suzuki and co-workers defined the role of this channel in RVD with the aid of patch clamp and $[Cl^-]_i$ with SPQ fluorescence. RVD occurred

1 min after osmotic shock; $[Cl^-]_i$ decreased simultaneously (Ogawa *et al.*, 1992). Phorbol ester also reduced $[Cl^-]_i$, and both phorbol ester and osmotic shock were not additive, suggesting a contribution of PKC to Cl^- exit during RVD. Patch-clamp studies revealed an increase in NP_o (aggregate open probability of multiple channels in a patch) of 23/33 pS Cl^- channels approximately 1 min after osmotic shock, which was concurrent with the observed RVD. When H-9, an inhibitor of protein kinases, was included, NP_o decreased. These observations indicate that this channel, present in rabbit proximal tubule cells, is activated during RVD through phosphorylation by PKC (Ogawa *et al.*, 1992).

Finally, a study showed that cytochalasin D, an F-actin-disrupting agent, decreases the activity of the 23–33 pS Cl^- channel described above (Suzuki *et al.*, 1993). The authors concluded that cytochalasin D, by disrupting F-actin, inhibits active Cl^- channel as well as whole cell Cl^- currents in rabbit proximal tubule cells (Suzuki *et al.*, 1993). The fact that the F-actin-based cytoskeleton may regulate ion channels is an exciting observation and, indeed, several groups have provided evidence for cytoskeletal regulation of ion channels (Schwiebert *et al.*, 1991a,b,1994; Mills *et al.*, 1993,1994; Cantiello *et al.*, 1993; Guharay and Sachs, 1984; Stanton *et al.*, 1991). However, in these other studies (see below in Section IV,C) disruption of F-actin led to activation of the channel. Moreover, the 23/33 pS Cl^- channel has been implicated in RVD in rabbit proximal tubule cells (Ogawa *et al.*, 1992; Suzuki *et al.*, 1993). The fact that a swelling-activated Cl^- channel is inhibited by F-actin fragmentation is difficult to reconcile, since cell swelling disrupts the membrane-associated actin network. Further studies are necessary to link the protein kinase-associated signaling mechanisms, the swelling-associated signaling mechanisms, and the F-actin-associated regulation into an integrated system of ion channel regulation. Nevertheless, this channel has provided an excellent end point for the study of cell signaling and ion channel regulation.

Figure 2a illustrates a model of the mechanisms of Cl^- transport in the proximal tubule and depicts the stimuli which activate the Cl^- channels described in this section. Cl^- channels in the apical and basolateral membranes of rabbit proximal tubules contribute to RVD. Therefore, it is possible that each channel or both channels may respond differently or simultaneously to specific stimuli that cause RVD. Cl^- channels in both apical and basolateral membranes are also stimulated by the cAMP/PKA signaling pathway. Weinstein has suggested that under normal conditions of transport, apical Cl^- channels could be a pathway for Cl^- secretion. However, the role of Cl^- secretion on transepithelial transport or in cell homeostasis remains to be delineated (Weinstein, 1992). Clearly, further studies will be necessary to define the function of Cl^- channels in the proximal tubule.

B. *Chloride Transport in Thin Descending (DLH) and Thin Ascending Limbs (ALH) of Henle's Loop*

The cells of the thin descending limb of Henle's loop are both morphologically and functionally heterogeneous (Imai *et al.*, 1988b; Imai, 1984). The transepithelial permeability of Cl^- measured either in the DLH of cortical nephrons or the upper portion of DLH of juxtamedullary nephrons of hamster kidney varies from low to moderately permeable to Cl^- (Imai, 1984). Lopes *et al.* (1988) through videooptical techniques found a Cl^- conductance that was activated by hypotonicity in the basolateral membrane of cells from the lower portion of DLH dissected from juxtamedullary nephrons. This Cl^- conductance is believed to be important for Cl^- efflux during RVD. Patch-clamp studies have not been performed on cells derived from the thin descending limb.

Kondo *et al.* (1987a,b,1988) established that the thin ascending limb of Henle's loop in juxtamedullary nephrons displayed the highest permeability to Cl^- of any nephron segment. A large portion of Cl^- transport across the ALH in hamster is transcellular, inhibited by furosemide and by DPC or NPPB, and dependent on intracellular pH and $[Ca^{2+}]_i$. Yoshitomi *et al.* (1988) measured, with intracellular microelectrodes, the transepithelial and basolateral membrane potentials during changes in luminal or basolateral Cl^- concentrations. The results showed that both luminal and basolateral membranes are conductive to Cl^- (Yoshitomi *et al.*, 1988), although the basolateral membrane Cl^- conductance was larger. The basolateral Cl^- conductance is stimulated by an increase in Ca^{2+}_i and inhibited by low pH. Imai *et al.* (1988a) also analyzed the effects of sulfhydryl reagents on Cl^- conductance in this nephron segment. They postulated that modulation of Cl^- conductance in this segment occurs by two distinct pathways, one stimulatory and another inhibitory both of which contain a SH groups (Imai *et al.*, 1988a). The exact nature of these sites is unknown.

In another study, Onuchic *et al.* (1992), using videooptical techniques, detected a Cl^- conductive pathway sensitive to anthracene-9-carboxylate (9-AC), a Cl^- channel blocker, in the basolateral membrane of epithelia from the ALH that was activated during RVD. Additional experiments showed that depletion of K^+, Cl^-, or Ca^{2+} or addition of barium, 9-AC, and EGTA abolished RVD. Since RVD is abolished in the absence of Ca^{2+}, it might be possible that a Ca^{2+}-activated Cl^- channel is the pathway for Cl^- loss during this process.

Figures 2b and 2c depict models of Cl^- transport across apical and basolateral membranes of the DLH and ALH of Henle's loop and how these conductances appear to be regulated. It is unclear whether the Ca^{2+}-activated Cl^- conductance found by Yoshitomi *et al.* and the swelling- and Ca^{2+}-activated Cl^- conductance characterized by Onuchic *et al.* are carried by a single Cl^- channel or represent multiple Cl^- channel subtypes. Of interest, a relative of the ClC-2 chloride channel gene, ClC-Kl, has been cloned and localized to this nephron segment. These results are discussed below.

C. Chloride Transport in the Thick Ascending Limb of Henle's Loop

The TAL of Henle's loop is responsible for 20–25% of NaCl reabsorption along the nephron (Berry and Rector, 1991; Reeves and Andreoli, 1992a). Of that quantity, 50% is absorbed via the paracellular pathway, whereas 50% is transported actively via transcellular mechanisms (Hebert and Andreoli, 1986). In studies by Andreoli and colleagues, Cl^- was shown to account for 90% of the equivalent short-circuit current in TAL (Hebert and Andreoli, 1984; Hebert *et al.*, 1984). By utilizing a variety of models for the TAL, several investigators describe Cl^- absorption as a two-step process (Reeves and Andreoli, 1992b; Greger, 1985; Oberleitner *et al.*, 1982; Yoshitomi *et al.*, 1987). Cl^- is absorbed across the apical membrane by a furasemide-sensitive, Na^+,K^+,$2Cl^-$ co-transporter. Once inside the cell, Cl^- exits across the basolateral membrane through a Cl^- channel down a favorable electrochemical gradient. A model of Cl^- reabsorption across the TAL is shown in Fig. 2d. The Cl^- channel blocker, DPC, was characterized initially as an inhibitor of Cl^- channels in the basolateral membrane of TAL (Distefano *et al.*, 1985; Wangemann *et al.*, 1986). Also involved in basolateral Cl^- exit is a KCl co-transporter which is thought to play only a minor role in net Cl^- reabsorption (Greger and Schlatter, 1983; Oberleitner *et al.*, 1983).

Cl^- reabsorption in the TAL is hormonally regulated. AVP, via its second-messenger cAMP, stimulates Cl^- reabsorption, in part, by activating the Na^+,K^+,$2Cl^-$ co-transporter in the apical membrane (Hebert and Andreoli, 1984; Hebert *et al.*, 1984; Molony *et al.*, 1987). Cl^- absorption by the TAL is inhibited by atrial natriuretic peptide (ANP) via its second messenger, cGMP (Bailly *et al.*, 1992). It is unknown whether cGMP regulates the entry step or the exit step of Cl^- transport in the TAL. Again, both stimulatory and inhibitory pathways modulate Cl^- reabsorption in this segment.

With the knowledge that the basolateral membrane of TAL contained Cl^- channels, studies were initiated to characterize the channels responsible for Cl^- absorption across the basolateral membrane. Despite the relative difficulty of forming gigaohm seals on the basolateral membrane of TALs due to a thick basement membrane resistant to enzymatic digestion, a handful of studies have successfully characterized Cl^- channels by patch clamping cells from this nephron segment.

Schlatter and Greger (1991), in intact TALs, and Greger *et al.* (1990), in isolated cells from TAL, characterized the same 31/42 pS Cl^- channel in the basolateral membrane. The activity of this channel increases with depolarization. Paulais and Teulon (1990) characterized the properties and regulation of a 40-pS Cl^- channel (P_{Cl}–P_{Na^+} = 20:1) in the basolateral membrane of mouse TAL. The activity of this channel was unaffected by Ca^{2+} or pH, but was increased with depolarization. The anion permeability sequence was $Cl^- > Br^- > NO_3 >> F^-$ (not permeable to gluconate). Interestingly, I^- blocked the channel, a property shared by other Cl^- channels including CFTR (Cliff *et al.*, 1992). In 8% of cell-attached and inside-out recordings from the basolateral membrane of untreated TALs, this channel was active. However, upon treatment with a cocktail containing forskolin (10 μM), 8-bromo-cAMP (100 μM), and IBMX (10 μM), an inhibitor of cyclic nucleotide phosphodiesterases, 24% of cell-attached recordings and 67% of inside-out recordings contained active, 40-pS Cl^- channels (Paulais and Teulon, 1990). Except for a minor difference in conductance, this channel and the one described above may represent the same channel. Nevertheless, these data show that cAMP increases Cl^- reabsorption across the TAL by activating the Cl^- channel in the basolateral membrane as well as the $Na^+,K^+,2Cl^-$ co-transporter in the apical membrane.

Because the basolateral membrane of intact TALs was difficult to study by conventional patch-clamp techniques, Andreoli and co-workers developed a method of isolating basolateral membrane vesicles from rabbit outer medulla (Bayliss *et al.*, 1990). The authors point out that this is not a pure preparation of basolateral membranes from rabbit TAL, which makes it difficult to pinpoint the precise location of the channels *in vivo*. Despite this caviat, this group characterized extensively the properties of Cl^- channels in isolated vesicles (Reeves and Andreoli, 1992b). Using a $^{36}Cl^-$ efflux assay, Bayliss *et al.* (1990) showed that this channel was inhibited by DPC with an IC_{50} of 154 μM. The anion permeability sequence was $I^- > Cl^- \geq NO_3{}^- >> Glu^-$. Similar results were obtained by basolateral membrane vesicles from porcine outer medulla (Breuer, 1989). In a series of papers, Reeves, Andreoli, and colleagues characterized Cl^- channels from membranes vesicles incorporated into lipid bilayers. The single-channel conductance in symmetrical 270 m*M* Cl^--containing solutions is 70–90 pS. Normalization of the single-channel conductance with other studies using symmetrical 135–145 m*M* Cl^--containing solutions yields a conductance of 35–40 pS, a conductance similar to that of channels described above in the basolateral membrane of intact TALs or isolated cells from TALs (Paulais and Teulon, 1990; Hebert and Andreoli, 1984; Greger *et al.*, 1990). Increasing Cl^- concentration on the outside of the membrane increases Cl^- channel activity (Winters *et al.*, 1991). The catalytic subunit of PKA and ATP activates these 70- to 90-pS Cl^- channels in a manner analogous to that observed when extracellular Cl^- concentration is increased (Winters *et al.*, 1991). Winters *et al.* (1990) also found that agents which disrupt arginine and lysine residues within the channel decrease channel activity. In additional studies, they show that Cl^- ions themselves interact with the channel by binding to arginine and lysine residues on their positively charged side chains (Winters *et al.*, 1992). Finally, this group has taken initial steps to obtain a cDNA for this Cl^- channel (see Section VII, Renal Chloride Channels Biochemistry and Molecular Biology).

IV. CHLORIDE TRANSPORT IN THE DISTAL NEPHRON: DISTAL CONVOLUTED TUBULE THROUGH THE COLLECTING DUCT

A. *Chloride Transport in the Distal Convoluted Tubule (DCT)*

The DCT was originally defined as the segment which extends from the macula densa to the first confluence of nephrons (Koeppen and Stanton, 1992). However, it is clear that three distinct segments exist between the macula densa and the crotical collecting duct: (1) the distal convoluted tubule; (2) the connecting tubule (CNT); and (3) the initial portion of the cortical collecting duct (Koeppen and Stanton, 1992). These segments together are responsible for the reabsorption of approximately 5% of the filtered load of Cl^-. The separation between DCT and CNT is abrupt in the rabbit but gradual in the rat and the human (Madsen and Tisher, 1986).

The entry mechanism for Cl^- across the apical membrane of the rat DCT involves a thiazide-sensitive NaCl co-transporter (Beaumont *et al.*, 1988; Beck *et al.*, 1988; Costanzo, 1985). Friedman and colleagues, using an immortalized cell line derived from mouse DCT, have demonstrated that Cl^- reabsorption is a two-step process involving entry across the apical membrane via an electroneutral, thiazide-sensitive NaCl co-transporter and exit across the basolateral membrane via an NPPB-sensitive Cl^- channel (Friedman and Gesek, 1993; Gesek and Friedman, 1993). In constrast to the rat or mouse DCT, the thiazide-sensitive NaCl co-transport system is absent from the luminal membrane of rabbit DCT (Koeppen and Stanton, 1992). A preliminary report suggests that Cl^- entry could also occur through a thiazide-insensitive co-transport mechanism which involves formate and oxalate-coupled Cl^- movement (Wang *et al.*, 1992,1993). Cl^- exit across the basolateral membrane occurs via a small Cl^- conductive pathway (Taniguchi *et al.*, 1989; Yoshitomi *et al.*, 1989b). In only one study of the single-channel properties of Cl^- channels in this segment, an 11-pS Cl^--selective channel was found in the apical membrane but was not fully characterized (Taniguchi *et al.*, 1989).

A recent study confirmed the fact that a low-conductance, 9-pS chloride channel was present in primary cultures of rabbit distal convoluted bright tubule (Poncet *et al.*, 1994). In unstimulated cells, these channels were present in 9% of cell-attached patches. However, pretreatment of cells with 5 μM forskolin or 1 mM 8-Br-cAMP increased the incidence of the low-conductance channels to 26 and 37%, respectively (Poncet *et al.*, 1994). Upon excision to the inside-out configuration, only 1 of 47 channels remained active, whereas addition of the catalytic subunit of protein kinase A with 2 mM ATP was necessary to restore channel activity (Poncet *et al.*, 1994). These 9-pS channels displayed a linear I–V relationship, appeared in patches as clusters of channels, were unaffected by depolarization or hyperpolarization, were sensitive to DPC at 1 mM but not at 100 μM, were insensitive to DNDS or NPPB, and exhibited a halide permeability sequence of $Br^- > Cl^- > I^-$ (I^- actually appears to block the channel)(Poncet *et al.*, 1994). In a low percentage of patches (<3%), outwardly rectifying, 30-pS Cl^- channels were observed, but their regulation or biophysical properties were difficult to characterize due to a low incidence of activity (Poncet *et al.*, 1994). These results suggest strongly that the apical surface of rabbit DCT cells contains cAMP-activated CFTR Cl^- channels. A summary of the studies that support the presence of CFTR Cl^- channels is presented at the end of this part of the review. Further studies are necessary to characterize further the functional role of CFTR in this segment. A cell model illustrating what is known about Cl^- transport in the rat DCT cell is shown in Fig. 3A.

B. Chloride Transport in the Connecting Tubule (CNT)

The CNT is composed of a heterogeneous population of cells (Berry and Rector, 1991; Koeppen and Stanton, 1992). Two cell types are present, CNT cells, involved primarily in Na^+ reabsorption, and intercalated cells, responsible for acid-base transport. Both cell types express specific mechanisms of Cl^- transport. Studies have identified the rabbit CNT cell as the thiazide-sensitive segment. For example addition of the thiazide diuretic, trichloromethiazide, to the lumen reduces NaCl reabsorption by 30% and hyperpolarizes V_{bl} of rabbit CNT cells, while rabbit DCT cells and intercalated cells were not affected by thiazides (Yoshitomi *et al.*, 1989a). Similar results have been found in the urinary bladder of the winter flounder (Stokes, 1984) and the late distal tubule from *Amphiuma* (Stanton, 1988, 1990). As in DCT cells, a Cl^- channel plays a role in the movement of

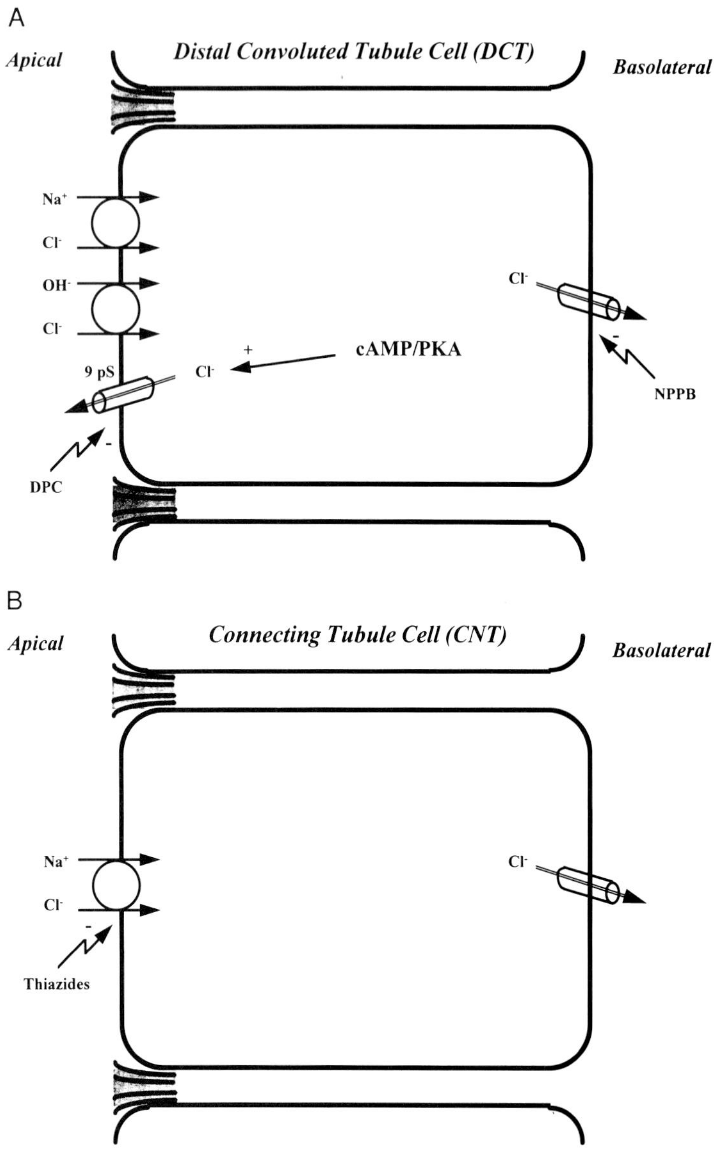

FIGURE 3 Cell models illustrating the mechanisms of Cl^- transport within the distal nephron of the kidney. For simplicity and clarity, only Cl^- transport mechanisms are shown.

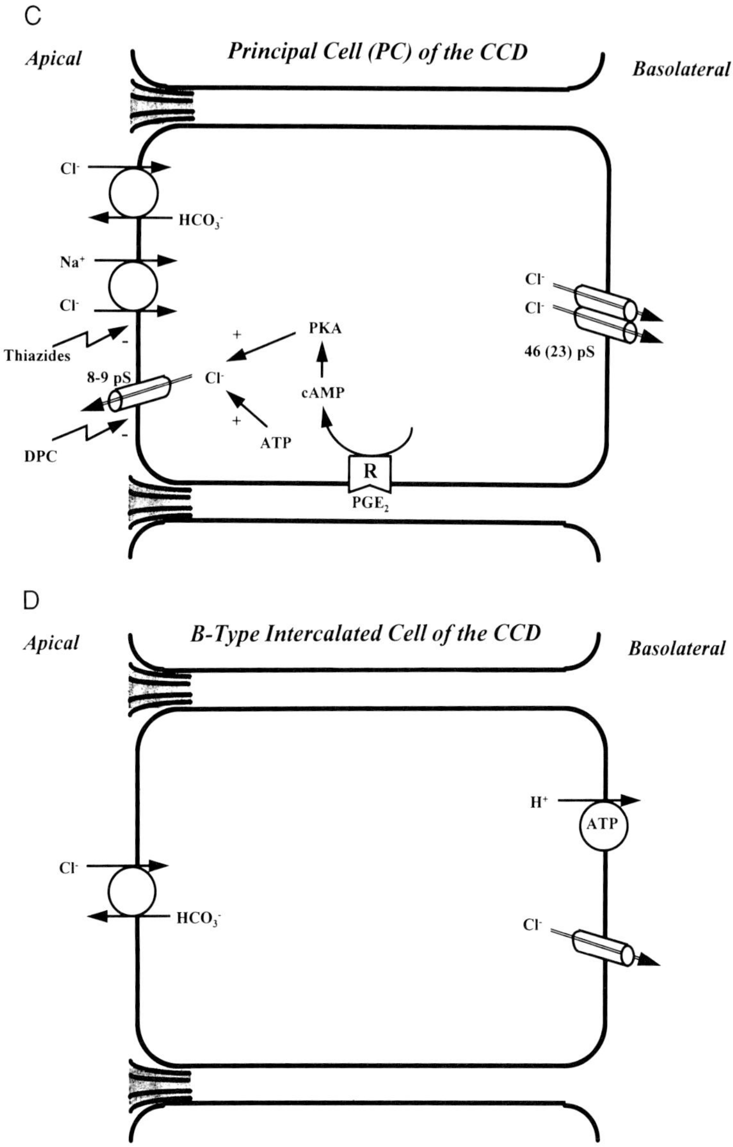

FIGURE 3 *Continued*

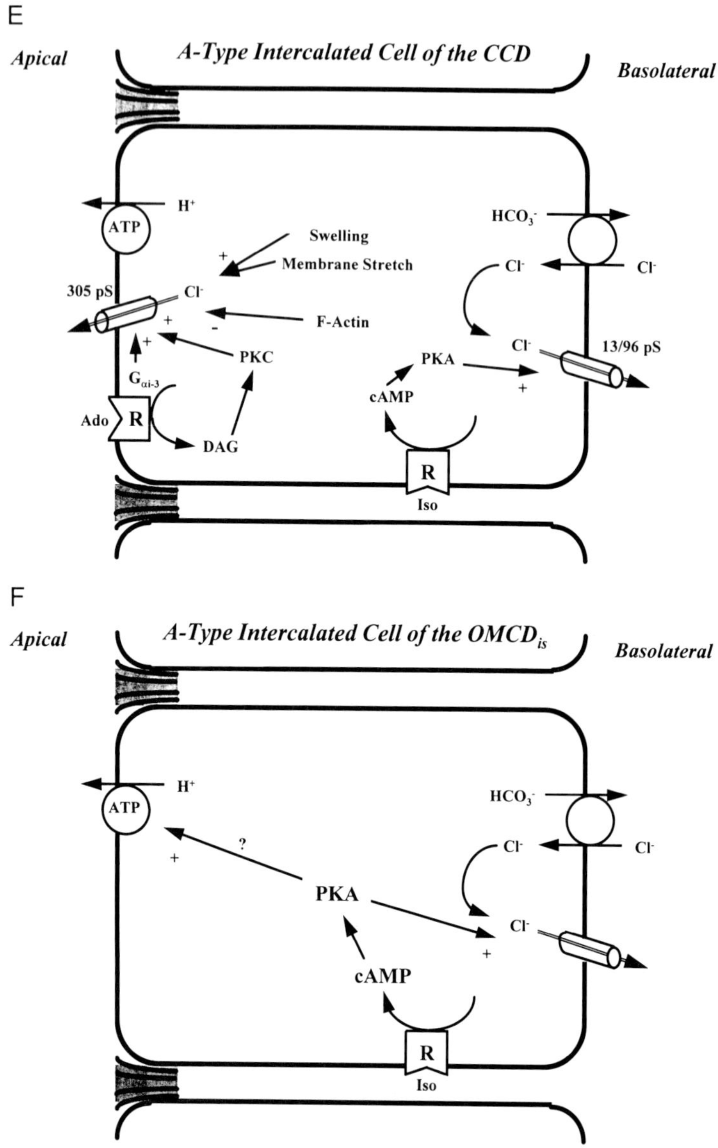

FIGURE 3 *Continued*

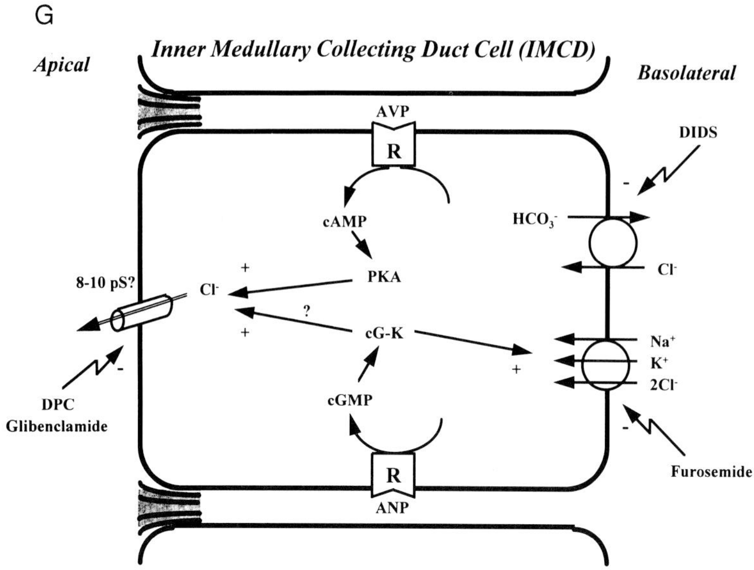

FIGURE 3 *Continued*

Cl^- out of the cell across the basolateral cell membrane (Koeppen and Stanton, 1992). The single-channel characteristics of the basolateral Cl^- channel have not been determined. A cell model of Cl^- transport across the rabbit CNT cell is shown in Fig. 3B.

C. Chloride Transport in the Cortical Collecting Duct (CCD)

The CCD is composed of two cell types: principal cells (PC) and intercalated cells (IC) (Kaissling and Kriz, 1979; Madsen and Tisher, 1986; Muto *et al.*, 1990; Koeppen *et al.*, 1983). In the CCD, principal cells comprise 60–70% of the cells, whereas intercalated cells constitute the remaining 30–40% (Lonnerholm and Ridderstrale, 1980; Muto *et al.*, 1990; Koeppen *et al.*, 1983). PCs are also the predominant cell type in the CNT and the CCD (Muto *et al.*, 1990). PC cells reabsorb Na^+ and secrete K^+ and have Cl^- channels in both the apical and basolateral membranes (see below). IC cells play a role in acid base as well as Cl^- transport. ICs can be further subdivided into A-type ICs, cells that secrete acid, and B-type ICs which secrete bicarbonate. Cell models illustrating Cl^- transport pathways in PCs, A-type ICs, and B-type ICs in the rabbit are shown in Figs. 3c–3e.

The paracellular pathway in the CCD is permeable to Cl^-. Consequently, as much as 80% of transepithelial Cl^- reabsorption takes place via the paracellular pathway driven by the lumen-negative transepithelial voltage (Schuster and Stokes, 1987; Sansom *et al.*, 1984). The remainder of Cl^- reabsorption is transcellular involving specific Cl^- transport pathways.

1. Principal Cells (Fig. 3C)

It is known that a basolateral Cl^- conductance participates in cell volume regulation in rabbit PCs (Strange, 1989). In contrast, an entry step for Cl^- in PCs is not fully understood. Evidence has been presented for the presence of a Cl^-/HCO_3^- exchanger in the apical membrane in rabbit PCs (Strange, 1991; Sansom *et al.*, 1984,1990; Koeppen *et al.*, 1983; Muto *et al.*, 1990). However, in the rat, one study identified a thiazide-sensitive NaCl cotransporter in the apical membrane (Terada and Knepper, 1990). Another study concluded that rat PCs express no mechanisms for Cl^- transport (Schlatter *et al.*, 1990). Finally, another study (see below) has found a Cl^- channel in the apical membrane of rabbit PCs (Ling and Kokko, 1992). Obviously more studies are necessary to define the precise mechanism(s) of Cl^- entry across the apical membrane of PCs.

Several patch-clamp studies succesfully characterized Cl^- channels in the basolateral membrane of PCs. Sansom *et al.* (1990) identified a "double-barreled" Cl^- channel in rabbit PCs that was similar to a Cl^- channel recorded from the electric organ of *Torpedo* (Miller, 1982). This channel displayed two open states (O_1 and O_2) and, therefore, two distinct conductances of 23 pS (O_1) and 46 pS (O_2)(Sansom *et al.*, 1990). The channel was not dependent upon voltage, although the open probability increased somewhat with hyperpolarization (Sansom *et al.*, 1990). The precise physiological role of this channel or its regulation has not been elucidated.

Other studies have focused on Cl^- conductances in the apical membrane of rabbit CCD cells grown in primary culture which displayed a PC-like phenotype. In cells treated with aldosterone, 9% of cell-attached patches contained a 9-pS Cl^-channel with a linear *I–V* relationship (Ling and Kokko, 1992). In cell-attached patches pretreated with PGE_2, however, the incidence of the 9 pS Cl^- channel increased to 32% (Ling and Kokko, 1992). Forskolin or permeable analogues of cAMP mimicked the effect of PGE_2 in increasing incidence and open probability of the channel (Ling and Kokko, 1992). Of interest, this channel is insensitive to DIDS, indicating that this channel shares many properties with CFTR Cl^- channels including single-channel conductance, a linear *I–V* relationahip, time and voltage independence, and resistance to DIDS blockade (Ling and Kokko,

1992; Cliff *et al.*, 1992; Egan *et al.*, 1992; Gray *et al.*, 1989; Tabcharani *et al.*, 1990). Breyer and co-workers in tubules isolated from rabbit CCD found a similar 8-pS Cl^- channel in the apical membrane (K. Superdock and M. Breyer, unpublished observations). Moreover, Breyer and co-workers performed RT-PCR on RNA isolated from rabbit kidney, RT-PCR and RNase protection assays on message isolated from immunodissected rabbit CCD cells, and immunoprecipitation of CFTR protein from CCD; all sets of experiments are consistent with the presence of CFTR mRNA and protein in the kidney and, specifically, in the CCD (M. Breyer, personal communication). In another preliminary report, Breyer and co-workers found an inwardly rectifying 54/22 pS Cl^- channel in the apical membrane of immunodissected rabbit CCD cells that was inhibited by ATP (Superdock *et al.*, 1992,1993). This channel type is novel and has not been found in any other renal cell type. Finally, in a preliminary report, Korbmacher *et al.*, (1992) found that forskolin increased the whole-cell Cl^- conductance in a cell line from mouse CCD (M1-CCD) approximately five-fold, consistent with more specific single-channel studies described above. However, due to the heterogeneous population of cells in the preparations used by Superdock *et al.* and Korbmacher *et al.*, it is not possible to localize these channels or conductances to a specific cell type in the CCD.

Taken together, these results demonstrate that specific chloride channels may be expressed on both the apical and basolateral membranes of PC cells in the CCD and are activated by cAMP/PKA signaling pathways. A similar distribution and regulation of Cl^- channels has been demonstrated in the proximal tubule. Further studies are necessary to define the physiological significance of cAMP/PKA-activated Cl^--conductive pathways in both membranes of PC cells.

2. Intercalated Cells

In the B-type ICs in the CCD from the rabbit (Fig. 3D), both Cl^- reabsorption and HCO_3^- secretion across the apical cell membrane occurs via an electroneutral pathway (Schuster and Stokes, 1987; Muto *et al.*, 1990; Star *et al.*, 1985; Schuster, 1985,1986; Schwartz *et al.*, 1985). In constrast, Cl^- is reabsorbed across the basolateral membrane, via a Cl^-conductance (Muto *et al.*, 1990; Sansom *et al.*, 1984). Inhibition of this Cl^- conductance reduces Cl^- absorption and HCO_3^- secretion in rabbit CCD by eliminating the ability of Cl^- to exit the cell across the basolateral cell membrane (Schuster, 1985). Currently, only one patch-clamp study has been performed in a preparation of B-type IC cells to define the properties and regulation of whole-cell Cl^- currents. Isoproterenol stimulates Cl^- currents via a β-adrenergic receptor-mediated process, and atenolol, a β-receptor antagonist, blocks the response (Ikeda *et al.*, 1992). Regulation of the whole-cell currents does not occur via cAMP or Ca^{2+}, but the currents are activated by GTPγS (Ikeda *et al.*, 1992). In addition, because the effects of isoproterenol and GTPγS are additive, the authors speculate that β-adrenergic receptors stimulate basolateral Cl^- conductance by a direct, membrane-delimited mechanism involving a G protein (Ikeda *et al.*, 1992).

In A-type ICs from the rabbit (Fig. 3E), the H^+-ATPase is located in the apical membrane and the Cl^-/HCO_3^- exchanger is expressed in the basolateral membrane. Under steady-state conditions, B-type ICs constitute most of the 30–40% of ICs in the rabbit CCD (Schwartz *et al.*, 1985). However, when the animal is subjected to an acid load, A-type ICs increase in number at the expense of B-type ICs (Schwartz *et al.*, 1985). In A-type ICs, there is no net transcellular Cl^- transport, but movement of Cl^- in and out of the cell across both apical and basolateral membranes does occur. Removal of Cl^- from the basolateral solution or inhibition of the basolateral Cl^- conductance attenuates electrogenic H^+ secretion across the apical cell membrane CCD (Hanley *et al.*, 1980). Thus, the basolateral Cl^- conductance is important for H^+ secretion and HCO_3^- reabsorption to occur by allowing for Cl^- to recycle across the basolateral cell membrane and by providing a counter ion to balance electrogenic H^+ secretion.

Patch-clamp studies were performed on RCCT-28A cells, an immortalized cell line derived from rabbit CCD, with properties most similar to A-type ICs (Arend *et al.*, 1989; Schwiebert *et al.*, 1992). In whole-cell patch recordings, the conductance of RCCT-28A cells was determined to be attributable mostly to Cl^- (Dietl *et al.*, 1992). Isoproterenol, as well as cAMP analogues, stimulate an outwardly rectifying, whole-cell Cl^- conductance (Dietl *et al.*, 1992). In single-channel recordings of the basolateral membrane of RCCT-28A cells, 13/96 pS Cl^- channels were observed with an outwardly rectifying *I–V* relationship (Dietl and Stanton, 1992). These channels are highly permeable to Cl^- compared with Na^+ ($P_{Cl^-}:P_{Na^+} = 64:1$) and were inhibited by DIDS and NPPB (Dietl and Stanton, 1992). A subset of these channels that displayed a low open probability was activated by the catalytic subunit of PKA, an effect that was reversed by alkaline phosphatase (Dietl and Stanton, 1992). In contrast, channels already activated by excision or a large depolarizing voltage pulse failed to respond to PKA (Dietl and Stanton, 1992). The authors concluded that this channel was the one regulated by isoproterenol via a second-messenger pathway involving cAMP and PKA and that this channel may be involved in Cl^- recycling across the basolateral membrane, by removing the Cl^- which enters the cell via the Cl^-/HCO_3^- exchanger in the same membrane.

Even though intracellular microelectrode recording of ICs displays no measurable conductance for any ion in the apical membrane under steady-state conditions (Muto *et al.*, 1990; Sansom *et al.*, 1984; Koeppen *et al.*, 1983), Cl^- channels are present. Stanton and co-workers identified a large-conductance (305-pS) Cl^- channel in the apical membrane of isolated rabbit CCDs, in rabbit IC cells in primary culture, and also in RCCT-28A cells (Light *et al.*, 1990; Schwiebert *et al.*, 1990,1992). These channels have the following characteristics: a linear *I–V* relationship; inactivation at voltages more negative than −30 mV and more positive than +30 mV; activation upon patch excision, activation by large depolarizing voltage pulses (>+50 mV); inhibition by DIDS, DPC, or NPPB; and greater permeability for Cl^- than HCO_3^-, gluconate, or Na^+ ($P_{Cl^-}:P_{HCO3^-} = 2:1$; $P_{Cl^-}:P_{Glu^-} = 9:1$;$P_{Cl^-}:P_{Na^+} = 14:1$). This channel is only observed in 3% of cell-attached patches under isotonic conditions in RCCT-28A cells. However, the 305-pS channel can be observed more frequently if the bath osmolality is reduced only by about 10% (Schwiebert *et al.*, 1994; Mills *et al.*, 1993,1994). After 2–5 min in hypotonic solutions, the channel becomes active attaining a high open probability (>0.50) for approximately 5–10 min before closing (Schwiebert *et al.*, 1994; Mills *et al.*, 1993,1994). At 15 min following hypotonic challenge, channel open probability is essentially 0.00 (Schwiebert *et al.*, 1994; Mills *et al.*, 1993,1994). The increase in open probability correlated with RVD, suggesting that this channel may be required for RVD.

Negative pressure (−2 to −15 cm H_2O pressure) applied to excised, inside-out patches of the apical membrane from these cells activates quiescent 305-pS channels. The negative pressure is also effective in increasing the open probability of channels spontaneously activated by patch excision (Schwiebert *et al.*, 1994; Mills *et al.*, 1993,1994). Both swelling- and negative pressure-induced activation of 305-pS Cl^- channels occurs via the filamentous actin (F-actin) cytoskeleton. Evidence for an involvement of filamentous actin comes from experiments which show that dihydrocytochalasin B (DHCB) as well as other cytochalasins (cytochalasins B and D) can activate single channels in inside-out patches. Moreover, swelling or stretch fails to stimulate single channels or whole-cell conductance in the presence of DHCB activation, indicating that disruption of F-actin was essential for mechanical activation (Schwiebert *et al.*, 1994; Mills *et al.*, 1993,1994). In support of the cytochalasin data, phalloidin, a mushroom toxin and F-actin-stabilizing agent, blocked the ability of swelling or stretch to stimulate the Cl^- channel or whole-cell Cl^- conductance. Preliminary studies utilizing rhodamine-phalloidin to visualize F-actin indicate that swelling or activation of protein kinase C depolymerize the F-actin cytoskeletal network in RCCT-28A cells (J. W. Mills, E. M. Schwiebert, and B. A. Stanton, unpublished observations). Interestingly, swelling stimulation of whole-cell Cl^- conductance also required protein kinase C and a pertussis toxin (PTX)-sensitive G protein, since swelling-induced stimulation was inhibited or abolished by calphostin C, a PKC inhibitor, and PTX, an inhibitor of a subset of heterotrimeric G proteins, respectively (Schwiebert *et al.*, 1994; Mills *et al.*, 1993,1994). Taken together, these results suggest that the 305-pS Cl^- channel in RCCT-28A cells mediates, at least in part, Cl^- efflux during RVD and that this channel is regulated during cell swelling by a signaling pathway that involves a PTX-sensitive G protein, protein kinase C, and the actin-based cytoskeleton. It is not clear whether autocoids such as adenosine (see below) are released during cell swelling and bind to receptors in an autocrine or paracrine fashion to trigger this pathway or whether cell swelling and the corresponding increase in membrane tension activate second messengers directly.

The 305-pS Cl^- channel is also regulated by autocoids. Adenosine, via its A_1 receptor subtype, stimulates the 305-pS Cl^- channel in cell-attached patches (Schwiebert *et al.*, 1992). Adenosine activates this channel both in cell-attached and inside-out patches via the following second-messenger pathway: binding of an agonist to an A_1 receptor coupled to phospholipase C, release of diacylglycerol, activation of protein kinase C, and, finally, stimulation of a PTX-sensitive G protein (Schwiebert *et al.*, 1990,1992). Moreover, in inside-out patches treated with PTX to inhibit 305-pS Cl^- channels, PKC failed to activate the channels, indicating that PKC required a G protein to activate the channel (Schwiebert *et al.*, 1990,1992).

These studies provide another example of agonist-activated conductances in the plasma membranes of renal epithelial cells where there is no macroscopic conductance for Cl^- or any ion detected under isotonic or steady-state conditions. Often a stimulant such as hypotonicity will activate "silent" Cl^- channels in cell membrane-generating macroscopic Cl^- conductances which are used by the cell when needed.

D. Chloride Transport in the Outer Medullary Collecting Duct (OMCD)

The outer stripe of the OMCD is comprised both of PCs and B-type ICs (Berry and Rector, 1991; Koeppen and Stanton, 1992; Koeppen, 1987). In constrast, the inner stripe of the OMCD ($OMCD_{is}$), is composed of A-type ICs (Koeppen and Stanton, 1992; Koeppen, 1987; Ridderstrale *et al.*, 1988). The $OMCD_{is}$ secretes Cl^- at rates equal to that of H^+ secretion (Stone *et al.*, 1983). Cl^- secretion across the A-type ICs in $OMCD_{is}$ occurs via the paracellular pathway. Microelectrode studies have failed to demonstrate the presence of a macroscopic Cl^- conductance in the apical membrane under steady-state conditions (Koeppen, 1986). However, given that a pattern has emerged in other nephron segments where macroscopic conductances for Cl^- cannot be detected by approaches utilizing intracellular microelectrodes but ion channels are identified by patch clamping, it is certainly possible that the movement of Cl^- across the apical cell membrane could involve an apical Cl^- channel.

A series of studies indicates that Cl^- channels are present in these cells. For example, Koeppen and co-workers (1991) have shown that isoproterenol, cAMP, and PKA stimulate an outwardly rectifying Cl^- current in whole-cell recordings from cells immunodissected from rabbit $OMCD_{is}$. Concomitant operation of basolateral Cl^- channels, basolateral Cl^-/HCO_3^- exchangers, and apical H^+-ATPase is important for properly maintaining acid-base transport across this nephron segment (Koeppen *et al.*, 1991). These currents may be generated by outwardly rectifying Cl^- channels similar to the ones found in the basolateral membrane of RCCT-28A cells, A-type ICs derived from rabbit CCD (Dietl and Stanton, 1992). Obviously, single-channel studies will be necessary to characterize the outwardly rectifying Cl^- channels in $OMCD_{is}$ cells and to determine their role in acid secretion.

E. Chloride Transport in the Inner Medullary Collecting Duct (IMCD)

Compared to other nephron segments, comparatively less information is available about Cl^- transport in the IMCD. The IMCD is composed primarily of IMCD cells; only the initial portion of the IMCD ($IMCD_1$ or $IMCD_i$) is believed to contain a low percentage of ICs (Berry and Rector, 1991; Koeppen and Stanton, 1992). As described above, PCs in the CCD are primarily responsible for Na^+ reabsorption, although Cl^- channels have been described in both membranes. Rocha and Kudo showed initially that IMCD cells possess the capacity to reabsorb Na^+ and secrete Cl^-. They showed that low concentrations of ANP reduced Na^+ reabsorption, while as little as 60 p*M* ANP increased bath-to-lumen Cl^- secretion and generated a lumen-negative potential (Rocha and Kudo, 1990a,b). Rocha and Kudo concluded that ANP, via its second-messenger cGMP, stimulates Cl^- secretion by activating a furosemide- and bumetanide-sensitive Na^+, K^+, $2Cl^-$ co-transporter in the basolateral membrane. Lumen-negative voltages were measured in the $IMCD_1$ of Brattleboro rats with the use of isolated perfused tubules and microelectrode impalements (Stanton, 1989).

Preliminary studies utilizing measurement of transepithelial short-circuit current (I_{SC}) generated by monolayers of immortalized mouse IMCD cells indicate that these cells are capable of Na^+ reabsorption and Cl^- secretion

(Kizer *et al.*, 1993; Vandorpe *et al.*, 1993). In the presence of amiloride (10 μM) to block Na^+ transport, vasopressin (1 nM), and permeable cyclic AMP analogues (100 μM CPT-cAMP) increased short-circuit current that was attributable to Cl^- secretion (Kizer *et al.*, 1993). Of interest, both Na^+ and Cl^- current were enhanced by aldosterone (1 nM)(Kizer *et al.*, 1993). Characterization of transepithelial Cl^- secretion across mouse IMCD cell monolayers showed that basolateral Cl^- entry was inhibited by furosemide, DIDS, or removal of HCO_3^- from the bathing solutions, suggesting roles for $Na^+,K^+,2Cl^-$ co-transport and Cl^-/HCO_3^- exchange in the basolateral entry mechanism (Kizer *et al.*, 1993). Apical Cl^- secretion was inhibited by 1–10 mM DPC or 100 μM glibenclamide but was unaffected by 100 μM DIDS or 100 μM NPPB, a result which suggests that Cl^- secretion may be mediated by CFTR Cl^- channel (Kizer *et al.*, 1993; Vandorpe *et al.*, 1993). Indeed, parallel, preliminary whole-cell patch-clamp studies indicate that CPT-cAMP increases whole-cell chloride conductance by 100% (Vandorpe *et al.*, 1993). These chloride currents displayed a linear I–V relationship, sensitivity to DPC, and time- and voltage-independent kinetics (Vandorpe *et al.*, 1993). Finally, CFTR was immonolocalized to the apical region of IMCD cells in mouse kidney (Vandorpe *et al.*, 1993). Taken together, these results are consistent with the observation by Rocha and Kudo that the IMCD reabsorbs Na^+ and secretes Cl^-. Moreover, these results indicate that vasopressin or cyclic AMP activates secretory Cl^- currents that may be carried by CFTR chloride channels in the apical membrane.

More studies will be necessary to examine the capacity of the IMCD cell both to absorb Na^+ and to secrete Cl^- and the cellular mechanisms which regulate each process. It is interesting to speculate that the cells which constitute the last segment of the nephron may also be the most diversified in their transport capabilities in order to define more precisely the ionic contents of the final urine.

The results described in the IMCD as well as results described above in earlier nephron segments (PCT, DCT, CCD) are consistent with a published study and a preliminary study by Guggino and co-workers on the distribution of CFTR along the nephron. Crawford *et al.*, (1991) found that CFTR was detected by immunocytochemistry in proximal and distal tubules in the outer cortex and in intracellular vesicles derived from proximal tubule. More recent preliminary studies, utilizing nested RT-PCR with two pairs of primers generated against exons 13 and 14 of the CF gene (these exons encode portions of the R domain of CFTR), indicate that CFTR is present in rat kidney, specifically within the cortex, outer medulla, and inner medulla (Morales *et al.*, 1993). CFTR was most abundant in the cortex. RT-PCR of individually dissected segments of the rat nephron indicate that CFTR is present throughout the proximal tubule (S_1,$S_{2,3}$), in the thin limbs of Henle's loop, in cortical and medullary TAL, in the CCD, and in the IMCD. CFTR was absent from glomeruli, a result consistent with previous studies (Morales *et al.*, 1993). In addition to a band of the expected size for full-length CFTR, a smaller band was also found using nested RT-PCR under identical conditions and may represent a splice variant of CFTR. This smaller band of CFTR was most abundant in outer medulla and inner medulla and, specifically, in OMCD, in descending thin limbs of Henle's loop, and in IMCD (Morales *et al.*, 1993). Taken together, functional studies as well as immunocytochemical and molecular genetic analysis along the nephron have revealed an unexpected abundance and a broad distribution of CFTR in the kidney. Future studies are critical to defining the distribution, regulation, and functional role of CFTR in the kidney. A perplexing question lies in the knowledge that CFTR is abundant in kidney, while the pathophysiology associated with cystic fibrosis is subtle and not readily observed in the kidneys of CF patients.

V. CHLORIDE TRANSPORT IN CULTURED CELL MODELS OF RENAL ION TRANSPORT

Several tissue culture cell models have been developed over the past several years. The ones described below have been used extensively to study renal transport mechanisms. Although the exact location in the kidney from which they were derived is not known, these models have provided much information regarding general mechanisms and types of regulation of Cl^- transporters and, thus, are a valuable tool.

A. *Xenopus Kidney Cell Line (A6)*

The A6 cell line has been used extensively by electrophysiologists as a model for renal "distal nephron" transport. The cells are particularly useful because they form tight, confluent monolayers. Several studies have described mechanisms of Cl^- transport in or across these cells. For example, two groups have characterized Cl^- secretion across A6 cell monolayers as a two-step process involving an Na^+,K^+,$2Cl^-$ cotransporter in the basolateral membrane and an NPPB-sensitive Cl^- channel in the apical membrane (Chalfant *et al.*, 1992; Middleton *et al.*, 1992). Chalfant *et al.* (1992) showed that AVP stimulated Cl^- secretion across monolayers of A6 cells, whereas Middleton *et al.* (1992) showed that adenosine and ATP via A2 adenosine and P2 purinergic receptors, respectively, stimulated Cl^- secretion by increasing intracellular cAMP.

Several laboratories have characterized Cl^- channels in the apical membrane of A6 epithelial cells. Marunaka and Eaton (1990a) found a 1- to 2-pS Cl^- channel with a linear *I–V* relationship and a dependency on bath calcium. The incidence of this channel in cell-attached patches was increased by insulin, and the channel was inhibited by alkaline phosphatases in inside-out patches (Marunaka and Eaton, 1990b). The precise physiological role of this channel is unknown.

Ling and co-workers (1992a) described an 8-pS Cl^- channel in the apical membrane of A6 cells that was similar to the one in the apical membrane of rabbit PC cells from the CCD. This channel was stimulated within 1 min by PGE_2, forskolin, and arginine vasotocin, agonists that increase intracellular cAMP (Ling *et al.*, 1992a). The channels remained active after stimulation for as long as 45 min. The authors concluded that this channel is the one which generates the amiloride-insensitive short-circuit current observed in Ussing chamber experiments (Ling *et al.*, 1992a).

Finally, Nelson *et al.* (1984) described a 360-pS Cl^- channel in the plasma membrane of nonconfluent A6 cells. This channel is inhibited by SITS, has a linear *I–V* relationship, is more permeable to Cl^- than Na^+ ($P_{Cl^-}:P_{Na^+} = 9:1$), and inactivates at a holding potential more negative than -20 mV and more positive than $+20$ mV (Nelson *et al.*, 1984). This channel has similar properties to one described in the apical membrane of RCCT-28A cells (Light *et al.*, 1990). Taken together, these data indicate that Cl^- channels in A6 epithelial cells have properties and mechanisms of regulation similar to those in native renal epithelia. Moreover, the type of transporters and channels described suggest that the A6 cells may be a "hybrid cell" which has transporters and channels found in multiple collecting duct cell types including prinicpal cells, intercalated cells, and IMCD cells.

B. Madin–Darby Canine Kidney Cell Line (MDCK)

Several studies have examined the mechanisms of Cl^- transport in another commonly used distal nephron cell model, the MDCK cell line. MDCK cell express an array of ion channels, including Cl^- channels (Lang *et al.*, 1990). Several years ago, investigators examined Cl^- conductances involved in cell volume regulation in this cell line. Kolb *et al.* (1985) identified a 456-pS Cl^- channel in the apical membrane of confluent MDCK cells at 37°C. This channel was rarely observed in cell-attached patches but its incidence was increased upon excision to the inside-out configuration (Kolb *et al.*, 1985). It was hypothesized then that this channel could mediate Cl^- efflux during cell volume regulation. Indeed, Schwiebert *et al.* (1994; Mills *et al.*, 1993,1994) in RCCT-28A cells, demonstrated that a similar large conductance Cl^- channel was activated by cell swelling. In a report, Winters *et al.* showed that mastoparan, a peptide from bee and wasp venom which activates heterotrimeric G proteins by mimicking the sequence in G protein, coupled hormone receptors for G protein interaction and activation, and GTP activated Cl^- and K^+ conductances in MDCK cells. The subsequent increase in Cl^- and K^+ permeability led to a decrease in cell volume. Similar results were found by Mills and Lubin, but these investigators used cytochalasins to activate Cl^- and K^+ channels in MDCK cells (Mills, 1987; Mills and Lubin, 1986).

In a review, Simmons (1993) describes the MDCK cell as an important cellular model for Cl^- secretion. Cl^- secretion across MDCK cell monolayers is a two-step process reminiscent of that which occurs across the IMCD cell (Simmons, 1993). Cl^- enters across the basolateral membrane via a Na^+, K^+,$2Cl^-$ co-transporter and exits across the apical membrane via a Cl^- channel (Simmons, 1993). The exact nature of the channel responsible for Cl^- secretion is not known; however, cAMP-activated Cl^- secretion is inhibited by NPPB and by DPC but not by DIDS or other stilbene derivatives (Simmons, 1993). Volume-activated Cl^- channels are present in the basolateral membrane of MDCK cells and are inhibited by DIDS (Simmons, 1993). Indeed, a novel Cl^- channel gene was cloned from MDCK cell messenger RNA (see below).

Taken together, these studies show that MDCK cells are an important model for many mechanisms of ion transport, specifically Cl^- secretion, and for mechanisms of regulatory volume decrease which involve Cl^- and K^+ channels. Like the A6 cell, the MDCK cell is a useful model since it grows into confluent, tight monolayers and because these cells display a "hybrid collecting duct" cell-like complement of ion transporters and channels.

VI. CHLORIDE CHANNELS IN INTRACELLULAR COMPARTMENTS

Cl^- channels play an important role in Cl^- movement across intracellular compartments such as endosomes, lysosomes, and the trans-Golgi network. In these organelles, Cl^- channels are important to compensate for positive charge produced by the electrogenic H^+-ATPase pump during proton influx and to allow the intraorganelle compartment to be acidified by the net influx of HCl (Mellman *et al.*, 1986; Rudnick, 1986). Several studies have confirmed the presence of a Cl^- conductance in renal endosomes. For example, Hilden showed that a small Cl^- conductance is present in rabbit renal cortex endosomes and is increased after addition of ATP (Hilden *et al.*, 1988). These results support the idea that ATP hydrolysis induces both H^+ translocation and Cl^- channel activation. Similarly, Bae and Verkman (1990) found a Cl^- conductance in endosomes from rabbit proximal tubules. This conductance is activated by protein kinase A, inhibited by phosphatases, and inhibited by DNDS (Bae and Verkman, 1990). Taken together, these studies established the presence and the importance of Cl^- channels in intraorganelle compartments.

To determine the characteristics of the Cl^- channels that underlie the Cl^- conductance found in renal endosomes, Schmid *et al.* (1989) used the patch-clamp technique to describe a Cl^- channel in an enriched endosomal vesicle preparation from rat kidney cortex. They found a channel with a linear *I–V* relationship and a single-channel conductance of 73 pS in symmetrical 140 m*M* KCl solutions. In contrast to the ATP-sensitive channel described above, the activity of this channel is independent of ATP or changes in pH. This channel was selective for anions over cations with an anion permeability sequence of $Cl^- = Br^- = I^- > SO_4^{2-} = F$. DIDS and NPPB completely inhibits channel activity. The authors speculate that the channel is responsible for the maintenance of electroneutrality during H^+ uptake into endocytic vesicles.

A novel Cl^- channel with a conductance of 115 pS in 0.41:0.15 *M* KCl solutions was characterized in vesicles from renal cortex (Reeves and Gurich, 1992). The authors identified three characteristics of their channel that differ from those involved in endosomal acidification: the 115-pS channel (1) is sensitive to the Cl^- channel blocker IAA-94; (2) is Ca^{2+} activated; and (3) is not activated by PKA or inactivated by alkaline phosphatase. Thus, the newly described channel is not the channel responsible for charge compensation during endosomal acidification. Its physiological role, so far, is unknown.

Taken together with the results above, multiple types of Cl^- channels clearly exist within vesicles and acidified organelles. However, it is possible that Cl^- channels normally present and functional in the plasma membrane may contaminate these vesicle preparations, particularly in endosomal vesicle fractions, and may cloud the study of vesiclar Cl^- channels which participate in vesicle acidification. Alternatively, since all plasma membrane Cl^- channel proteins reside in intracellular vesicles for a portion of their existence within epithelial cells, some Cl^- channels may participate in vesicle acidification when located intracellularly and in plasma membrane Cl^- movement when located on the cell surface. Future studies and better fractionation techniques need to be developed to address these questions.

VII. RENAL CHLORIDE CHANNEL BIOCHEMISTRY AND MOLECULAR BIOLOGY

Three groups have used expression and homology cloning approaches to isolate genes which encode Cl^- channels (Thiemann *et al.*, 1992; Paulmichl *et al.*, 1992). Jentsch and colleagues, by using a homology cloning approach with cDNA probes from ClC-0 and ClC-1 (cDNAs cloned from *Torpedo* electric organ and skeletal muscle, respectively))Jentsch *et al.*, 1990; Steinmeyer *et al.*, 1991), cloned a third type of Cl^- channel cDNA (Thiemann *et al.*, 1992). This channel gene appears to be ubiquitously expressed; 3.3-kb mRNA was detected in several tissues which encoded for a 99-kDa protein, and a larger 4.5-kb species was observed in an African green monkey kidney cell line (BSC-1), an SV-40 transformed monkey kidney cell line (COS-7), and a porcine kidney cell line ($LLC-PK_1$)(Thiemann *et al.*, 1992). When this cDNA was expressed in *Xenopus* oocytes, a whole-cell current was recorded that was quite low and was time and voltage independent between holding potentials of -100 to $+100$ mV. However, the current was markedly activated at hyperpolarizing potentials more negative than -100 mV. This current was inhibited by 9-AC and by DPC and was insensitive to DIDS, and the anion selectivity sequence was calculated to be $Cl^- = Br^- > I^-$ (Thiemann *et al.*, 1992). More recent studies have shown that this channel, when expressed in *Xenopus* oocytes, is activated by cell swelling (Grunder *et al.*, 1992) (See also Chapter 2 by Jentsch and colleagues in this volume).

Unfortunately, single-channel analysis has also not been performed on cells expressing this cDNA. *I–V* plots in excised, inside-out patches are normally constructed between -100 and $+100$ mV. In order to study this channel electrophysiologically, the range of holding potentials should be widened to record ClC-2 channel activity. These studies will require very high resistance seals (in excess of 10 GΩ or more) to maintain the inside-out configuration at such large potentials, Since this channel is expressed in several cell lines derived from kidney, future studies on Cl^- channels in renal epithelial cells should be performed in order to determine whether this channel is expressed. So far, studies have not been undertaken to define the relative distribution and expression of ClC-2 along the nephron. Moreover, the role of this channel in renal Cl^- transport is not known.

Clapham and co-workers utilized expression cloning in *Xenopus* oocytes with mRNA derived from MDCK cells to clone a Cl^- channel cDNA (Paulmichl *et al.*, 1992). When this cDNA was used to probe Northern blots, a 1.8-kb mRNA was recognized that encodes for a 26-kDa protein (Paulmichl *et al.*, 1992). Cl^- currents generated by expressing the RNA encoding this channel in *Xenopus* oocytes have the following characteristics: (1) inhibition by NPPB and DIDS; (2) insensitivity to [Ca^{2+}]; and (3) an outwardly rectifying *I–V* relationship with inactivation at depolarizing voltages (Paulmichl *et al.*, 1992). Of interest, this channel was inhibited by varying amounts by nucleotides such as AMP, ADP, ATP, GTP, cAMP, ITP, and cGMP (cAMP and cGMP were the most potent) (Paulmichl *et al.*, 1992). Since this channel was cloned with message derived from a canine kidney cell line, future studies are necessary to determine which cell types along the nephron express this new Cl^- channel. Like ClC-2, the contribution of this channel to renal Cl^- transport is not understood. A new study by this group has since revealed that this gene encodes a swelling-activated chloride channel *regulator* which is linked to the actin cytoskeleton.

Uchida *et al.* (1993) used RT-PCR techniques and obtained a probe based on conserved regions of the genes cloned by Jentsch and co-workers to screen the rat whole kidney cDNA library. This group isolated a cDNA clone (ClC-K1) of approximately 2.4 kb in length with a single long open reading frame of 1965 bp encoding 686 amino acids. They found 40% amino acid identity with the Cl^- channel cloned from *Torpedo marmorata*. Hydropathy analysis demonstrated a profile very similar to those of the previously cloned Cl^- channels by Jentsch and colleagues, suggesting that ClC-K1 is a member of the Cl^- channel family. RT-PCR using microdissected nephron segments revealed that the main site of expression in the medulla was the thin ascending limb of Henle's loop, the nephron segment which displays the highest permeability to Cl^-; however, expression of ClC-K1 mRNA was observed in thin descending limb, cortical collecting duct, and inner medullary collecting duct as well. As for ClC-1 and ClC-2, cRNA from ClC-K1 failed to express a Cl^- current in *Xenopus* oocytes, unless the 5′-untranslated region was replaced by the sequence found in ClC-0. With this modification, large, time-dependent Cl^- currents were observed, but their characteristics were quite different from those found in the other ClC channels. ClC-K1-expressed currents displayed outward rectification, time-dependent activation at depolarizing voltages, and inhibition by DIDS and DPC which was complete at a concentration of 1 m*M* for each blocker. Finally, the mRNA for ClC-K1 was upregulated by dehydration, suggesting that this channel's principal functional role may be in the renal urinary concentration mechanisms.

Several investigators have isolated putative Cl^- channel proteins or partial cDNAs thought to encode for Cl^- channels (Redhead *et al.*, 1992; Tsai *et al.*, 1991; Zimniak *et al.*, 1992; Uchida *et al.*, 1992; Schuster and Bao, 1992). Using a Cl^- channel blocker, IAA-94, as a ligand in an affinity column, four major proteins were purified from membranes isolated from membrane vesicles derived from renal cortex and trachea (Landry *et al.*, 1987,1989). Antibodies were generated to the 64-kDa protein, and this antibody depleted IAA-94-inhibitable Cl^- channel activity in solbilized kidney membranes (Redhead *et al.*, 1992). As described above, Bae and Verkman (1990) showed that endosomes from proximal tubule contain a chloride conductance that is activated by PKA. More recent studies by Reenstra *et al.* (1992) indicate that a single protein is a substrate for PKA in these endosomes and that has a molecular weight of 66 kDa. The antibody to the 64-kDa protein labels the apical surface of CFPAC-1 cells, cells isolated from the pancreas of a CF patient (Redhead *et al.*, 1992). The authors conclude from these studies that this 64-kDa protein is a component or subunit of Cl^- channels present in epithelial plasma membrane and the membranes of intracellular organelles.

Finn and colleagues generated a monoclonal antibody (E12) which inhibits Cl^- conductances in several cell types including *Necturus* gallbladder, toad gallbladder, rat colon, and human nasal epithelia (Tsai *et al.*, 1991). In Western blots, this antibody recognizes a 200- 220-kDa protein in whole cell lysates from *Necturus* gallbladder, toad gallbladder, urinary bladder, small intestine, A6 cells, rat colon, rabbit gastric mucosa, human lymphocytes, and human nasal epithelia (Tsai *et al.*, 1991). Following extensive purification and reconstitution into planar lipid bilayers suggests that it is a Cl^- channel with a unitary conductance of 62 pS in assymetrical 150/50 m*M* NaCl solutions (Tsai *et al.*, 1991). The gating characteristics show that the channel behaves as an independent channel or as aggregates of several channels (Tsai *et al.*, 1991). As with the 64-kDa protein described above, the authors conclude that the 200- 220-kDa protein, expressed in a wide variety of epithelial tissues, is a component of an epithelial Cl^- channel. In order to determine definitively whether these proteins are subunits of Cl^- channels, future studies are necessary to show that these proteins affect directly the function of other proteins known to form the conductive pore of a Cl^- channel.

Several initial preliminary reports also indicate that progress is being made toward cloning novel Cl^- channel genes. The polymerase chain reaction was employed to identify Cl^- channel genes homologous to the channels identified by Jentsch and colleagues. A PCR product was identified which has 88% nucleotide identify and 50–75% amino acid identity with ClC-2 and ClC-0, genes identified by Jentsch and co-workers (1990). A riboprobe generated from the PCR product recognized mRNA species of ~5 kb in cortex and inner medulla and ~5 and 9 kb in outer medulla and a cell line from rabbit cortical collecting duct (RCSV3).

Using expression cloning in *Xenopus* oocytes, Zimniak *et al.* (1992) have identified a specific fraction of messenger RNA (mRNA) isolated

TABLE I
Chloride Channels Characterized in the Kidney

Cell type (species)	Membrane	Conductance (γ)	I–V	Blockers	Anion selectivity	Regulation
Mesangial cell (rat)	Cell	4 pS	OR	—	—	Insulin, angiotensin II, Ca^{2+}_i (+)
PT (*Ambystoma*)	BLM	10 pS	L	DPC	I > Br > Cl >> F > Asp	cAMP/PKA (+)
PT (rabbit or *Rana*)	BLM	25–30 pS	L	DPC	Cl = Na = K	Depolarization, stretch (+)
PT (rabbit)	APM	23/33 pS	OR	—	Cl > Na	PTH,PKA,PKC,F-actin (+)
TAL (rabbit)	BLM	31/42 pS	OR	DPC, 9-AC	Cl > Na	Depolarization (+)
TAL (mouse)	BLM	40 pS	L	DPC	Cl > Br > NO_3 > F >> I,Na	cAMP (+)
TAL (rabbit)	BLMV	70–90 pS (sym. 270 Cl)	L	DPC	I > Cl ≥ NO_3 > Glu	cAMP/PKA (+)
DCT (rabbit)	APM	9 pS	L	DPC	Cl > I >> Glu, Na	cAMP/PKA (+)

CCD,PC (rabbit)	BLM	23 and 46 pS	L	DIDS	Cl > Na	Hyperpolarization (+)
CCD, PC (rabbit)	APM	9 pS	L	DPC	Cl > Na	PGE_2, cAMP (+)
CCD, A-IC (rabbit)	BLM	13/96 pS	OR	DIDS, NPPB	Cl >>> Na	Iso, cAMP, PKA (+)
CCD, A-IC (rabbit)	APM	305 pS	L	DPC, DIDS, NPPB	Cl > HCO_3 > Glu > Na	Swelling, stretch (+) Ado, PKC, $G_{\alpha i}$ (+) F-actin (−) PKA (NE)
CCD (rabbit)	APM	54–22 pS	IR	—	Cl > NA	ATP,GTP,ATPγS,GTPγS (−)
A6 (*Xenopus*)	APM	1–2 pS	L	DPC, DIDS	Cl > Na	PKA (+), Phosphatases (−)
A6 (*Xenopus*)	APM	8 pS	L	DPC	Cl > I >> Na	PGE_2, cAMP, AVT (+)
MDCK (Dog)	Cell	360 pS	L	SITS	Cl > Na	—
MDCK (Dog)	APM	456 pS	L	DIDS	Cl > Na	—

Each channel is discussed in detail in the text and appropriate references are cited. All studies used approximately 130–150 m*M* symmetrical Cl^--containing solutions unless otherwise noted. L, linear I–V relationship; OR, outwardly rectifying I–V relationship; IR, inwardly rectifying I–V relationship; BLM, basolateral membrane; and APM, apical membrane. (+) activation, (−) inhibition, (NE) "no effect." Single-channel conductance (γ) is included.

from medullary TALs that express a DPC-inhibitable Cl^- conductance. In water-injected oocytes, the Cl^- conductance was 1.43 ± 0.15 μS, whereas injection of 1.8 through 3.2 kb mRNA induced a Cl^- conductance of 9.05 ± 1.56 μS. This fraction of mRNA was used to make a cDNA library with which to clone a full-length cDNA that encodes this Cl^- channel (Zimniak *et al.*, 1992).

Finally, a group has succeeded in cloning a gene for the thiazide-sensitive NaCl co-transporter from winter flounder urinary bladder (Gamba *et al.*, 1993). The gene predicts a 112-kDa protein with 12 membrane-spanning segments (Gamba *et al.*, 1993). Two species of mRNA were observed routinely, a 3.7-kb mRNA in urinary bladder and a 3.0-kb mRNA in nonbladder tissues including distal tubule of the kidney (Gamba *et al.*, 1993). This gene is the first of several genes to be isolated in the near future that encode for Cl^- transporters.

Taken together, these published studies and studies in progress are identifying several Cl^- channel genes that are expressed along the nephron. In the next few years, a family or multile families of Cl^- channel and Cl^- transporter genes are likely to emerge from these projects. With this information, much analysis will be needed to elucidate the expression and functional role of these channels along the nephron.

VIII. SUMMARY

Table I provides a summary of Cl^- channels characterized by single-channel patch-clamp methods in specialized cells associated with the glomerulus, in epithelial cells along the entire length of the nephron, and in cell lines used as models of renal epithelial Cl^- transport. It is clearly evident from this review that although many studies have characterized the properties of individual channels along the nephron, the distribution, the functional roles, the regulation, and the molecular properties of Cl^- channels in the kidney are still not understood. A possible reason for this is that Cl^--conductive pathways are often silent under basal conditions and require specific stimuli to become active. Moreover, more attention has been paid to cation transport in the kidney, specifically sodium and potassium. Analysis of ionic currents in renal epithelial cells along the nephron in the presence of hormones which elevate cyclic AMP, cyclic GMP, or Ca^{2+}, or during cell swelling, should reveal novel Cl^--conductive pathways. The recent explosion of Cl^- channel molecular biology should also accelerate our knowledge of Cl^- channel biology in renal epithelial cells. Many exciting answers should be discovered in the near future, some of which are (1) the functional role(s) and distribution of CFTR in the kidney; (2) the function and distribution of the ClC-2 channel, the novel Cl^- channel gene cloned by Clapham and co-workers (Paulmichl *et al.*, 1992), and the most recent channel gene (ClC-Kl) cloned by Uchida *et al.* in the kidney; (3) the physiology and pathophysiology of Cl^- secretion along the nephron; and (4) the identification and characterization of as yet unidentified Cl^- channels which underlie Cl^- conductances in macula densa cells, cells of the thin loops of Henle, distal convoluted tubule cells, and connecting tubule cells.

Acknowledgments

We thank Dr. Matthew Breyer and Dr. Bruce Stanton for valuable comments and criticism of this review.

References

Abboud, H. E., Luke, R. G., Galla, J. H., and Kotchen, T. A. (1979). Stimulation of renin by acute selective chloride depletion in the rat. *Circ. Res.* **44,** 815–821.

Alpern, R. J., and Chambers, M. (1987). Basolateral membrane Cl/HCO_3 exchange in the rat proximal convoluted tubule. Na-dependent and -independent modes. *J. Gen. Physiol.* **89,** 581–598.

Arend, L. J., Haramati, A., Thompson, C. I., and Spielman, W. S. (1984). Adenosine-induced decreases in renin release: Dissociation from hemodynamic effects. *Am. J. Physiol.* **247,** F447–F452.

Arend, L. J., Handler, J. S., Rhim, J. S., Gusovsky, F., and Spielman, W. S. (1989). Adenosine-sensitive phosphoinositide turnover in a newly established renal cell line. *Am. J. Physiol.* **256,** F1067–F1074.

Bae, H.-R., and Verkman, A. S. (1990). Protein kinase A regulates chloride conductance in endocytic vesicles from proximal tubule. *Nature (London)* **348,** 637–639.

Bailly, C., Neant, F., and Amiel, C. (1992). Cyclic GMP inhibits Cl reabsorption in the thick ascending limb (TAL) from mouse kidney. *JASN* **3,** 802 (abstr.).

Bayliss, J. M., Reeves, W. B., and Andreoli, T. E. (1990). Cl transport in basolateral renal medullary vesicles. I. Cl transport in intact vesicles. *J. Membr. Biol.* **113,** 49–56.

Beaumont, K., Vaughn, D. A., and Healy, D. P. (1988). Thiazide diuretic receptors: Autoradiographic localization in rat kidney with [^{3}H] metolazone. *J. Pharmacol. Exp. Ther.* **250,** 414–419.

Beck, F. X., Dorge, A., Giebisch, G., and Thurau, K. (1988). Effects of diuretics on cell potassium transport: An electron microprobe study. *Kidney Int.* **37,** 1423–1428.

Bell, P. D., Lapointe, J.-Y., and Cardinal, J. (1989). Direct measurement of basolateral potentials from cells of the macula densa. *Am. J. Physiol.* **257,** F463–F468.

Berry, C. A., and Rector, F. C. J. (1991). Renal transport of glucose, amino acids, sodium, chloride, and water. *In* "The Kidney" (B. M. Brenner and F. C. J. Rector, eds.), 4th ed., pp. 245–282. Saunders, Philadelphia.

Blantz, R. C., Konnen, K. S., and Tucker, B. J. (1976). Angiotensin II effects upon the glomerular microcirculation and ultrafiltration coefficient of the rat. *J. Clin. Invest.* **57,** 419–434.

Blantz, R. C., Gabbai, F. B., Tucker, B. J., Yamamoto, T., and Wilson, C. B. (1993). Role of mesangial cell in glomerular response to volume and angiotensin II. *Am. J. Physiol.* **264,** F158–F165.

Bonventre, J. V., Skorecki, K. L., Kreisberg, J. I., and Cheung, J. Y. (1986). Vasopressin increases cytosolic free calcium concentration in glomerular mesangial cells. *Am. J. Physiol.* **25,** F94–F102.

Breuer, W. (1989). Characterization of chloride channels in membrane vesicles from the kidney outer medulla. *J. Membr. Biol.* **107,** 35–42.

Burckhardt, B.-C., Cassola, A. C., and Fromter, E. (1984). Electrophysiological analysis of bicarbonate permeation across the peritubular cell membrane of rat kidney proximal tubule. II. Exclusion of HCO_3 effects on other ion permeabilities and of coupled electroneutral HCO_3 transport. *Pfuegers Arch.* **401,** 43–51.

Burnham, C., Munzesheimer, C., Rabon, E., and Sachs, G. (1982). Ion pathways in renal brush border membranes. *Biochim. Biophys. Acta* **685,** 260–272.

Cantiello, H. F., Stow, J. L., Prat, A. G., and Ausiello, D. A. (1991). Actin filaments regulate epithelial Na^+ channel activity. *Am. J. Physiol.* **261,** C882–C888.

Cantiello, H. F., Prat, A. G., Bonventre, J. V., Cunningham, C. C., Hartwig, J. H., and Ausiello, D. A. (1993). Actin-binding protein contributes to cell volume regulatory ion channel activation in melanoma cells. *J. Biol. Chem.* **268,** 4596–4599.

Cardinal, J., Lapointe, J.-Y., and Laprade, R. (1984). Luminal and peritubular ionic substitutions and intracellular potential of the rabbit proximal convoluted tubule. *Am. J. Physiol.* **247,** F352–F364.

Cassola, A. C., Mollenhauer, M., and Fromter, E. (1983). The intracellular chloride channel activity of rat kidney proximal tubular cells. *Pfluegers Arch.* **399,** 259–265.

Chalfant, M. L., Coupaye-Gerard, B., Civan, M. M., and Kleyman, T. R. (1992). Rapid activation of Cl secretion by arginine vasopressin in the epithelial cell line A6. *JASN* **3,** 805 (abstr.).

Chantrelle, B. M., Cogan, M. G., and Rector, F. C. J. (1985). Active and passive components of NaCl absorption in the proximal tubule of the rat kidney. *Min. Electr. Metab.* **11,** 209–214.

Chen, P.-Y., Illsley, N. P., and Verkman, A. S. (1988). Renal brush-border chloride transport mechanisms characterized using a fluorescent indicator. *Am. J. Physiol.* **254,** F114–F120.

Churchill, P. C., and Churchill, M. C. (1985). A_1 and A_2 adenosine receptor activation inhibits and stimulates renin secretion of rat renal cortical slices. *J. Pharmacol. Exp. Ther.* **232,** 589–594.

Churchill, P. C., Churchill, M. C., and McDonald, F. D. (1979). Effects of saline and mannitol on renin and distal tubule Na in rats. *Circ. Res.* **45,** 786–792.

Cliff, W. H., Schoumacher, R. A., and Frizzell, R. A. (1992). Cyclic AMP-activated chloride channels in CFTR-transfected cystic fibrosis pancreatic epithelial cells. *Am. J. Physiol.* **262,** C1154–C1160.

Cook, C. B., and Churchill, P. C. (1984). Effects of renal denervation on the renal responses of anesthetized rats to cyclohexyladenosine. *Can. J. Physiol. Pharmacol.* **62,** 934–938.

Costanzo, L. S. (1985). Localization of diuretic action in microperfused rat tubules: Ca and Na transport. *Am. J. Physiol.* **248,** F527–F535.

Crawford, I., Maloney, P. C., Zeitlin, P. L., Guggino, W. B., Hyde, S. C., Turley, H., Gatter, K. C., Harris, A., and Higgins, C. F. (1991). Immunocytochemical localization of the cystic fibrosis gene product CFTR. *Proc. Natl. Acad. Sci. U.S.A.* **88,** 9262–9266.

Dietl, P., and Stanton, B. A. (1992). Chloride channels in apical and basolateral membranes of CCD cells (RCCT-28A) in culture. *Am. J. Physiol.* **263,** F243–F250.

Dietl, P., Kizer, N., and Stanton, B. A. (1992). Conductive properties of a rabbit cortical collecting duct cell line: Regulation by isoproterenol. *Am. J. Physiol.* **262,** F578–F582.

Distefano, A., Wittner, M., Schlatter, E., Lang, H. J., Englert, H., and Greger, R. (1985). Diphenylamine-2-carboxylate, a blocker of the Cl conductive pathway in Cl-transporting epithelia. *Pfluegers Arch.* **405,** S95–S100.

Egan, M., Flotte, T., Afione, S., Solow, R., Zeitlin, P. L., Carter, B. J., and Guggino,

W. B. (1992). Defective regulation of outwardly rectifying chloride channels by protein kinase A corrected by insertion of CFTR. *Nature (London)* **358,** 581–584.

Eveloff, J., and Warnock, D. G. (1987). KCl cotransport systems in rabbit renal basolateral membrane vesicles. *Am. J. Physiol.* **252,** F883–F889.

Frederiksen, O., Leyssac, P. P., and Skinner, S. L. (1975). Sensitive osmometer function of juxtaglomerular cells *in vitro*. *J. Physiol. (London)* **252,** 669–679.

Friedman, P. A., and Gesek, F. A. (1993). Calcium transport in renal epithelial cells. *Am. J. Physiol.* **264,** F181–F198.

Gamba, G., Saltzberg, S. N., Lombardi, M., Miyanoshita, A., Lytton, J., Hediger, M. A., Brenner, B. M., and Hebert, S. C. (1993). Primary structure and functional expression of a cDNA encoding the thiazide-sensitive, electroneutral sodium-chloride cotransporter. *Proc. Natl. Acad. Sci. U.S.A.* **90,** 2749–2753.

Gesek, F. A., and Friedman, P. A. (1993). Calcitonin stimulates calcium transport in distal convoluted tubule cells. *Am. J. Physiol.* **264,** F744–F751.

Gogelein, H., and Greger, R. (1986). A voltage-dependent ionic channel in the basolateral membrane of late proximal tubules of the rabbit kidney. *Pfluegers Arch.* **407,** S142–S148.

Goormaghtigh, N. (1937). L'appareil neuro-myo-artériel juxtaglomerulaire du rein: Ses réactions en pathologie et ses rapports avec le tube urinifère. *C. R. Seances Soc. Biol. Ses Fil.* **124,** 293–296.

Goormaghtigh, N. (1939). Une glande endocrine dans la paroi des artérioles renales. *Bruxelles-Med.* **19,** 1541–1549.

Grassl, S. M., and Aronson, P. S. (1986). Na/HCO_3 cotransport in basolateral membrane vesicles isolated from rabbit renal cortex. *J. Biol. Chem.* **261,** 8778–8783.

Grassl, S. M., Holohan, P. D., and Ross, C. R. (1987). HCO_3 transport in basolateral membrane vesicles isolated from rat renal cortex. *J. Biol. Chem.* **262,** 2682–2687.

Gray, M. A., Harris, A., Coleman, L., Greenwell, J. R., and Argent, B. E. (1989). Two types of chloride channels on duct cells cultured from human fetal pancreas. *Am. J. Physiol.* **257,** C240–C251.

Greger, R. (1985). Ion transport mechanisms in thick ascending limb of Henle's loop of mammalian nephron. *Physiol. Rev.* **65,** 760–767.

Greger, R., and Schlatter, E. (1983). Properties of the basolateral membrane of the cortical thick ascending limb of Henle's loop of rabbit kidney. A model of secondary active chloride transport. *Pfluegers Arch.* **396,** 325–334.

Greger, R., Bleich, M., and Schlatter, E. (1990). Ion channels in the thick ascending limb of Henle's loop. *Pfluegers Arch.* **13,** 37–50.

Grunder, S., Thiemann, A., Pusch, M., and Jentsch, T. J. (1992). Regions involved in the opening of ClC-2 chloride channel by voltage and cell volume. *Nature (London)* **360,** 759–762.

Guggino, W. B., Boulpaep, E. L., and Giebisch, G. (1982). Electrical properties of Cl transport across the *Necturus* proximal tubule. *J. Membr. Biol.* **65,** 185–196.

Guggino, W. B., London, R., Boulpaep, E. L., and Giebisch, G. (1983). Chloride transport across the basolateral cell membrane of the *Necturus* proximal tubule: Dependence on bicarbonate and sodium. *J. Membr. Biol.* **71,** 227–240.

Guharay, F., and Sachs, F. (1984). Stretch-activated single ion channel currents in tissue-cultured embryonic chick skeletal muscle. *J. Physiol. (London)* **352,** 685–701.

Hanley, M. J., Kokko, J. P., Gross, J. B., and Jacobson, H. R. (1980). Electrophysiologic study of the cortical collecting tubule of the rabbit. *Kidney Int.* **17,** 74–81.

Hassid, A., Pidikiti, N., and Gamero, D. (1986). Effects of vasopressin peptides on cytosolic calcium in cultured mesangial cells. *Am. J. Physiol.* **251,** F1018–F1028.

Hebert, S. C., and Andreoli, T. E. (1984). Effects of antidiuretic hormone on cellular conductance pathways in mouse meduallary thick ascending limbs of Henle. II. Determinants of the ADH-mediated increases in transepithelial voltage and in net Cl absorption. *J. Membr. Biol.* **80,** 221–233.

Hebert, S. C., and Andreoli, T. E. (1986). Ionic conductance pathways in the mouse medullary thick ascending limb of Henle. *J. Gen. Physiol.* **87,** 567–590.

Hebert, S. C., Friedman, P. A., and Andreoli, T. E. (1984). Effects of antidiuretic hormone on cellular conductive pathways in mouse medullary thick ascending limbs of Henle. I. ADH increase transcellular conductance pathways. *J. Membr. Biol.* **80,** 201–219.

Hilden, S. A., Johns, C. A., and Madias, N. E. (1988). Cl^--dependent ATP-driven H^+ transport in rabbit renal cortical endosomes. *Am. J. Physiol.* **255,** F885–F897.

Hunter, M. (1990). Stretch-activated channels in the basolateral membrane of single proximal cell of frog kidney. *Pfluegers Arch.* **416,** 448–453.

Ikeda, M., Iyori, M., Yoshitomi, K., Hayashi, M., Imai, M., Kurokawa, K., and Saruta, T. (1992). Isoproterenol (Iso) stimulates Cl currents in b-intercalated cells via G protein-associated processes. *JASN* **3,** 809 (abstr.).

Imai, M. (1984). Functional heterogeneity of the descending limbs of Henle's loop. II. Interspecies difference among rabbits, rats, and hamsters. *Pfluegers Arch.* **402,** 313–401.

Imai, M., Kondo, Y., Koseki, C., and Yoshitomi, K. (1988a). Dual effect of N-ethylmaleimide on Cl transport across the thin ascending limb of Henle's loop. *Pfluegers Arch.* **411,** 520–528.

Imai, M., Taniguchi, J., and Yoshitomi, K. (1988b). Transition of permeability properties along the descending limb of long-loop nephrons. *Am. J. Physiol.* **254,** F323–F328.

Ishibashi, K., Rector, F. C. J., and Berry, C. A. (1990). Chloride transport across the basolateral membrane of rabbit proximal convoluted tubules. *Am. J. Physiol.* **258,** F1569–F1578.

Ishibashi, K., Rector, F. C., Jr., and Berry C. A. (1993). Role of Na-dependent Cl/HCO_3 exchange in basolateral Cl transport of rabbit proximal tubules. *Am. J. Physiol.* **264,** F251–F258.

Itoh, S., Carretero, O. A., and Murray, R. D. (1985). Possible role of adenosine in the macula densa mechanism of renin release in rabbits. *J. Clin. Invest.* **76,** 1412–1417.

Jentsch, T. J., Steinmeyer, K., and Schwarz, G. (1990). Primary structure of *Torpedo marmorata* chloride channel isolated by expression cloning in *Xenopus* oocytes. *Nature (London)* **348,** 510–514.

Kaissling, B., and Kirz, W. (1979). Structural analysis of the rabbit kidney. *Adv. Anat. Embryol. Cell Biol.* **56,** 1–123.

Karniski, L. P., and Aronson, P. S. (1985). Chloride/formate exchange with formic acid recycling: A mechanism of active chloride transport across epithelial membranes. *Proc. Natl. Acad. Sci. U.S.A.* **82,** 6362–6365.

Karniski, L. P., and Aronson, P. S. (1987). Anion exchange pathways for Cl transport in rabbit renal microvillus membranes. *Am. J. Physiol.* **253,** F513–F521.

Kashgarian, M., Biemesderfer, D., Caplan, M., and Forbush, B. (1985). Monoclonal antibodies to Na-K-ATPase: Immunocytochemical localization along nephron segments. *Kidney Int.* **28,** 899–913.

Kirchner, K. A., Kotchen, T. A., Galla, J. H., and Luke, R. G. (1978). Importance of chloride for acute inhibition of renin by sodium chloride. *Am. J. Physiol.* **235,** F444–F450.

Kizer, N. L., Lewis, B., Vandorpe, D., Guggino, W. B., and Stanton, B. A. (1993). Vasopressin and cyclic AMP stimulate electrogenic chloride secretion by an inner medullary collecting duct (IMCD) cell line isolated from a transgenic mouse. *JASN.* **4,** 869 (abstr.).

Koeppen, B. M. (1986). Conductive properties of the rabbit outer medullary collecting duct: outer stripe. *Am. J. Physiol.* **250,** F70–F76.

Koeppen, B. M. (1987). Electrophysiological identification of principal and intercalated cells in the rabbit outer medullary collecting duct. *Pfluegers Arch.* **409,** 138–141.

Koeppen, B. M., and Stanton, B. A. (1992). Sodium chloride transport: Distal nephron. *In* "The Kidney: Physiology and Pathophysiology" (D. W. Seldin and G. Giebisch, eds.), 2nd ed., pp. 2003–2040. Raven Press, New York.

Koeppen, B. M., Biagi, B. A., and Giebisch, G. H. (1983). Intracellular microelectrode characterization of the rabbit cortical collecting duct. *Am. J. Physiol.* **244,** F35–F47.

Koeppen, B. M., Pappas, C., Manger, T., and Oyler, A. (1991). Cellular mechanisms of Cl transport in outer medullary collecting duct. *Kidney Int.* **40,** S131–S135.

Kolb, H. A., Brown, C. D. A., and Murer, H. (1985). Identification of a voltage-dependent anion channel in the apical membrane of a Cl^--secretory epithelium (MDCK). *Pfluegers Arch.* **403,** 262–265.

Kondo, Y., and Fromter, E. (1990). Evidence for chloride/bicarbonate exchange mediating bicarbonate efflux from S3 segments of rabbit renal proximal tubule. *Pfluegers Arch.* **415,** 726–733.

Kondo, Y., Yoshitomi, K., and Imai, M. (1987a). Effect of anion transport inhibitors and ion substitution on Cl transport in the thin ascending limb of Henle's loop. *Am. J. Physiol.* **253,** F1206–F1215.

Kondo, Y., Yoshitomi, K., and Imai, M. (1987b). Effect of pH on Cl^- transport in the thin ascending limb of Henle's loop. *Am. J. Physiol.* **253,** F1216–F1222.

Kondo, Y., Yoshitomi, K., and Imai, M. (1988). Effect of Ca^{2+} on Cl^- transport in the thin ascending limb of Henle's loop. *Am. J. Physiol.* **254,** F232–F239.

Korbmacher, C., Segal, A. S., Boulpaep, E. L., Schroder, U. H., and Fromter, E. (1992). Forskolin stimulates a chloride conductance in M-1 mouse cortical collecting duct cells. *JASN* **3,** 812 (abstr.).

Kremer, S. G., Breuer, W. V., and Skorecki, K. L. (1989). Vasoconstrictor hormones depolarize renal glomerular mesangial cells by activating chloride channels. *J. Cell. Physiol.* **138,** 97–105.

Kurtz, A., and Penner, R. (1989). Angiotensin II induces oscillations of intracellular calcium and blocks anomalous inward rectifying potassium current in mouse renal juxtaglomerular cells. *Proc. Natl. Acad. Sci. U.S.A.* **86,** 3423–3427.

Kurtz, I. (1989). Basolateral membrane Na/H antiport, Na/base cotransport, and Na-independent Cl/base exchange in the rabbit S3 proximal tubule. *J. Clin. Invest.* **83,** 616–622.

Kuwahara, M., Rector, F. C. J., and Berry, C. A. (1988). SITS-sensitive basolateral anion current in rabbit proximal convoluted tubules. *Am. J. Physiol.* **254,** F828–F836.

Landry, D. W., Reitman, M., Cragoe, E. J., Jr., and Al-Awqati, Q. (1987). Epithelial chloride channels: Development of inhibitory ligands. *J. Gen. Physiol.* **90,** 779–798.

Landry, D. W., Akabas, M. H., Redhead, C., Edelman, A., Cragoe, E., J., Jr., and Al-Awqati, Q. (1989). Purification and reconstitution of chloride channel from kidney and trachea. *Science* **244,** 1469–1472.

Lang, F., Friedrich, F., Paulmichl, M., Schobersberger, W., and Jungwirth, A. (1990). Ion channels in Madin-Darby canine kidney cells. *Renal Physiol. Biochem.* **13,** 82–93.

Lapointe, J.-Y., Bell, P. D., Hurst, A. M., and Cardinal, J. (1991). Basolateral ionic permeabilities of macula densa cells. *Am. J. Physiol.* **260,** F856–F860.

Light, D. B., Schwiebert, E. M., Fejes-Toth, G., Naray-Fejes-Toth, A., Karlson, K. H., McCann, F. V., and Stanton, B. A. (1990). Chloride channels in the apical membrane of cortical collecting duct cells. *Am. J. Physiol.* **258,** F273–F280.

Ling, B. N., and Kokko, K. E. (1992). PGE_2 activates apical chloride channels in rabbit

cortical collecting tubule (RCCT) principal cells via a cAMP-dependent pathway. *JASN* **3,** 813 (abstr.).
Ling, B. N., Kokko, K. E., Matsumoto, P. S., and Eaton, D. C. (1992a). Activation of a chloride conductive pathway in a distal nephron cell line (A6) by PGE_2, forskolin, and arginine vasotocin. *JASN* **3,** 813 (abstr.).
Ling, B. N., Matsunaga, H., and Eaton, D. C. (1992b). Regulation of chloride and non-selective cation channels (NSCC) by insulin and angiotensin (AG) II in glomerular mesangial cells (GMC). *JASN* **3,** 813 (abstr.).
Lipkowitz, M. S., and Abramson, R. G. (1987). Ionic permeabilities of rat renal cortical brush-border membrane vesicles. *Am. J. Physiol.* **252,** F700–F711.
Lonnerholm, G., and Ridderstrale, Y. (1980). Intracellular distribution of carbonic anhydrase in the rat kidney. *Kidney Int.* **17,** 162–174.
Lopes, A. G., Amzel, L. M., Markakis, D., and Guggino, W. B. (1988). Cell volume regulation by the thin descending limb of Henle's loop. *Proc. Natl. Acad. Sci. U.S.A.* **85,** 2873–2877.
Lorenz, J. N., Kotchen, T. A., and Ott, C. E. (1990). Effect of Na and Cl infusion on loop function and plasma renin activity in rats. *Am. J. Physiol.* **258,** F1328–F1335.
Lorenz, J. N., Weihprecht, H., Schnermann, J., Skott, O., and Briggs, J. P. (1991). Renin release from isolated juxtaglomerular apparatus depends on macula densa chloride transport. *Am. J. Physiol.* **260,** F486–F493.
Macri, P., Breton, S., Cardinal, J., and Laprade, R. (1990). Ionic basis for volume regulation in the rabbit proximal convoluted tubule (PCT). *JASN* **1,** 689 (abstr.).
Macri, P., Breton, S., Beck, J. S., Cardinal, J., and Laprade, R. (1993). Basolateral K^+, Cl^-, and HCO_3^- conductances and cell volume regulation in rabbit PCT. *Am. J. Physiol.* **264,** F365–F376.
Madsen, K. M., and Tisher, C. C. (1986). Structural-functional relationships along the distal nephron. *Am. J. Physiol.* **250,** F1–F15.
Marunaka, Y., and Eaton, D. C. (1990a). Chloride channels in the apical membrane of a distal nephron A6 cell line. *Am. J. Physiol.* **258,** C352–C368.
Marunaka, Y., and Eaton, D. C. (1990b). Effects of insulin and phosphatases on the Ca^{2+}-dependent Cl channel in a distal nephron cell line (A6). *J. Gen. Physiol.* **95,** 773–789.
Mellman, I., Fuchs, R., and Helenius, A. (1986). Acidification of the endocytic and exocytic pathways. *Annu. Rev. Biochem.* **55,** 663–700.
Middleton, J. P., Basavappa, S., Mangel, A. W., and Fitz, J. G. (1992). Potential autacoid regulation of Na,K,Cl cotransport and chloride secretion by purinergic receptors. *JASN* **3,** 815 (abstr.).
Miller, C. (1982). Open-state substructure of single chloride channels from *Torpedo* electroplax. *Philos. Trans. R. Soc. London* **299,** 401–411.
Mills, J. W. (1987). The cell cytoskeleton: Possible role in volume control. *Curr. Top. Membr. Trans.* **30,** 75–101.
Mills, J. W., and Lubin, M. (1986). Effect of adenosine 3′,5′-cyclic monophosphate on volume and cytoskeleton of MDCK cells. *Am. J. Physiol.* **250,** C319–C324.
Mills, J. W., Schwiebert, E. M., and Stanton, B. A. (1993). The cytoskeleton and cell volume regulation. *In* "Cellular and Molecular Physiology of Cell Volume Regulation" (K. Strange, ed.) CRC Critical Reviews, pp. 241–258. Boca Raton, Florida.
Mills, J. W., Schwiebert, E. M., and Stanton, B. A. (1994). Evidence for a role of actin filaments in regulating cell swelling. *J. Exp. Zool.* **268,** 111–120.
Molony, D. A., Reeves, W. B., Hebert, S. C., and Andreoli, T. E. (1987). ADH increases apical Na,K,2Cl entry in mouse medullary thick ascending limbs of Henle. *Am. J. Physiol.* **252,** F177–F187.

Morales, M. M., Morita, T., Stanton, B. A., Lopes, A. G., and Guggino, W. B. (1993). Localization of the cystic fibrosis transmembrane conductance regulator (CFTR) along the rat nephron using the polymerase chain reaction. *JASN* **4,** 875 (abstr.).

Muto, S., Yasoshima, K., Yoshitomi, K., Imai, M., and Asano, Y. (1990). Electrophysiological identification of alpha- and beta-intercalated cells and their distribution along the rabbit distal nephron segments. *J. Clin. Invest.* **86,** 1829–1839.

Nelson, D. J., Tang, J. M., and Palmer, L. G. (1984). Single-channel recordings of apical membrane chloride conductance in A6 epithelial cells. *J. Membr. Biol.* **80,** 81–89.

Oberleitner, H., Guggino, W. B., and Giebisch, G. (1982). Mechanism of distal tubular chloride transport in *Amphiuma* kidney. Am. J. Physiol. **242,** F331–F339.

Oberleitner, H., Ritter, M., Lang, F., and Guggino, W. (1983). Anthracene-9-carboxylic acid inhibits renal chloride reabsorption. *Pfluegers Arch.* **398,** 172.

Ogawa, A., Suzuki, M., Kawaguchi, Y., and Sakai, O. (1992). An apical chloride channel is activated during regulatory volume decrease of rabbit proximal tubule cells via protein kinase C (PKC). *JASN* **3,** 817 (abstr.).

Okuda, T., Yamashita, N., and Kurokawa, K. (1986). Angiotensin II and vasopressin stimulate calcium-activated chloride conductance in rat mesangial cells. *J. Clin. Invest.* **78,** 1443–1448.

Okuda, T., Kojima, I., Ogata, E., and Kurokawa, K. (1989). Ambient Cl^- ions modify rat mesangial contraction by modulating cell inositol trisphosphate and Ca^{2+} via enhanced prostaglandin E_2. *J. Clin. Invest.* **84,** 1866–1872.

Onuchic, L. F., Arenstein, I. R., and Lopes, A. G. (1992). Cell volume regulation in rat thin ascending limb of Henle's loop. *Am. J. Physiol.* **263,** F353–F362.

Osswald, H., Schmitz, H.-J., and Heidenreich, O. (1975). Adenosine response of the rat kidney after saline loading, sodium restriction and hemorrhagia. *Pfluegers Arch.* **357,** 323–333.

Paulais, M., and Teulon, J. (1990). Cyclic AMP-activated chloride channel in the basolateral membrane of the thick ascending limb of the mouse kidney. *J. Membr. Biol.* **113,** 253–260.

Paulmichl, M., Li, Y., Wickman, K., Ackermann, M., Peralta, E., and Clapham, D. (1992). New mammalian chloride channel identified by expression cloning. *Nature (London)* **356,** 238–241.

Persson, A. E. G., Salomonsson, M., Westerlund, P., Greger, R., Schlatter, E., and Gonzalez, E. (1991). Macula densa cell function. *Kidney Int.* **39,** S39–S44.

Persson, B.-E., Sakai, T., and Marsh, D. J. (1988). Juxtaglomerular interstitial hypertonicity *in Amphiuma:* Tubular origin-TGF signal. *Am. J. Physiol.* **254,** F445–F449.

Pfeilschifter, J. (1986). Tumor promotor 12-o-tetradecanoyl-phorbol 13-acetate inhibits angiotensin II-induced phosphate production and cytosolic Ca^{2+} rise in rate renal mesangial cells. *FEBS Lett.* **203,** 262–266.

Pfeilschifter, J., and Bauer, C. (1986). Pertussis toxin abolishes angiotensin II-induced phosphoinositde hydrolysis and prostaglandin synthesis in rat renal mesangial cells. *Biochem. J.* **236,** 289–294.

Poncet, V., Tauc, M., Bidet, M., and Poujeol, P. (1994). Chloride channels in the apical membrane of primary cultures of the rabbit distal bright convoluted tubule. *Am. J. Physiol.* (in press).

Redhead, C. R., Edelman, A. E., Brown, D., Landry, D. W., and Al-Awqati, Q. (1992). A ubiquitous 64-kDa protein is a component of a chloride channel of plasma and intracellular membranes. *Proc. Natl. Acad. Sci. U.S.A.* **89,** 3716–3720.

Reenstra, W. W., Sabolic, I., Bae, H. R., and Verkman, A. S. (1992). Protein kinase A dependent membrane protein phosphorylation and chloride conductance in endosomal vesicles from kidney cortex. *Biochemistry* **31(1),** 175–81.

Reeves, W. B., and Andreoli, T. E. (1992a). Sodium chloride transport in the loop of Henle. *In* "The Kidney: Physiology and Pathophysiology" (D. W. Seldin and G. Giebisch, eds.), 2nd ed., pp. 1975–2002. Raven Press, New York.

Reeves, W. B., and Andreoli, T. E. (1992b). Renal epithelial chloride channels. *Annu. Rev. Physiol.* **54,** 29–50.

Reeves, W. B., and Gurich, R. W. (1992). Apparent heterogeneity of chloride channels in endosomal membranes. *JASN* **3,** 818 (abstr.)

Ridderstrale, Y., Kashgarian, M., Koeppen, B. M., Giebisch, G., Stetson, D., Ardito, T., and Stanton, B. A. (1988). Morphological heterogeneity of the rabbit collecting duct. *Kidney Int.* **34,** 655–670.

Riordan, J. R., Rommens, J. M., Kerem, B., Alon, N., Rozmahel, R., Grzelczak, Z., Zielenski, J., Lok, S., Plavsic, N., Chou, J.-L., Drumm, M. L., Iannuzzi, M. C., Collins, F. S., and Tsui, L.-C. (1989). Identification of the cystic fibrosis gene: Cloning and characterization of complementary DNA. *Science* **245,** 1066–1073.

Rocha, A. S., and Kudo, L. H. (1990a). Atrial peptide and cyclic GMP effects on NaCl transport in inner medullary collecting duct. *Am. J. Physiol.* **259,** F258–F268.

Rocha, A. S., and Kudo, L. H. (1990b). Factors governing sodium and chloride transport across the inner medullary collecting duct. *Kidney Int.* **38,** 654–657.

Rostand, S. G., Work, J., and Luke, R. G. (1985). Effect of reduced chloride reabsorption on renin release in the isolated rat kidney. *Pfluegers Arch.* **405,** 46–51.

Rudnick, G. (1986). ATP-driven H^+ pumping into intracellular organelles. *Annu. Rev. Physiol.* **48,** 403–413.

Salomonsson, M., Gonzalez, E., Westerlund, P., and Persson, A. E. G. (1991). Chloride concentration in macula densa and cortical thick ascending limb cells. *Kidney Int.* **39,** S51–S54.

Sansom, S. C., Weinman, E. J., and O'Neil, R. G. (1984). Microelectrode assessment of chloride conductive properties of cortical collecting duct. *Am. J. Physiol.* **247,** F291–F302.

Sansom, S. C., La, B.-Q., and Carosi, S. L. (1990). Double-barreled chloride channels of the collecting duct basolateral membrane. *Am. J. Physiol.* **259,** F46–F52.

Sasaki, S., and Yoshiyama, N. (1988). Interaction of chloride and bicarbonate transport across the basolateral membrane of rabbit proximal straight tubule. Evidence for sodium-coupled chloride/bicarbonate exchange. *J. Clin. Invest.* **81,** 1004–1011.

Sasaki, S., Ishibashi, K., Yoshiyama, N., and Shiigai, T. (1988). KCI cotransport across the basolateral membrane of rabbit proximal straight tubule. *J. Clin. Invest.* **81,** 194–199.

Schild, L., Aronson, P. S., and Giebisch, G. (1991). Basolateral transport pathways for K^+ and Cl^- in rabbit proximal tubule: effects on cell volume. *Am. J. Physiol.* **260,** F101–F109.

Schlatter, E., and Greger, R. (1991). Patch clamp experiments with freshly isolated renal tubular cells. *Pfluegers Arch.* **418,** R75.

Schlatter, E., Salomonsson, M., Persson, A. E. G., and Greger, R. (1989). Macula densa cells sense luminal NaCl concentration via furosemide-sensitive Na,K,2Cl cotransport. *Pfluegers Arch.* **414,** 286–290.

Schlatter, E., Greger, R., and Schafer, J. A. (1990). Principal cells of the cortical collecting ducts of the rat are not a route of transepithelial Cl transport. *Pfluegers Arch.* **417,** 317–323.

Schlondorff, D. (1987). The glomerular mesangial cell: An expanding role for a specialized pericyte. *FASEB J* **1,** 272–281.

Schmid, A., Burckhardt, G., and Gogelein, H. (1989). Single chloride channels in endosomal vesicle preparations from rat kidney cortex. *J. Membr. Biol.* **111,** 265–275.

Schnermann, J., and Briggs, J. P. (1992). Function of the juxtaglomerular apparatus: Control of glomerular hemodynamics and renin secretion. *In* "The Kidney: Physiology and Pathophysiology" D. W. Seldin and G. Giebisch, eds.), 2nd ed., pp 1249–1289. Raven Press, New York.

Schnermann, J., and Marver, D. (1986). ATPase activity in macula densa cells of the rabbit kidney. *Pfluegers Arch.* **407,** 82–86.

Schuster, V. L. (1985). Cyclic adenosine monophosphate-stimulated bicarbonate secretion in rabbit cortical collecting tubules. *J. Clin. Invest.* **75,** 2056–2064.

Schuster, V. L. (1986). Cyclic adenosine monophosphate-stimulated anion transport in rabbit cortical collecting duct. Kinetics, stoichiometry, conductive pathway. *J. Clin. Invest.* **78,** 1621–1630.

Schuster, V. L., and Bao, Y. (1992). PCR cloning of a rabbit kidney C1 channel cDNA. *JASN* **3,** 819 (abstr.).

Schuster, V. L., and Stokes, J. B. (1987). Chloride transport by the cortical and outer medullary collecting duct. *Am. J. Physiol.* **253,** F203–F212.

Schwartz, G. J., Barasch, J., and Al-Awqati, Q. (1985). Plasticity of functional epithelial polarity. *Nature (London)* **318,** 368–371.

Schwiebert, E. M., Light, D. B., Fejes-Toth, G., Naray-Fejes-Toth, A., and Stanton, B. A. (1990). A GTP-binding protein activates chloride channels in a renal epithelium. *J. Biol. Chem.* **265,** 7725–7728.

Schwiebert, E. M., Mills, J. W., and Stanton, B. A. (1991a). The cytoskeleton regulates chloride channels in the apical membrane of cortical collecting duct cells. *FASEB J.* **5,** 739a (abstr.).

Schwiebert, E. M., Mills, J. W., and Stanton, B. A. (1991b). The cytoskeleton regulates a volume regulatory chloride channel in cortical collecting duct cells in culture. *JASN* **2,** 750a (abstr.).

Schwiebert, E. M., Karlson, K. H., Friedman, P. A., Dietl, P., Spielman, W. S., and Stanton, B. A. (1992). Adenosine regulate a chloride channel via protein kinase C and a G protein in a rabbit cortical collecting duct. *J. Clin. Invest.* **89,** 834–841.

Schwiebert, E. M., Mills, J. W., and Stanton, B. A. (1994). The actin-based cytoskeleton regulates a chloride channel and cell volume in renal cortical collecting duct cell line. *J. Biol. Chem.* **269,** 7081–7089.

Segal, A. S., and Boulpaep, E. L. (1990). Basolateral chloride channels in single isolated proximal tubule cells of *Ambystoma*. *JASN* **1,** 691 (abstr.).

Segal, A. S., and Boulpaep, E. L. (1992). cAMP-activated chloride channel on the basolateral membrane of renal proximal tubule. *JASN* **3,** 819 (abstr.).

Shindo, T., and Spring, K. R. (1981). Chloride movement across the basolateral membrane of proximal tubule cells. *J. Membr. Biol.* **58,** 35–42.

Shiuan, D., and Weinstein, S. W. (1984). Evidence for electroneutral chloride transport in rabbit renal cortical brush border membrane vesicles. *Am. J. Physiol.* **247,** F837–F847.

Simmons, N. L (1993). Renal epithelial chloride secretion. *Exp. Physiol.* **78,** 117–137.

Skott, O. (1988). Do osmotic forces play a role in renin secretion? *Am. J. Physiol.* **255,** F1–F10.

Skott, O., and Briggs, J. P. (1987). Direct demonstration of macula densa-mediated renin secretion. *Science* **237,** 1618–1620.

Spring, K. R., and Kimura, G. (1978). Chloride reabsorption by renal proximal tubules of *Necturus*. *J. Membr. Biol.* **38,** 233–254.

Stanton, B. A. (1988). Electroneutral NaCl transport by distal tubule: Evidence for Na/H-Cl/HCO_3 exchange. *Am. J. Physiol.* **254,** F80–F86.

Stanton, B. A. (1989). Characterization of apical and basolateral membrane conductances of rat inner medullary collecting duct. *Am. J. Physiol.* **256,** F862–F868.

Stanton, B. A. (1990). Cellular actions of thiazide diuretics in the distal tubule. *JASN* **1,** 832–836.
Stanton, B. A., Mills, J. W., and Schwiebert, E. M. (1991). Role of the actin-based cytoskeleton in regulatory volume decrease in cortical collecting duct cells in culture. *JASN* **2,** 751a (abstr.).
Star, R. A., Burg, M. B., and Knepper, M. A. (1985). Bicarbonate secretion and chloride absorption by rabbit cortical collecting ducts: Role of chloride-bicarbonate exchange. *J. Clin. Invest.* **76,** 1123–1130.
Steinmeyer, K., Ortland, C., and Jentsch, T. J. (1991). Primary structure and functional expression of a developmentally regulated skeletal muscle chloride channel. *Nature (London)* **354,** 301–308.
Stokes, J. B. (1984). Sodium chloride reabsorption by the urinary bladder of the winter flounder. A thiazide-sensitive, electrically neutral transport system. *J. Clin. Invest.* **74,** 7–16.
Stone, D. K., Seldin, D. W., Kokko, J. P., and Jacobson, H. R. (1983). Anion dependence of rabbit medullary collecting duct acidification. *J. Clin. Invest.* **72,** 77–83.
Strange, K. (1989). Ouabain-induced cell swelling on rabbit cortical collecting tubule: NaCl transport by principal cells. *J. Membr. Biol.* **107,** 249–261.
Strange, K. (1991). Volume regulatory Cl loss after Na pump inhibition in CCD principal cells. *Am. J. Physiol.* **260,** F225–F234.
Superdock, K. R., Snyders, D. J., and Breyer, M. D. (1992). A novel ATP-inhibitable Cl channel in apical membranes of cultured rabbit cortical collecting duct (CCD) cells. *JASN* **3,** 821 (abstr.).
Superdock, K. R., Snyders, D. J., and Breyer, M. D. (1993). ATP-inhibitable Cl^- channel in apical membranes of cultured rabbit cortical collecting duct cells. *Am. J. Physiol.* **265,** C957–C965.
Suzuki, M., Morita, T., Hanaoka, K., Kawaguchi, Y., and Sakai, O. (1991). A chloride channel activated by parathyroid hormone in rabbit renal proximal tubule cells. *J. Clin. Invest.* **88,** 735–742.
Suzuki, M., Miyazaki, K., Ikeda, M., Kawaguchi, Y., and Sakai, O. (1993). F-actin network may regulate a Cl^- channel in renal proximal tubule cells. *J. Membr. Biol.* **134,** 31–39.
Tabcharani, J. A., Low, W. Elie, D., and Hanrahan, J. W. (1990). Low-conductance chloride channel activated by cAMP in the epithelial cell line T84. *FEBS Lett.* **270,** 157–164.
Tagawa, A., and Vander, A. J. (1970). Effects of adenosine compounds on renal function and renin secretion in dogs. *Circ. Res.* **26,** 327–338.
Taniguchi, J., Yoshitomi, K., and Imai, M. (1989). K and Cl channels in the basolateral membrane of rabbit distal convoluted tubule (DCT). *Kidney Int.* **256,** F246–F254.
Terada, Y., and Knepper, M. A. (1990). Thiazide-sensitive NaCl absorption in rat cortical collecting duct. *Am. J. Physiol.* **259,** F519–F528.
Thiemann, A., Grunder, S., Pusch, M., and Jentsch, T. J. (1992). A chloride channel widely expressed in epithelial and non-epithelial cells. *Nature (London)* **356,** 57–60.
Tasi, L.-M., Dillard, M., Rosenberg, R. L., Falk, R. J., Gaido, M. L., and Finn, A. L. (1991). Reconstitution of an epithelial chloride channel: Conservation of the channel from mudpuppy to man. *J. Gen. Physiol.* **98,** 723–750.
Uchida, S., Sasaki, S., and Marumo, F. (1992). Molecular cloning of a putative chloride channel cDNA expressed in kidney. *JASN* **3,** 822 (abstr.).
Uchida, S., Sasaki, S., Furukawa, T., Hiraoka, M., Imai, T., Hirata, Y., and Marumo, F. (1993). Molecular cloning of a chloride channel that is regulated by dehydration and expressed predominantly in kidney medulla. *J. Biol. Chem.* **268,** 3821–3824.
Vandorpe, D., Kizer, N. L., Ciampolillo, F., Memoli, V. A., Guggino, W. B., and Stanton,

B. A. (1993). Cyclic AMP stimulates CFTR (cystic fibrosis transmembrane conductance regulator) chloride channels in inner medullary collecting duct. *JASN*, **4,** 882 (abstr.).

Wang, T., Giebisch, G., and Aronson, P. S. (1992). Formate and oxalate facilitate chloride transport in the rat distal convoluted tubule. *JASN* **3,** 823.

Wang, T., Agulian, S. K., Giebisch, G., and Aronson, P. S. (1993). Effects of formate and oxalate on chloride absorption in rat distal tubule. *Am. J. Physiol.* **264,** F730–F736.

Wangemann, P., Wittner, M., Distefano, A., Englert, H. C., Lang, H. J., Schlatter, E., and Greger, R. (1986). Cl channel blockers in the thick ascending limb of the loop of Henle. Structure-activity relationship. *Pfluegers Arch.* **407,** S128–S141.

Warnock, D. G., and Yee, V. J. (1981). Chloride uptake by brush border membrane vesicles isolated from rabbit and renal cortex. *J. Clin. Invest.* **67,** 103–115.

Weihprecht, H., Lorenz, J. N., Schnermann, J., Skott, O., and Briggs, J. P. (1990). Effect of adenosine$_1$-receptor blockade on renin release from rabbit perfused juxtaglomerular apparatus. *J. Clin. Invest.* **85,** 1622–1628.

Weinstein, A. M. (1992). Sodium and chloride transport: Proximal nephron. *In* "The Kidney: Physiology and Pathophysiology" D. W. Seldin and G. Giebisch, eds.), 2nd ed., pp. 1925–1974. Raven Press, New York.

Welling, P. A., and Linshaw, M. A. (1988). Importance of anion in hypotonic volume regulation of rabbit proximal straight tubule. *Am. J. Physiol.* **255,** F853–F860.

Welling, P.A., and O'Neil, R. G. (1990a). Ionic conductive properties of rabbit proximal straight tubule basolateral membrane. *Am. J. Physiol.* **258,** F940–F950.

Welling, P. A., and O'Neil, R. G. (1990b). Cell swelling activates basolateral membrane Cl and K conductances in rabbit proximal tubule. *Am. J. Physiol.* **258,** F951–F962.

Winters, C. J., Reeves, W. B., and Andreoli, T. E. (1990). Cl channels in basolateral renal medullary membrane vesicles. III. Determinants of single channel activity. *J. Membr. Biol.* **118,** 269–278.

Winters, C. J., Reeves, W. B., and Andreoli, T. E. (1991). Cl channels in basolateral renal medullary membrane vesicles. IV. Analogous channel activation by Cl or cAMP-dependent protein kinase. *J. Membr. Biol.* **122,** 89–95.

Winters, C. J., Reeves, W. B., and Andreoli, T. E. (1992). Intracellular anions modulate basolateral mTALH Cl channels by interacting with lysine-or arginine-rich domains. *JASN* **3,** 823 (abstr.).

Yoshitomi, K., Kondo, Y., and Imai, M. (1987). Functional heterogeneity in the hamster medullary thick ascending limb of Henle's loop. *Pfluegers Arch.* **408,** 600–608.

Yoshitomi, K., Kondo, Y., and Imai, M. (1988). Evidence for conductive Cl pathways across the cell membranes of the thin ascending limb of Henle's loop. *J. Clin. Invest.* **82,** 866–871.

Yoshitomi, K., Shimizu, T., and Imai, M. (1989a). Electrophysiological evidence that tri-chlormethiazide (TCM) acts on the luminal membrane of the connecting tubule (CNT) cell. *Kidney Int.* **35,** 492.

Yoshitomi, K., Shimizu, T., Taniguchi, J., and Imai, M. (1989b). Electrophysiological characterization of rabbit distal convoluted tubule cells. *Pfluegers Arch.* **414,** 457–463.

Zimniak, L., Reeves, W. B., and Andreoli, T. E. (1992). Cl channels in basolateral renal medullary membranes. VI. Cl conductance expression in *Xenopus* oocytes. *Am. J. Physiol.* **263,** F979–F984.

CHAPTER 11

Chloride Ion Channels in Mammalian Heart Cells

Tzyh-Chang Hwang and David C. Gadsby
Laboratory of Cardiac/Membrane Physiology, The Rockefeller University, New York, New York 10021

I. INTRODUCTION AND OVERVIEW

For the past 3 decades, cellular cardiac electrophysiological studies have been focused mainly on voltage-gated cation channels, so that our understanding of cardiac anion channels has been left lagging far behind. Fortunately, an explosion of interest in, and information about, cardiac Cl^- channels over the past 3 years now promises to redress this imbalance. Whereas, until recently, the very existence of any cardiac Cl^- channel was controversial, there is now strong electrophysiological evidence for at least three kinds of cardiac Cl^- channels, and molecular biological evidence suggests that there could be at least two more.

The earliest investigations of anion-conductive pathways were carried out on multicellular preparations of sheep Purkinje fibers by Carmeliet (1961) and dog and sheep Purkinje fibers by Hutter and Noble (1961), and both demonstrated that replacement of external Cl^- ions with larger, less-permeant anions (like acetylglycinate, pyroglutamate, or methylsulfate) prolonged the action potential while anions like I^-, NO_3^-, and Br^- reduced the action potential duration and slowed spontaneous activity, consistent with their being more permeant. These studies indicated that Cl^- ions make a significant contribution to action potential repolarization, and play some small role in diastolic depolarization, and suggested an anion permeability sequence for the latter effect of $I^- > NO_3^- > Br^- > Cl^-$ (the reverse of that for skeletal muscle; Hutter and Padsha, 1959). Later work, using Purkinje fibers voltage clamped with two microelectrodes, demonstrated that a transient outward current elicited by strong depolarization was diminished by external Cl^- replacement with acetylglycinate, implying that Cl^- ions normally carry that current (Dudel *et al.*, 1967). Although interpretation of these early observations on Cl^- ion replacement was later found to be complicated by Ca^{2+}-chelating properties of some of the substitute anions, and hence inadvertent modulation of the extracellular free $[Ca^{2+}]$ (Kenyon and Gibbons, 1977), careful measurements did eventually provide strong support for a $[Cl^-]$-sensitive component of the transient outward current as well as for an additional, substantial portion carried by K^+ (Kenyon and Gibbons, 1979). Cardiac Cl^- currents subsequently attracted relatively little attention until recently when three groups (Harvey and Hume, 1989; Bahinski *et al.*, 1989a; Matsuoka *et al.*, 1990), all using the whole-cell current-recording technique in mammalian cardiac myocytes, independently obtained convincing evidence for the existence of a Cl^- current regulated by cAMP-dependent protein kinase (PKA). Ehara and Ishihara (1990) soon recorded the underlying anion-selective single-channel currents in intact myocytes exposed to epinephrine, and unitary channel currents have since been studied in excised inside-out (Nagel *et al.*, 1992) and outside-out Ehara and Matsuura, 1993) patches.

As a result of the ensuing flurry of investigation, three different types of Cl^- conductance in cardiac myocytes have now been reasonably well characterized electrophysiologically, namely, that regulated by PKA (Harvey and Hume, 1989; Bahinski *et al.*, 1989a; Matsuoka *et al.*, 1990), another induced by cell swelling (Tseng, 1992; Sorota, 1992; Hagiwara *et al.*, 1992b), and a third activated by a rise in intracellular free $[Ca^{2+}]$ (Zygmunt and Gibbons, 1991, 1992). There are already preliminary reports of unitary currents for the latter two types of Cl^- channel (Hagiwara *et al.*, 1992a; Zhang *et al.*, 1992; Zygmunt and Gibbons, 1993). Stimulation of purinergic receptors has also been reported to activate a cardiac Cl^-

conductance (Matsuura and Ehara, 1992; cf. Scamps and Vassort, 1990) as has presumed activation of protein kinase C (Walsh, 1991; Zhang and Ten Eick, 1993), but these Cl^- conductances are not yet fully characterized; nor are the extremely large conductance anion channels found in neonatal rat cardiac myocytes (Coulombe *et al.*, 1987). Molecular cloning technology has similarly begun to uncover potential cardiac Cl^- channel molecules (e.g., Paulmichl *et al.*, 1992; Thiemann *et al.*, 1992; Moorman *et al.*, 1992), although their electrophysiological properties when expressed in myocytes remain uncertain.

In this review, we emphasize the latest developments contributing to our understanding of the distribution, regulation, biophysical characteristics, and functions of the various cardiac Cl^- channel types, in an attempt to minimize duplication with recent comprehensive reviews of this subject (Hume and Harvey, 1992; Ackerman and Clapham, 1993).

II. PKA-REGULATED Cl^- CHANNELS (CFTR)

Given the years of relative uncertainty surrounding cardiac Cl^- currents, it may seem ironic that the PKA-regulated Cl^- conductance, now the most completely characterized of the Cl^- conductances in heart, was first identified in guinea pig ventricular myocytes as a Na^+-sensitive pathway (Egan *et al.*, 1987,1988): in myocytes impaled with single, high-resistance, KCl-filled microelectrodes connected to a switch-clamp amplifier, isoproterenol (Iso) was found to activate an inward current near the normal resting potential (~80 mV) that was absent when all external Na^+ ions were replaced by tetramethylammonium (TMA^+) ions. This, coupled with insensitivity to standard maneuvers for reducing K^+ or Ca^{2+} channel, hyperpolarization-activated or Na^+/K^+ pump currents, and the negligible influence on the inward current of Cl^-_o replacement with isethionate$^-$ ions, led to the suggestion that Na^+ ions might carry the inward current (Egan *et al.*, 1987,1988). Convincing evidence was presented for activation of the inward current via the adenylyl cyclase/cAMP pathway (Egan *et al.*, 1987,1988). Subsequent experiments, such as that in Fig. 1, using the tight-seal whole-cell current-recording technique have clearly established Cl^- ions as the charge carrier for this current (Harvey and Hume, 1989; Bahinski *et al.*, 1989a; Matsuoka *et al.*, 1990), accounting for its insensitivity to most of the challenges tested by Egan *et al.* (1987,1988; the sensitivity to external Na^+ ions is discussed below); since *inward* Cl^- current reflects Cl^- *efflux,* at sufficiently negative potentials Cl^- current is not expected to be diminished (and could even be enhanced) by substitution of isethionate$^-$ for Cl^-_o.

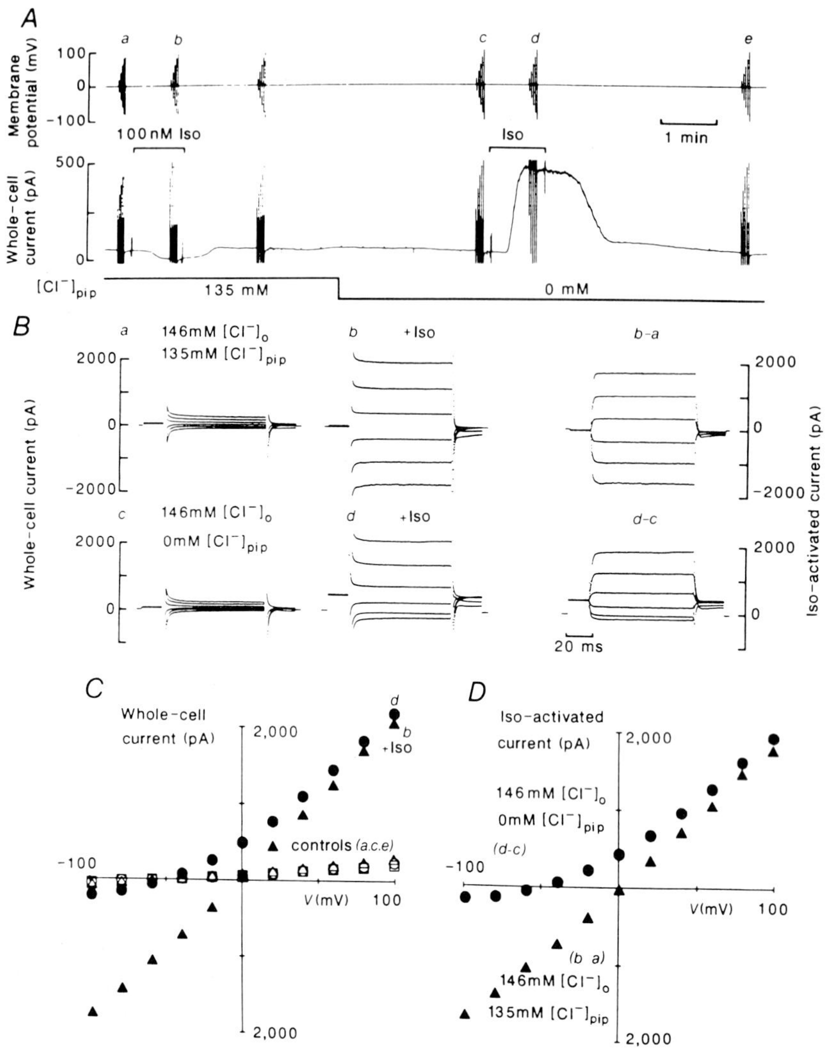

FIGURE 1 A time- and voltage-independent Cl^- conductance activated by isoproterenol (Iso). Chart recording of membrane potential (top) and whole-cell current (A), superimposed records of whole-cell currents for pulses to ±100, ±60, and ±20 mV, from the 0-mV holding potential, in the presence or absence of isoproterenol under conditions of approximately symmetrical or asymmetrical Cl^- concentrations (B), and steady-state *I–V* relationships of whole-cell current (C) or isoproterenol-activated difference current (D). (From Bahinski *et al.*, 1989a, with permission.)

Dialysis of the cell contents against the pipette solution during whole-cell current recording has been exploited to demonstrate that, like the cystic fibrosis (CF) transmembrane conductance regulator (CFTR, the epithelial Cl^- channels whose dysfunction is believed responsible for CF; e.g., Collins, 1992; Welsh *et al.*, 1992), this Cl^- conductance is activated by PKA-dependent phosphorylation following activation of β-adrenoceptors, G_s proteins, and adenylyl cyclase (evidence reviewed below). Identity of these cardiac PKA-regulated Cl^- channels and CFTR is strongly suggested by their common biochemical and biophysical characteristics, determined in both whole-cell and single-channel current experiments (below). Further evidence comes from the demonstration, by Northern analysis (Fig. 2), that CFTR mRNA exists in guinea pig and rabbit ventricle and human atrium (Levesque *et al.*, 1992; Nagel *et al.*, 1992). Although the complete amino acid sequence of cardiac CFTR is not yet available, preliminary results using the polymerase chain reaction technique indicate extremely high homology between rabbit cardiac CFTR and human epithelial CFTR (e.g., 98% identity over a 180-amino acid stretch corresponding to the first nucleotide binding domain, NBD1; Levesque *et al.*, 1992), at least throughout the 12 putative membrane-spanning regions (Horowitz *et al.*, 1993). The most recent evidence suggests that the major difference between rabbit cardiac (Horowitz *et al.*, 1993) and human epithelial (Riordan *et al.*, 1989) CFTR is that the former lacks 30 amino acids from the predicted first cytoplasmic loop; those residues are encoded by exon 5 of the CFTR gene. The implication is that cardiac CFTR might be an alternatively spliced isoform (Horowitz *et al.*, 1993).

A. Regional Localization within the Heart

Mechanically the heart behaves as a syncytium, but cardiac myocytes are differentiated into several types each displaying a particular set of excitability properties, determined by the relative densities of the various ion channels. A striking example is the virtually exclusive localization to atrial and nodal cells of the K^+ channels that are opened by stimulation of muscarinic receptors. As a result, parasympathetic stimulation affects the electrical activity of atrial and ventricular cells quite differently, strongly hyperpolarizing sinoatrial (SA) nodal cells and thereby lowering the heart rate (negative chronotropic effect), but having little influence on the membrane potential of ventricular cells during diastole. Ventricular myocytes do not lack muscarinic receptors, however, and under appropriate conditions their stimulation can inhibit adenylyl cyclase, lowering the cellular [cAMP] and hence PKA activity which, in turn, decreases

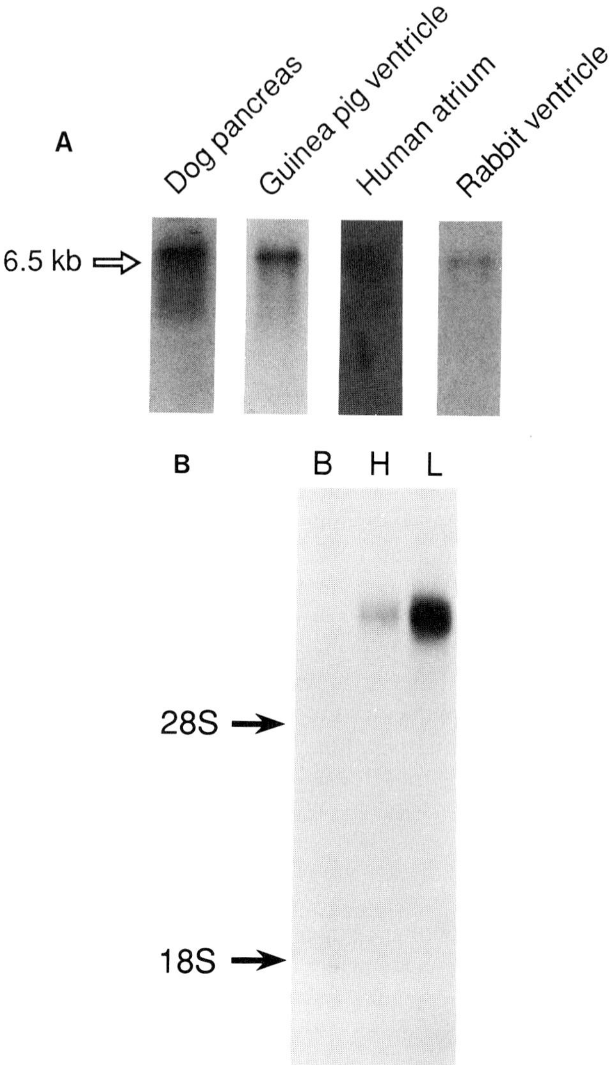

FIGURE 2 Northern analysis of mRNA isolated from four different species (A), or of total RNA from guinea pig brain, heart, and lung (B). Blots were probed with the polymerase chain reaction amplification product corresponding to 550 bp encoding the first nucleotide binding domain of CFTR from rabbit ventricle (A) (from Levesque *et al.*, 1992, with permission) or with the antisense RNA transcribed from the full-length human CFTR clone (B) (from Nagel *et al.*, 1992, with permission).

Ca^{2+} current, leading to a reduction of action potential amplitude and duration. By analogy, a full understanding of the physiological role of PKA-regulated Cl^- channels therefore requires, among other things, information on their regional distribution within the heart.

Takano and Noma (1992) examined the density of the catecholamine-induced Cl^- current in cells isolated from the SA node, atrium, and ventricle of rabbit heart: in none of the 20 nodal cells or 20 atrial cells they examined did Iso activate any Cl^- current, despite the fact that it consistently enhanced hyperpolarization-activated current in SA nodal cells and L-type Ca^{2+} current in atrial cells. Hagiwara *et al.* (1992b) similarly reported no activation of Cl^- current by Iso or forskolin in either SA nodal or atrial cells of the rabbit. On the other hand, Harvey and Hume (1989) mentioned observing the catecholamine-induced Cl^- current in guinea pig atrial cells, whereas Matsuura and Ehara (1992) found that β-adrenergic stimulation activated the Cl^- conductance in only ~10% of guinea pig atrial cells, and Horowitz *et al.* (1993) found no Iso-induced Cl^- current in either guinea pig or dog atrial myocytes.

In ventricular myocytes from guinea pig (e.g., Egan *et al.*, 1987, 1988; Harvey and Hume, 1989; Bahinski *et al.*, 1989a; Matsuoka *et al.*, 1990), rabbit (Harvey and Hume, 1989; Takano and Noma, 1992), and cat (Zhang and Ten Eick, 1993), but not dog (Sorota *et al.*, 1991) or rat (Dukes *et al.*, 1990), Iso does activate robust Cl^- currents which, in rabbit ventricle, are twofold larger in myocytes isolated from the epicardial side than in those from the endocardial side (Takano and Noma, 1992). Takano and Noma (1992) noted activation of the Cl^- conductance by Iso in every (≥40) rabbit ventricular myocyte examined, and Hwang *et al.* (1992a) found that 1 μM Iso elicited PKA-regulated Cl^- current (~1 $\mu A/\mu F$ at 0 mV) in 79 of 95 (83%) guinea pig ventricular myocytes (with 100 μM pipette [GTP]). Correspondingly, Northern analysis reveals expression of CFTR in guinea pig and rabbit ventricle, but not in dog ventricle or in dog, rabbit, or guinea pig atrium (Horowitz *et al.*, 1993).

The questionable presence of this Cl^- conductance in atrial cells (Matsuura and Ehara, 1992; Takano and Noma, 1992; Hagiwara *et al.*, 1992b) and its absence from SA nodal cells (Takano and Noma, 1992; Hagiwara *et al.*, 1992b) argue that it can play little role in the positive chronotropic effect of catecholamines (Takano and Noma, 1992), whereas the higher density of activated Cl^- current in epicardial than in endocardial myocytes might be able to account, at least partly, for the shorter action potential duration of epicardial cells (Takano and Noma, 1992). However, that would require significant *basal* activation of the PKA-regulated Cl^- conductance (in the absence of adrenergic stimulation), which has not been examined in intact cardiac tissue but which has been found lacking in voltage-clamped internally dialyzed ventricular myocytes (Fig. 1; Bahinski *et al.*, 1989a; Harvey *et al.*, 1990; Hwang *et al.*, 1992a), possibly because diffusion from cell to pipette lowers the intracellular [cAMP] (Hwang *et al.*, 1992a). Given the relative scarcity of PKA-regulated Cl^- conductance in guinea pig, rabbit, and dog atrial myocytes, and yet the positive Northern blot of human atrial mRNA (Fig. 2A; Levesque *et al.*, 1992), it will be of interest to see whether catecholamines elicit PKA-regulated Cl^- current in human atrial myocytes.

B. Whole-Cell Studies

1. Regulatory Pathways

a. Stimulatory Pathway. In myocytes, β-adrenoceptor activation modulates the activity of several ion channels through the adenylyl cyclase–PKA phosphorylation pathway (reviewed by Hartzell, 1988; Gadsby, 1990). The evidence that the same pathway regulates this catecholamine-induced Cl^- conductance is now overwhelming and includes such facts as (1) β-adrenergic activation of this current can be mimicked by forskolin (Harvey and Hume, 1989; Matsuoka *et al.*, 1990; Hwang *et al.*, 1992a) or histamine (Harvey and Hume, 1990; Hwang *et al.*, 1992a) or by intracellular application of cAMP (Bahinski *et al.*, 1989a,b; Tareen *et al.*, 1991,1992; Hwang *et al.*, 1992a); (2) intrapipette application of the catalytic subunit of PKA can activate the current (Bahinski *et al.*, 1989a; Harvey *et al.*, 1991); (3) the Cl^- conductance induced by cAMP (Bahinski *et al.*, 1989b), Iso (Hwang *et al.*, 1992a), or forskolin (Hwang *et al.*, 1993a) can be completely shut down by intrapipette application of a small peptide (PKI) that specifically inhibits PKA. The PKA regulatory pathway is also influenced by the activity of the phosphodiesterases that hydrolyze cAMP. Thus, Ono *et al.* (1992) found that intrapipette application of cGMP enhanced the Cl^- conductance induced by submaximal concentrations of Iso, forskolin, or intracellular cAMP, and suggested that the cGMP might work by inhibiting phosphodiesterase, increasing intracellular [cAMP]. This idea was supported by the abolition of cGMP's effect by milrinone, a specific inhibitor of cGMP-sensitive phosphodiesterase. Because cGMP and β-adrenoceptor stimulation have a similar synergistic action on Ca^{2+} current in myocytes (Ono and Trautwein, 1991), cGMP-sensitive phosphodiesterase seems to play a general role in the regulation of cellular cAMP levels.

b. Inhibitory Pathway. Stimulation of muscarinic receptors by acetylcholine or carbachol inhibits Iso- or forskolin-induced Cl^- currents (Harvey and Hume, 1989; Harvey *et al.*, 1990; Tareen *et al.*, 1991; Hwang *et al.*, 1992a), but not Cl^- currents activated by pipette cAMP (Tareen *et al.*, 1991; Hwang *et al.*, 1992a), suggesting that the inhibition is exerted before synthesis of cAMP by adenylyl cyclase. This inhibitory effect required pipette GTP (Horie *et al.*, 1992) and was abolished in cells incubated with pertussis toxin or in cells dialyzed with a high concentration of GDPβS, a nonspecific inhibitor of G protein turnover, demonstrating involvement of a pertussis toxin-sensitive G protein, presumably G_i (Hwang *et al.*, 1992a). So, as previously suggested for regulation of L-type Ca^{2+} current in guinea pig (Hescheler *et al.*, 1986) and frog (Fischmeister and Hartzell, 1986; Nakajima *et al.*, 1990; Parson *et al.*, 1991) cardiac myocytes, the stimulatory β-adrenergic and inhibitory muscarinic pathways appear to regulate the Cl^- current by impinging on the adenylyl cyclase via activated G_s and G_i proteins, respectively.

However, the precise mechanism of adenylyl cyclase inhibition by G_i seems to differ depending on whether the cyclase is stimulated by forskolin or by β-adrenoceptor activation (i.e., G_s). As demonstrated for Ca^{2+} currents in frog myocytes (Fischmeister and Shrier, 1989; Parson *et al.*, 1991), inhibition of forskolin-induced Cl^- currents by muscarinic receptor stimulation can be attenuated, and eventually overcome, by increasing the forskolin concentration (Tareen *et al.*, 1991; Hwang *et al.*, 1992b). In other words, the forskolin dose–response curve is simply shifted to higher forskolin concentrations by exposure to carbachol, the amplitude of the maximal response remaining unaltered; a similar effect is obtained by intrapipette application of the nonhydrolyzable GTP analogue, GTPγS, which, in the absence of agonists, preferentially activates G_i (Hwang *et al.*, 1992a,b). In contrast to this result with forskolin, however, Hwang *et al.* (1992b) found little change in the apparent affinity for Iso but a substantial reduction of the maximal response it elicited (even at 10 μM Iso) in the presence of carbachol (Hwang *et al.*, 1992b), a result again echoing findings on frog myocyte Ca^{2+} currents (Fischmeister and Shrier, 1989; Parson *et al.*, 1991). These results have led to the suggestion that G_i activation reduces the potency of forskolin but reduces the efficacy of Iso (Parson *et al.*, 1991; Hwang *et al.*, 1992b), implying that G_i might interfere with the action of G_s through a noncompetitive, allosteric, effect but might interfere in a competitive manner, perhaps sterically, with forskolin binding. Muscarinic inhibitory effects in guinea pig ventricular myocytes were also examined by Hescheler *et al.* (1986) and Tareen *et al.* (1991) who studied Ca^{2+} and Cl^- currents, respectively, but both these groups reported that acetylcholine simply shifted the Iso dose–response curves (as described above for forskolin). The apparent discrepancies between some of these reports clearly warrant further investigation. Nevertheless, the PKA-regulated Cl^- conductance provides a convenient monitor of cellular [cAMP] for examining the regulation of adenylyl cyclase via interactions between G_s, G_i, and forskolin in an intact mammalian cell.

c. Lack of Direct Coupling between G Proteins and Cl^- Channels. It is clear that Iso elicits PKA-regulated Cl^- current through activation of the G_s proteins. Thus, (i) withdrawal of GTP from pipette solutions reversibly abolished activation of the Cl^- current by 1 μM Iso or by 10 μM histamine, but activation by 1 μM forskolin or by pipette cAMP remained unaffected (Horie *et al.*, 1992); (ii) switching to pipette solution containing 1 mM GDPβS prevented Cl^- current activation by Iso but not by forskolin (Hwang *et al.*, 1992a); and (iii) with GTPγS (or GppNHp) in the pipette, a brief exposure to Iso or histamine, but not to forskolin, caused persistent activation of the Cl^- current (Hwang *et al.*, 1992a). However, the demonstration that muscarinic K^+ channels in atrial myocytes are opened directly by activated G_K proteins (Breitwieser and Szabo, 1985; Pfaffinger *et al.*, 1985) and evidence (reviewed by Brown and Birnbaumer, 1988) supporting direct modulation by G_s of cardiac L-type Ca^{2+} and Na^+ channels, both of which are also known to be regulated through the PKA phosphorylation pathway, raise the question of a possible involvement of such a direct, membrane-delimited pathway in the activation of PKA-regulated Cl^- channels. But this has been ruled out by the observations (Hwang *et al.*, 1992a) that (i) intrapipette application of PKI completely abolished Iso-induced Cl^- current with either GTP (Fig. 3) or GTPγS in the pipette, (ii) Iso had no effect in the presence of PKI, and (iii) Iso failed to increase Cl^- current activated by maximal stimulation of the PKA pathway with forskolin or intracellular cAMP (even though the current can still be increased by, e.g., phosphatase inhibition; Hwang *et al.*, 1993a,b).

Because PKI fully abolished Iso-induced Cl^- current even with GTPγS in the pipette, we can also rule out direct modulation of these Cl^- channels by any G proteins with a significant basal turnover rate, since GTPγS should have activated them all (Hwang *et al.*, 1992a). Any influence of G_i on activation of the Cl^- channels other than through its effect on adenylyl cyclase (e.g., by affecting kinase or phosphatase activity) is equally unlikely, because carbachol (or acetylcholine) did not inhibit Cl^- current elicited by either maximal (Tareen *et al.*, 1991; Hwang *et al.*, 1992a) or submaximal (Tareen *et al.*, 1991) levels of intracellular cAMP.

d. Influence of External Na^+ Ions. The surprising sensitivity of the Cl^- current induced by β-adrenoceptor stimulation to the external [Na^+] (cf. Egan *et al.*, 1987,1988) is also a consistent finding in the more recent experiments incorporating cell dialysis (Bahinski *et al.*, 1989a; Matsuoka

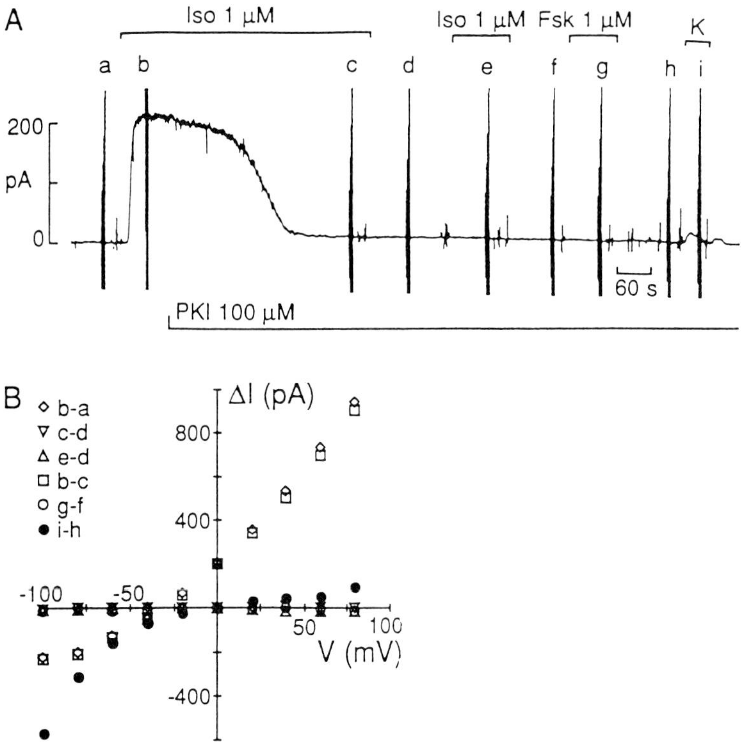

FIGURE 3 Abolition of Iso-induced Cl^- conductance by PKA inhibitor peptide, PKI. Chart record of current changes at 0 mV showing complete abolition of Iso- and forskolin (Fsk)-induced Cl^- current by intracellular dialysis with PKI (A) and steady-state difference *I–V* relationships obtained by subtraction as indicated (B). (From Hwang *et al.*, 1992a, by copyright permission of the Rockefeller University Press.)

et al., 1990; Harvey *et al.*, 1990), although its interpretation remains controversial (Harvey *et al.*, 1991; Tareen *et al.*, 1992). In the first systematic study of this phenomenon, Matsuoka *et al.* (1990) found no response to epinephrine when Na^+_o was replaced by TEA^+ or $Tris^+$ and an apparent facilitation of the response by external cations in the sequence $Na^+ > K^+$, $Rb^+ > Cs^+$, Li^+, divalent cations. The mechanism was subsequently clarified by Tareen *et al.* (1992), who concluded that the Na^+ substitutes competitively *inhibited* β-adrenoceptor activation of G_s because (i) they shifted rightward the dose–response curve for Cl^- conductance activation by Iso or epinephrine, but not by histamine, forskolin, or pipette cAMP; (ii) they similarly modified Iso-induced enhancement of Ca^{2+} current; and yet (iii) the apparent affinity for Iso activation of inward Cl^- current was unaffected by substitution of sucrose for external NaCl. These results corroborate the finding of Nagel *et al.* (1992) that Cl^- channels in excised inside-out patches could be directly activated by PKA catalytic subunit plus Mg-ATP in the complete absence of extracellular or cytoplasmic Na^+ [replaced by *N*-methyl-D-glucamine$^+$ (NMG), or Cs^+], implying that neither intra- nor extracellular Na^+ influence Cl^- channel phosphorylation, gating, or permeation. In contrast, Harvey *et al.* (1991) reported that replacement of Na_o^+ by TMA^+ (or choline, $Tris^+$, or NMG^+) reduced whole-cell Cl^- conductance activated by Iso, forskolin, 8-Br-cAMP (membrane-permeable cAMP derivative), or IBMX or by including PKA catalytic subunit in the pipette; substitution of TMA^+ for Na_o^+ also diminished β-adrenoceptor-mediated stimulation of Ca^{2+} and delayed K^+ currents. Since Harvey *et al.* (1991), unlike Egan *et al.* (1988) and Matsuoka *et al.* (1990), did not test for the influence of atropine, the inhibition of β-adrenoceptor-induced stimulation of Cl^-, Ca^{2+}, and K^+ currents by 140 m*M* TMA^+ (or choline$^+$) is presumably largely attributable to a maximal stimulation of muscarinic receptors (cf. Zakharov and Harvey, 1993); any remaining effect (cf. Zakharov and Harvey, 1993) can be accounted for by the reduced apparent affinity for β-agonists (Tareen *et al.*, 1992). Of course, this explanation is far less satisfactory for the effects of TMA^+ on Cl^- conductance activated by 8-Br-cAMP, IBMX, or PKA catalytic subunit. Harvey *et al.* (1991) also reported that TMA^+ replacement of Na_o^+ caused almost no diminution of forskolin-activated Cl^- conductance when 10 m*M* Na^+ was included in the usually Na^+-free pipette solution, or when pipette ATP was replaced by ATPγS (presumed to give rise to hydrolysis-resistant phosphoprotein), the latter finding being interpreted as suggesting that external Na^+ replacement affects Cl^- channel phosphorylation or dephosphorylation and the former as suggesting this effect is due to a secondary depletion of intracellular Na^+. However, Tareen *et al.* (1992) found the same ~70% reduction of Iso-activated Cl^- current on replacing Na_o^+ with TEA^+ whether the pipette solution contained 0, 15, or 115 m*M* Na^+. Although some of the apparent discrepancies between the Harvey *et al.* (1991) and Tareen *et al.* (1992) studies remain unexplained, three results suggest some impairment of intracellular dialysis in the former. First, the difference between results with 10 m*M* Na^+ and Na^+-free pipette solutions was interpreted as indicating considerable intracellular Na^+ accumulation during dialysis with Na^+-free, pipette solution (Harvey *et al.*, 1991). Second, β-adrenoceptor-mediated stimulation of membrane currents, a process known to require G protein turnover (see above), occurred in the absence of pipette GTP (Harvey *et al.*, 1991) implying uncontrolled availability of endogenous cytoplasmic GTP. Third, cardiac PKA-phosphorylated Cl^- channels, like overexpressed CFTR (Anderson *et al.*, 1991a), require hydrolyzable nucleoside triphosphate to open (see below; Nagel *et al.*, 1992) so that complete replacement of cellular ATP with ATPγS would be expected to result in phosphorylated but closed Cl^- channels and probably ATP-lack contracture of the cell (Hwang *et al.*, 1993a).

e. Modulation of Cl^- Channels by Phosphatases. Deactivation of PKA-regulated Cl^- conductance upon removal of agonists or a switch to pipette solution containing PKI is rapid, indicating the presence of highly active endogenous protein phosphatases (Hwang *et al.*, 1993a). Inhibition of that phosphatase activity should thus increase the steady-state level of phosphoprotein, and hence of the activated Cl^- conductance, and slow the subsequent deactivation of the conductance. Intrapipette application of okadaic acid or microcystin, both specific inhibitors of phosphatases 1 and 2A, causes both effects (Hwang *et al.*, 1993a). However, even maximally effective concentrations of okadaic acid and/or microcystin fail to prevent deactivation of the Cl^- current upon withdrawal of forskolin, implying that phosphatase(s) other than phosphatases 1 and 2A can also dephosphorylate the Cl^- channel. But, that (those) okadaic acid- or microcystin-insensitive phosphatase(s) could not fully deactivate the Cl^- conductance, and about half of it remained long after (≥40 min) washout of forskolin (Fig. 4); this residual component was insensitive to high [PKI], [Cl^-] sensitive, time independent, and insensitive to stilbene derivatives. A type 1 and/or 2A phosphatase is thus required for complete dephosphorylation of these Cl^- channels, although partial dephosphorylation can be achieved by some other phosphatase(s).

This implies two distinct forms of phosphorylated Cl^- channel: since the residual Cl^- conductance is ~40–50% of the fully activated Cl^- conductance, the partially phosphorylated Cl^- channels must contribute less to the whole-cell Cl^- conductance than the fully phosphorylated channels. In other words, the degree of phosphorylation must regulate either the conductance or open probability of a single channel (Hwang *et al.*, 1993a). This demonstration of the functional consequences of incremental phosphorylation is in keeping with the observation (Riordan *et al.*, 1989) of multiple consensus sites for PKA phosphorylation in the cytoplasmic regulatory domain of CFTR and with the finding that PKA does indeed phosphorylate four or five serines in CFTR both *in vivo* and *in vitro* (Cheng *et al.*, 1991; Picciotto *et al.*, 1992).

The okadaic acid- and microcystin-sensitive phosphatase is likely to be phosphatase 2A because preliminary results indicate that a phosphorylated inhibitor-1 peptide, which specifically inhibits phosphatase 1, fails to affect forskolin-induced Cl^- current (Hwang *et al.*, 1993b). The okadaic acid-insensitive phosphatase appears to be sensitive to orthovanadate, since

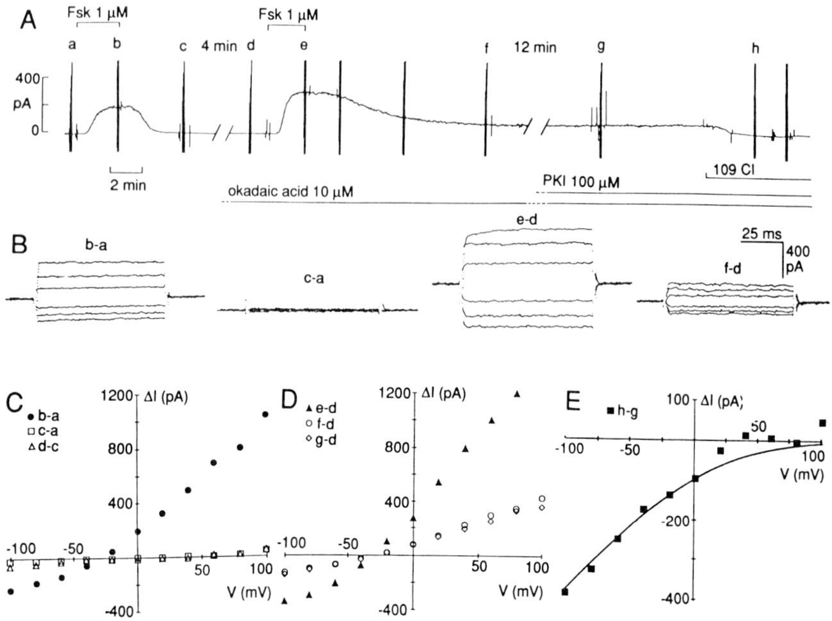

FIGURE 4 Okadaic acid prevents complete deactivation of the Fsk-activated Cl^- conductance. Whole-cell current at 0 mV showing enhancement of Fsk-induced Cl^- conductance by okadaic acid and persistence of a Cl-sensitive component after removal of Fsk (A), superimposed sample traces of difference currents from A, determined by subtracting digitized records of currents elicited by voltage pulses to ±20, ±40, and ±60 mV under one condition from corresponding records obtained under another, as indicated (B), and steady-state whole-cell difference *I–V* relationships as indicated (C, D, and E). The smooth curve shows a nonlinear least-squares fit of the constant field equation to the data points. (From Hwang *et al.*, 1993a, by copyright permission of the Rockefeller University Press.)

its inclusion with okadaic acid in pipette solutions completely prevents any deactivation of the Cl^- conductance following withdrawal of forskolin (Hwang *et al.*, 1993b). This vanadate-sensitive, okadaic acid-insensitive, phosphatase is probably phosphatase 2C, a Mg^{2+}-dependent protein phosphatase, because all pipette solutions contained ≥10 m*M* EGTA (pCa ~11) and so phosphatase 2B, which is Ca^{2+}/calmodulin-dependent, should have been completely inhibited (reviewed by Shenolikar and Nairn, 1991).

2. Biophysical and Pharmacological Characteristics

With approximately symmetrical Cl^- concentrations the whole-cell PKA-regulated Cl^- conductance is time- and voltage-independent (Bahinski *et al.*, 1989a; Harvey and Hume, 1989; Matsuoka *et al.*, 1990), one of the characteristics of epithelial CFTR Cl^- conductance. The activated currents reverse not far from estimated Cl^- equilibrium potentials over a range of pipette and extracellular Cl^- concentrations: measured reversal potentials vary roughly linearly with $\log_{10}$ $[Cl^-]_o$ with slopes of ~60 mV (Matsuoka *et al.*, 1990) or ~53 mV (Harvey *et al.*, 1990). Shifts of the reversal potential, caused by anion substitution under nominally biionic conditions, suggest a permeability sequence of $NO_3^- > Br^- > Cl^- \geq I^- > F^-$ for this Cl^- channel (Overholt and Harvey, 1992; Hwang *et al.*, 1992c). Although I^- and Cl^- appear almost equally permeant when I^- is outside and Cl^- is inside the cell, both the outward current at positive voltages and the inward current at negative potentials (mainly carried by Cl^-) are much smaller than with Cl^- alone (Overholt and Harvey, 1992), suggesting that I^- might act as a permeant blocker of the channel (Hwang *et al.*, 1992c; cf. Walsh and Long, 1992). Overexpressed CFTR Cl^- channels exhibit a similar permeability sequence (Anderson *et al.*, 1991b) and also seem to be blocked by I^- (Tabcharani *et al.*, 1992).

Although relatively high concentrations of the stilbene derivatives, 4,4′-dinitro-2,2′-stilbene disulfonic acid (DNDS) and SITS (4-acetamido-4′-isothiocyanatostilbene-2,2′-disulfonate), were found to inhibit Iso-induced Cl^- current irreversibly under certain conditions (Bahinski *et al.*, 1989a; Matsuoka *et al.*, 1990; Takano and Noma, 1992), there was no effect when pipette solutions contained high [Hepes] (40 m*M*) and [EGTA] (50 m*M*) (Bahinski *et al.*, 1989a) or when pipettes with wider tips were used to facilitate intracellular dialysis (access resistances 1–4 MΩ; Nagel *et al.*, 1992; Hwang *et al.*, 1992c), indicating that stilbenes do not block the channel itself; nor do they block epithelial CFTR Cl^- channels (Cliff *et al.*, 1992). Indeed, Harvey and Hume (1991) noted that 0.1–1 m*M* SITS appeared to enhance Iso-activated Cl^- current and that SITS alone sometimes elicited Cl^- current, results also interpreted in terms of indirect effects on the regulatory pathway. Further preliminary tests (Hwang *et al.*, 1992c) on well-dialyzed guinea pig ventricular myocytes show no inhibitory effects on forskolin-induced Cl^- conductance of up to 2-min exposures to diphenylamine-2-carboxylic acid (DPC, 200 μM), DIDS (100 μM), or DNDS (0.2–1 m*M*). Harvey *et al.* (1990) found ~40% inhibition of the Iso-induced Cl^- conductance by 100 μM anthracene-9-carboxylate (9-AC) and Levesque *et al.* (1993) reported almost complete block by 200 μM 9-AC. Bearing in mind that β-adrenoceptor-activated whole-cell Cl^- conductance is regulated via a complex metabolic pathway (see Fig. 6, below) and that these channel blockers are predominantly hydrophobic compounds capable of nonspecific effects at high concentrations, cautious interpretation of the results of pharmacological experiments seems warranted.

C. Single-Channel Current Studies

1. Cell-Attached Mode

Ehara and Ishihara (1990) made the first recordings of single anion channel currents in cell-attached patches on guinea pig myocytes stimulated by epinephrine. Channels could be detected in only <5% of the patches, indicating a low density of this channel in ventricular myocyte membranes. The single-channel *I–V* relationship showed outward rectification, presumably largely because of the lower cytoplasmic than extracellular [Cl^-], and a limiting slope conductance at positive potentials of ~13 pS. The channels showed the slow gating kinetics characteristic of CFTR, with mean open and closed times on the order of hundreds of milliseconds and an open probability (P_o) of 0.69 (SD ± 0.14; Ehara and Matsuura, 1993) which did not vary with the membrane potential (between −120 and +60 mV). Binomial analysis of amplitude histograms of current in multichannel patches before and after recruitment of new channels, during activation by intrapipette cAMP (via a second pipette in whole-cell current recording configuration), suggested that channels opened with this high P_o which then remained constant for several minutes (Ehara and Matsuura, 1993): the interpretation was that PKA phosphorylation activates each Cl^- channel in an all-or-none manner, effectively regulating the number of active channels. Interestingly, the open probability estimated by Ehara and Matsuura (1993) is similar to the higher (corresponding to full phosphorylation) of two P_o values observed in inside-out patch experiments (see below). It is possible that Ehara and Matsuura (1993) did not observe the lower P_o mode of channel gating (see below) because of their very low cytoplasmic free [Mg^{2+}] (~10^{-2} m*M*) which would limit phosphatase 2C activity.

2. Excised Inside-Out Mode

a. Regulation of Cl^- Channel by PKA and Hydrolyzable Nucleoside Triphosphates. To overcome the low channel density, the giant patch technique (Hilgemann, 1990) was used to examine activation by PKA at the single-channel level in excised inside-out patches (Nagel *et al.*, 1992). Once phosphorylated by PKA plus Mg-ATP, channel activity can be maintained for tens of minutes in the presence of Mg-ATP, but its removal causes prompt closure of the channel(s) (Fig. 5). Although GTP can substitute for ATP in opening phosphorylated channels, although with lower

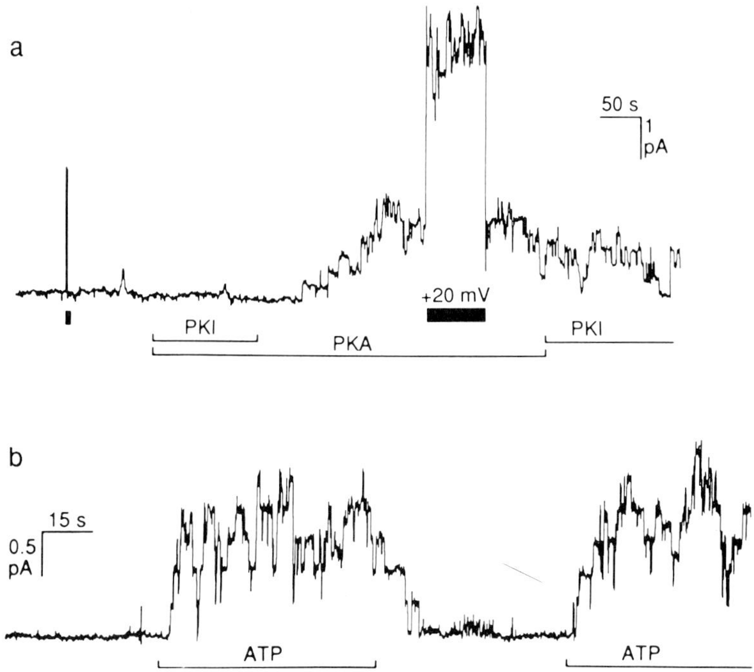

FIGURE 5 Phosphorylated Cl^- channels in excised inside-out patches need ATP to open. (a) PKI peptide (5-24-amide) prevented channel activation by PKA plus Mg-ATP, but failed to close the phosphorylated channels after withdrawal of PKA. (b) Reversible activation of phosphorylated Cl channels by ATP. The current record begins ~30 sec after the withdrawal of PKA and ATP had closed all channels; reexposure to 500 μM Mg-ATP reopened the channels. (From Nagel *et al.*, 1992, with permission.)

potency, neither ADP, which has no γ-phosphate, nor AMP-PNP, which cannot donate its γ-phosphate, can open phosphorylated channels that have been closed by withdrawal of ATP. This requirement of a hydrolyzable nucleoside triphosphate for opening PKA phosphorylated channels is believed to be a hallmark of CFTR Cl^- channels (Anderson *et al.*, 1991a).

In patches containing several channels the patch current generally declines after removal of PKA, usually with a concomitant reduction of channel open time, but channel activity can be partially (or completely) restored by reapplication of PKA. The implication is that even in excised patches channels can be partially dephosphorylated by membrane-bound phosphatases and that channel activity (i.e., channel open probability) depends on the degree of phosphorylation. Consistent with this suggestion, patches containing a single channel occasionally show, after addition of PKA, a jump of P_o from a relatively low value ($\leq$0.3) to a higher value ($\geq$0.6) despite the presence of a constant [ATP] (0.5 m*M*; Hwang *et al.*, 1993b). These observations provide additional support for the interpretation of the residual whole-cell Cl^- conductance in the presence of okadaic acid (or microcystin), suggesting that partially phosphorylated Cl^- channels are associated with a low P_o gating mode, whereas fully phosphorylated Cl^- channels have a high P_o. According to the whole-cell data, phosphatase 2C should be responsible for the conversion from fully to partially phosphorylated Cl^- channels and should therefore be the membrane-bound phosphatase.

D. Summary/Synthesis of Regulatory and Gating Mechanisms

Figure 6 summarizes the metabolic pathways for modulation of PKA-regulated Cl^- channels in cardiac myocytes. Agonist stimulation of receptors regulates adenylyl cyclase activity through stimulatory or inhibitory G proteins. The intracellular cAMP level that results from cyclase and phosphodiesterase activity determines the level of protein kinase A activa-

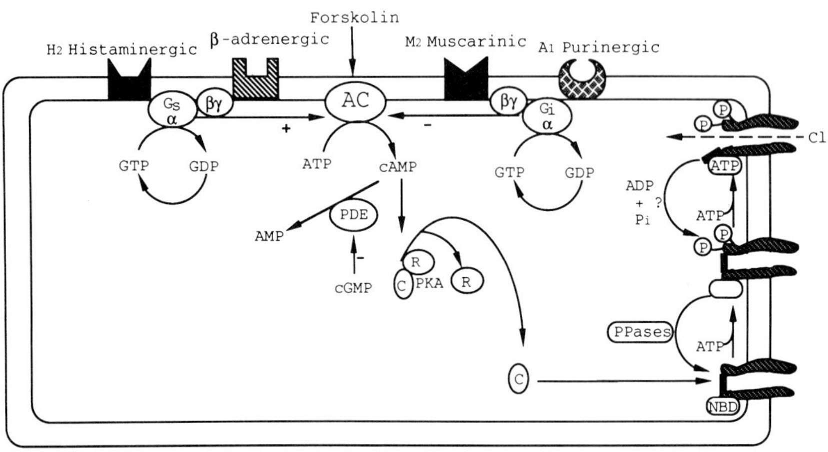

FIGURE 6 Schematic representation of the regulatory pathways of cardiac PKA-activated Cl^- channels. Various hormonal receptors regulate the CFTR Cl^- channel by impinging on the adenylyl cyclase through G proteins, and the resulting alteration of intracellular cAMP levels modulates PKA, which directly phosphorylates the channel molecule. At least two PKA phosphorylation sites can be differentiated by their different sensitivity to protein phosphatases and their different contribution to whole-cell Cl^- conductance (see text and Fig. 7).

tion, which in turn controls the rate of phosphorylation of Cl^- channel molecules. The dual phosphorylation model in Fig. 7 can account for all present results obtained both in whole cells and in excised membrane patches. Complete dephosphorylation of the channel requires phosphatase 2A and a second, vanadate-sensitive, phosphatase, likely phosphatase 2C. Because it is known that PKA-phosphorylated CFTR Cl^- channels are gated by Mg-ATP, it will be important to learn whether, and if so how, ATP control of channel gating differs depending on the degree of channel phosphorylation.

E. Functional Role and Cardiological Significance

Documentation of the molecular identity of the cardiac PKA-regulated Cl^- channel and elucidation of its regulatory mechanisms have preceded a clear understanding of its functional role in the normal human ventricle, let alone the potential pathophysiological impact of its absence or dysfunction in patients with cystic fibrosis. Upon β-adrenoceptor stimulation the normally steep Cl^- ion gradient is expected to drive a substantial outward, repolarizing, current through these channels during the plateau of the cardiac action potential. This outward current should counteract the influence of the simultaneously enhanced Ca^{2+} current to prolong the action potential duration, and, at the same time, it should increase Ca^{2+} entry into the cells during the action potential plateau by holding the membrane potential away from the Ca^{2+} equilibrium potential. In support of this suggestion, Harvey *et al.* (1990) demonstrated that activation of the Cl^- current did shorten the action potential duration, but the measurements

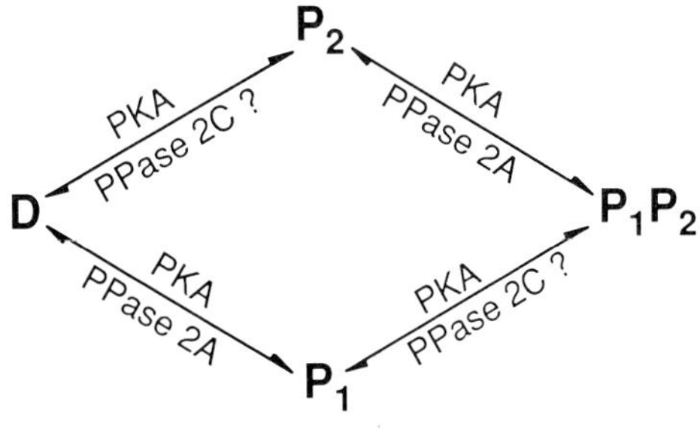

FIGURE 7 Suggested phosphorylation scheme for the CFTR Cl^- channel. PKA activates the inactive channel by phosphorylating the P_1 site, while additional phosphorylation of the P_2 site, to yield P_1P_2, increases the open probability of the active channel opened by ATP. Preliminary results suggest that channel molecules with only the P_2 site phosphorylated are likely to be nonconducting.

were made at room temperature to prevent simultaneous activation of delayed rectifier K^+ channels so that the relative contributions of these two current components to the shape of the action potential remains unclear. Measurements using 200 μM 9-AC to abolish the Cl^- current also demonstrated its ability to shorten action potential duration, but again precluded direct comparison of the relative roles of K^+ and Cl^- currents (Levesque *et al.*, 1993). On the other hand, Takano and Noma (1992) used the nystatin-perforated patch technique on rabbit ventricular myocytes, at 36°C, and found that whereas 0.3 μM Iso caused a slight reduction of the action potential duration (on average), subsequent exposure to 1 mM DNDS caused a marked prolongation. Because, in the same cells, DNDS did not alter action potential duration in the absence of Iso, implying lack of effect on Ca^{2+} or K^+ current, the conclusion was that (i) DNDS prolonged the action potential in Iso by inhibiting PKA-regulated Cl^- current and, hence, (ii) that this current does normally counter the tendency of the enhanced Ca^{2+} current to increase action potential duration (Takano and Noma, 1992). (As already mentioned, this effect of DNDS in poorly dialyzed myocytes is presumably not due to direct block of the Cl^- channel, but to some indirect effect on its regulation.)

Unopposed lengthening of the action potential duration by Ca^{2+} current enhancement could excessively prolong the refractory period, predisposing the heart to arrhythmias, as in the long Q-T syndrome (Vincent *et al.*, 1974). Despite the sparse electrocardiological literature on CF patients, ventricular dysfunction during stress (Benson *et al.*, 1984) and arrhythmias (Cheron *et al.*, 1984; Sullivan *et al.*, 1986) have been reported, but are usually assumed to be consequences of the pulmonary dysfunction (cor pulmonale) often associated with CF. However, if systematic electrocardiographic studies were to reveal ventricular repolarization abnormalities in CF patients, then a lack of PKA-dependent stimulation of cardiac Cl^- channels, but not of Ca^{2+} channels, would provide a reasonable explanation. Several kinds of studies are now warranted to address the question of the functional role of CFTR in human hearts. First, Northern analysis of atrial (Levesque *et al.*, 1992) and ventricular tissue should establish whether there are regional differences in the level of CFTR expression within the human heart. Second, electrophysiological studies of atrial and ventricular myocytes from normal and CF hearts should establish patterns of function of CFTR in those hearts. And finally, electrocardiographic measurements, with and without sympathetic stress, should be compared for normal hearts, for CF patients, for CF patients with lung transplants, for CF patients with heart and lung transplants, and for non-CF patients who have received CF hearts in "domino" transplant operations; such a systematic set of measurements should shed light both on the normal role of these Cl^- channels and on the pathophysiological significance of their possible dysfunction in CF patients. On the other hand, since one result of the enhanced Cl^- current is a facilitation of Ca^{2+} entry (above), the postulated failure of these channels in CF patients might afford them some degree of protection against the possibly dangerous consequences of excessive Ca^{2+} entry. Another, related, suggestion (Levesque *et al.*, 1993) is that the abbreviated action potential resulting from Cl^- current activation might be considered dangerous and that agents that diminish the Cl^- current, and so prolong the action potential, might have antiarrhythmic properties (like the Class III antiarrhythmics that inhibit K^+ channels). In this view, again, possible dysfunction of cardiac Cl^- channels in CF patients could be considered beneficial. It is even conceivable that a mechanism like one of these could contribute to the selection pressure that has managed to keep CF mutations so prevalent in the caucasian population. A careful analysis of appropriately recorded and selected electrocardiograms should begin to shed light on these kinds of questions.

III. STRETCH-ACTIVATED Cl^- CHANNELS

Stretch-activated ion channels have been found in a wide variety of cell types and are thought to play an important role in the regulation of cell volume (see review by Morris, 1990). A Cl^- current activated by the membrane stretch associated with cell swelling has been identified in cardiac myocytes isolated from rabbit atria and SA node (Hagiwara *et al.*, 1992a,b), dog atria (Sorota, 1992) and ventricles (Tseng, 1992), and cultured embryonic chick heart (Zhang *et al.*, 1992).

This Cl^- current can be elicited by hypertonic internal (Tseng, 1992) or hypotonic external solutions (Hagiwara *et al.*, 1992a; Tseng, 1992; Sorota, 1992; Zhang *et al.*, 1992), by direct mechanical inflation of the cell (Hagiwara *et al.*, 1992b), or by exposing ventricular myocytes to anion amphipaths, such as dipyridamole or 2,4,6-trinitrophenol, which are believed to preferentially insert into the outer leaflet of the membrane and so cause a distortion of the membrane comparable to that occurring during cell swelling (Tseng, 1992). Since induction of this stretch-activated Cl^- current was found to be affected by neither a specific PKA inhibitor nor the nonspecific kinase inhibitor, H7, it seems unlikely that phosphorylation is involved in the regulatory mechanism (Hagiwara *et al.*, 1992b). Nor is it likely that intracellular Ca^{2+} ions are important for its activation because of the high [EGTA] (e.g., 10 m*M*; Hagiwara *et al.*, 1992b) generally included in pipette solutions and because cell swelling can elicit the current in myocytes simultaneously exposed to Ca^{2+}-free external solutions containing ryanodine (to prevent Ca^{2+} release from the sarcoplasmic reticulum) and dialyzed with EGTA-containing pipette solutions (S. Sorota, personal communication).

Preliminary reports of single anion channel currents elicited by osmotic swelling demonstrate (i) in the cell-attached patches on chick myocytes, a strong outwardly rectifying *I-V* relationship and a single-channel conductance of 31 pS at 0 mV (Zhang *et al.*, 1992) and (ii) in inside-out patches excised from rabbit atrial myocytes, a single-channel conductance of 35 pS in symmetrical 150 m*M* [Cl^-] solutions (Hagiwara *et al.*, 1992a). Estimation of the anion selectivity from reversal potential shifts in whole-cell experiments gave a permeability sequence of $I^- > NO_3^- > Br^- > Cl^- > F^-$ (Hagiwara *et al.*, 1992b). And, this current can be reduced by some Cl^- channel blocking agents, including DNDS (1–5 m*M*), SITS (1 m*M*) (Hagiwara *et al.*, 1992b), 9-AC (1 m*M*; Hagiwara *et al.*, 1992b; Tseng, 1992; Sorota, 1992), niflumic acid (100 μM; Sorota, 1993), indanyloxyacetate (100 μM; Sorota, 1993), and nitrophenylpropylamino benzoate (NPPB, 40 μM; Tseng, 1992).

Like that of PKA-regulated Cl^- channels, the function of this stretch-activated Cl^- channel remains speculative, although, by permitting efflux of Cl^-ions together with K^+ ions, it could be involved in a regulatory volume decrease in response to swelling of cardiac myocytes. Activity of the channel could become important, for example, during the cardiac cell swelling known to accompany cardiac ischemia and reperfusion (Tranum-Jensen *et al.*, 1981).

IV. Ca^{2+}-ACTIVATED Cl^- CHANNELS

Early studies on cardiac Purkinje fibers by Kenyon and Gibbons (1979) showed that the transient outward current induced by step depolarization to positive potentials included two components: a K^+ current sensitive to 4-aminopyridine (4-AP) and a chloride-sensitive but 4-AP-insensitive component. From patch-clamp studies on rabbit ventricular and atrial myocytes the charge carrier for this latter component was concluded to be Cl^- ions, because the 4-AP-insensitive transient current disappeared upon reduction of the external [Cl^-] sufficient to make the Cl^- equilibrium potential equal to the pulse potential (Zygmunt and Gibbons, 1991), and reversal potentials of quasi-instantaneous *I–V* relations, determined with double-pulse protocols, shifted with the Cl^- equilibrium potential when $[Cl^-]_o$ was altered (Zygmunt and Gibbons, 1992).

The Ca^{2+}-dependent nature of this current was based on the observations that the current appeared only when the intracellular Ca^{2+} buffering was low (pCa = 7.7 or 8.3 with 0.2 or 0.8 m*M* EGTA) and that it was abolished by 10 m*M* intracellular EGTA or by inhibition of voltage-activated, L-type Ca^{2+} current with Cd^{2+} or nisoldipine (Zygmunt and Gibbons, 1991). Interestingly, this Cl^- current was also abolished by ryanodine or caffeine, both of which prevent sarcoplasmic reticulum Ca^{2+} release, indicating that the underlying Cl^- channels are activated by the Ca^{2+} ions that are released from the sarcoplasmic reticulum rather than by the trigger Ca^{2+} ions that enter through L-type Ca^{2+} channels (Zygmunt and Gibbons, 1991,1992). The Ca^{2+}-activated Cl^- current was larger in atrial than in ventricular cells, but it is not yet clear whether this reflects a difference in channel density or in factors influencing the localized intracellular [Ca^{2+}] near the channels.

So far, only preliminary single-channel data are available (Zygmunt and Gibbons, 1993); the on-cell single-channel conductance was reported to be 43 pS for outward current and 18 pS for inward current, not very different from the conductance of Cl^- channels activated by swelling in the same cells (Hagiwara *et al.*, 1992a) or even in cultured chick heart cells (Zhang et al., 1992). The Ca^{2+}-activated whole-cell Cl^- current was consistently found to be abolished by 0.1 m*M* DIDS or 2 m*M* SITS (Zygmunt and Gibbons, 1991, 1992). Although the Ca^{2+}-sensitive Cl^- channel is evidently distinct from the PKA-regulated Cl^- channel, it is not at all clear that it is distinct from either the stretch-activated Cl^- channel or Cl^- channels suggested to be activated by protein kinase C (Walsh, 1991; Zhang and Ten Eick, 1993).

Because of its transient nature, by virtue of its dependence on the brief, large rise in intracellular [Ca^{2+}], this Cl^- current is likely to contribute to whole-cell membrane conductance only early during the cardiac action potential. Together with the larger, 4-AP-sensitive, transient K^+ current, it probably contributes to the rapid phase 1 repolarization of the atrial action potential; indeed, Takano and Noma (1992) noted that application of 1 m*M* DNDS blunted that early repolarization phase in rabbit atrial myocytes, but had a much smaller (if any) effect on the far less extensive phase 1 repolarization of ventricular myocytes.

V. OTHER Cl^- CHANNELS

Walsh (1991) reported activation of a time-independent Cl^- conductance in guinea pig ventricular myocytes, at room temperature, by external application of the phorbol ester, PMA, or of norepinephrine in the presence of propranolol (to stimulate α-adrenoceptors) or by dialyzing the cell with partially purified protein kinase C. The whole-cell I–V relationships were described as approximately linear despite asymmetrical [Cl^-] and were shifted in a parallel fashion by lowering $[Cl^-]_o$. This PMA-activated current was reduced to about half by exposure to the monocarboxylic acid derivative, 8-chlorophenoxypropionic acid (CPPA). More recently, Zhang and Ten Eick (1993) described activation by PMA in feline ventricular myocytes of a time-independent Cl^- conductance that was sensitive to 9-AC. Maximally effective concentrations of PMA precluded activation of PKA-regulated Cl^- conductance by forskolin, but the effects of submaximally activating concentrations of PMA and forskolin seemed additive, leading to the suggestion that the same population of Cl^- channels subserves the Cl^- conductance activated by both signaling pathways. It will be interesting to see whether similar effects can be demonstrated in excised membrane patches by direct application of PKC and/or PKA.

Extracellular ATP, its nonhydrolyzable analogue ATPγS, ADP, AMP, and adenosine were all found to activate a Cl^- current in guinea pig atrial myocytes (Matsuura and Ehara, 1992; cf. Scamps and Vassort, 1990), but the subtype of the purinergic receptor involved and the detailed mechanism of channel activation are not yet clear.

In addition to the electrophysiological characterization of Cl^- currents described above, molecular expression studies of mRNA injected into *Xenopus* oocytes have begun to contribute to our understanding of the full spectrum of cardiac Cl^- channels. Thus, Jentsch and colleagues have already identified three Cl^- channel homologs: ClC-0, the voltage-dependent Cl^- channel cloned from the electric organ of *Torpedo marmorata* (Jentsch *et al.*, 1990); ClC-1, the principal Cl^- channel of skeletal muscle (Steinmeyer *et al.*, 1991); and ClC-2, a Cl^- channel activated by hyperpolarization (Thiemann et al., 1992) and probably by cell swelling (Gründer *et al.*, 1992) that appears to be ubiquitously expressed in, for example, heart, brain, pancreas, lung, and liver. Although oocytes injected with mRNA transcribed from ClC-2 generated a Cl^- conductance that was activated by strong hyperpolarization (to between −100 and −180 mV) and that could be partially blocked by 1 mM DPC or 1 mM 9-AC, an equivalent current has not yet been demonstrated in cardiac myocytes, probably because of the large negative voltages required to open the channels (Thiemann et al., 1992).

Another, quite different, suspected Cl^- channel clone has been isolated from rat heart (Paulmichl *et al.*, 1992). When expressed in oocytes, it induced a Cl^- current that could be diminished not only by DIDS (50% inhibition at 20 μM) and NPPB (2 μM) but, surprisingly, also by extracellular cAMP. The primary structure (235 amino acids) of this clone suggests no α-helical membrane-spanning segments, but is consistent with a β-barrel structure spanning the membrane of the kind shown for porin channels (Weiss *et al.*, 1991; cf. Blachly-Dyson *et al.*, 1990). Expression has been demonstrated in rat atria and ventricle, but it has yet to be seen whether this Cl$^-$ channel is present in the sarcolemmal membrane.

Finally, the least-characterized putative cardiac Cl$^-$ channel, a 72-amino acid sarcolemmal protein called phospholemman, is the major cardiac substrate for phosphorylation by PKA and PKC. Its expression in *Xenopus* oocytes is associated with the appearance of a hyperpolarization-activated Cl$^-$ current sensitive to 9-AC (Moorman *et al.*, 1992), a characteristic shared by the expression product of ClC-2. But it is not yet clear whether phospholemman itself forms a Cl$^-$ channel or modulates preexisting covert Cl$^-$ channels.

VI. SUMMARY AND OUTLOOK

The application of patch-clamp techniques to cardiac myocytes has provided convincing evidence for the existence of at least three kinds of cardiac Cl$^-$ channel. The differences in their regulatory mechanisms suggest that each might play a distinct role in modulating the resting membrane potential or the action potential of myocytes according to the circumstances, depending, for example, on whether there are circulating catecholamines or there is cell swelling or release of Ca^{2+} from the sarcoplasmic reticulum. Because the equilibrium potential for Cl$^-$ ions in heart cells normally lies between −65 and −45 mV (Vaughan-Jones, 1982), opening of Cl$^-$ channels at ventricular diastolic membrane potentials will tend to cause Cl$^-$ efflux and hence membrane depolarization, whereas Cl$^-$ channel opening during the action potential plateau will shorten its duration due to a repolarizing Cl$^-$ influx. Indeed, shortening of the action potential duration as a result of activation of PKA-regulated Cl$^-$ channels has been established (Takano and Noma, 1992; cf. Harvey *et al.*, 1990; Levesque *et al.*, 1993). Correspondingly, adrenergic agonists have been reported to cause a small variable depolarization of ventricular cells during diastole (Harvey *et al.*, 1990; Matsuoka *et al.*, 1990; but cf. Sorota *et al.*, 1991) which became marked when the intracellular Cl$^-$ concentration was artificially elevated (Harvey *et al.*, 1990; Sorota *et al.*, 1991; Yamawake *et al.*, 1992). However, thorough examination of the significance of effects like these has been hindered by the lack of specific blockers for any of the cardiac Cl$^-$ channels, leading to difficulty in evaluating the contributions of changes in Cl$^-$ current relative to those of other currents.

As demonstrated with great success for cation (notably K^+ ion) channels, the appropriate combination of molecular biological and electrophysiological tools should eventually allow us to understand the relationship between the structure and function of individual Cl^- channels. Thus, although the CFTR protein is now known to be responsible for PKA-regulated cardiac Cl^- currents, further work will be required to shed equivalent light on the molecular identities of the other cardiac Cl^- channels. Even more work, of a different kind, including clinical studies, will be needed before we understand the physiological roles of each of the cardiac Cl^- channels and the potential pathophysiological consequences of their dysfunction, for example, in patients with cystic fibrosis.

Acknowledgments

The preparation of this review and our research summarized in it were supported by NIH HL-14899 and HL-49907 and the American Heart Association New York City Affiliate. We thank Peter Hoff for invaluable technical assistance.

References

Ackerman, M. J., and Clapham, D. E. (1993). Cardiac chloride channels. *Trends Cardiov. Med.* **3,** 23–28.

Anderson, M. P., Berger, H. A., Rich, D. R., Gregory, R. J., Smith, A. E., and Welsh, M. J. (1991a). Nucleoside triphosphates are required to open the CFTR chloride channel. *Cell (Cambridge, Mass.)* **67,** 775–784.

Anderson, M. P., Gregory, R. J., Thompson, S., Souza, D. W., Paul, I., Muligan, R. C., Smith, A. E., and Welsh, M. J. (1991b). Demonstration that CFTR is a chloride channel by alteration of its anion selectivity. *Science* **253,** 202–254.

Bahinski, A., Nairn, A. C., Greengard, P., and Gadsby, D. C. (1989a). Chloride conductance regulated by cyclic AMP-dependent protein kinase in cardiac myocytes. *Nature (London)* **340,** 718–721.

Bahinski, A., Gadsby, D. C., Greengard, P., and Nairn, A. C. (1989b). Chloride conductance regulated by protein kinase A in isolated guinea-pig ventricular myocytes. *J. Physiol. (London)* **418,** 32P (abstr.).

Benson, L. N., Newth, C. J. L., Desouza, M., Lobraico, R., Kartodihardjo, W., Corkey, C., Gilday, D., and Olley, P. M. (1984). Radionuclide assessment of right and left ventricular function during bicycle exercise in young patients with cystic fibrosis. *Am. Rev. Respir. Dis.* **130,** 987–992.

Blachly-Dyson, E., Peng, S., Colombini, M., and Forte, M. (1990). Selectivity change in site-directed mutants of the VDAC ion channel: Structural implications. *Science* **247,** 1233–1236.

Breitwieser, G., and Szabo, G. (1985). Uncoupling of cardiac muscarinic and β-adrenergic receptors from ion channels by a guanine nucleotide analogue. *Nature (London)* **317,** 538–540.

Brown, A. M., and Birnbaumer, L. (1988). Direct G protein gating of ion channels. *Am. J. Physiol.* **254,** H401–H410.

Carmeliet, E. E. (1961). Chloride ions and the membrane potential of Purkinje fibers. *J. Physiol. (London)* **156,** 375–388.

Cheng, S. H., Rich, D. P., Marshall, J., Gregory, R. J., Welsh, M. J., and Smith, A. L.

(1991). Phosphorylation of R domain by cAMP-dependent protein kinase regulates the CFTR chloride channel. *Cell (Cambridge, Mass.)* **66,** 1027–1036.
Cheron, G., Paradis, K., Steru, D., Demay, G., and Lenoir, G. (1984). Cardiac involvement in cystic fibrosis revealed by a ventridular arrythmia. *Acta Paediatr. Scand.* **73,** 697–700.
Cliff, W. H., Schoumacher, R. A., and Frizzell, R. A. (1992). cAMP-activated Cl channels in CFTR-transfected cystic fibrosis pancreatic epithelial cells. *Am. J. Physiol.* **262,** C1154–C1160.
Collins, F. S. (1992). Cystic fibrosis: Molecular biology and therapeutic implications. *Science* **256,** 774–779.
Coulombe, A., Duclohier, H., Coraboeuf, E., and Touzet, N. (1987). Single chloride-permeable channels of large conductance in cultured cardiac cells of newborn rats. *Eur. Biophys. J.* **14,** 155–162.
Dudel, J., Peper, K., Rudel, R., and Trautwein, W. (1967). The dynamic chloride component of membrane current in Purkinje fibers. *Pfluegers Arch.* **295,** 197–212.
Dukes, I. D., Cleemann, L., and Morad, M. (1990). Tedisamil blocks the transient and delayed rectifier K^+ currents in mammalian cardiac and glial cells. *J. Pharmacol. Exp. Ther.* **254,** 560–569.
Egan, T. M., Noble, D., Noble, S. J., Powell, T., and Twist, V. W. (1987). An isoprenaline activated sodium-dependent inward current in ventricular myocytes. *Nature (London)* **328,** 634–637.
Egan, T. M., Noble, D., Noble, S. J., Powell, T., Twist, V. W., and Yamaoka, K. (1988). On the mechanism of isoprenaline- and forskolin induced depolarization of single guinea-pig ventricular myocytes. *J. Physiol. (London)* **400,** 299–320.
Ehara, T., and Ishihara, K. (1990). Anion channels activated by adrenaline in cardiac muscle. *Nature (London)* **347,** 284–286.
Ehara, T., and Matsuura, H. (1993). Single channel study of the cyclic AMP-regulated chloride current in guinea-pig ventricular myocytes. *J. Physiol. (London)* **464,** 307–320.
Fischmeister R., and Shrier, A. (1989). Interactive effects of isoprenaline, forskolin, and acetylcholine on Ca^{+2} current in frog ventricular myocytes. *J. Physiol. (London)* **417,** 231–239.
Fischmeister, R., and Hartzell, H. C. (1986). Mechanism of action of acetylcholine on calcium current in single cells from frog ventricle. *J. Physiol. (London)* **376,** 183–202.
Gadsby, D. C. (1990). Effects of β-adrenergic catecholamines on membrane currents in cardiac cells. *In* "Cardiac Electrophysiology" (M. R. Rosen, M. J. Janse, and A. L. Wit, (eds.), pp. 857–876. Futura Publ. Co., Mount Kisco, NY.
Gründer, S., Thiemann, A., Pusch, M., and Jentsch, T. J. (1992). Regions involved in the opening of ClC-2 chloride channel by voltage and cell volume. *Nature (London)* **360,** 759–762.
Hagiwara, N., Masuda, H., Shoda, M., and Irisawa, H. (1992a). Chloride conductance activated by membrane stretch in rabbit single cardiac myocytes. *J. Physiol. (London)* **446,** 330P.
Hagiwara, N., Masuda, H., Shoda, M., and Irisawa, H. (1992b). Stretch activated anion channels of rabbit cardiac myocytes. *J. Physiol. (London)* **456,** 285–302.
Hartzell, H. C. (1988). Regulation of cardiac ion channels by catecholamines, acetylcholine and second messenger systems. *Prog. Biophy. Mol. Biol.* **52,** 165–247.
Harvey, R. D., and Hume, J. R. (1989). Autonomic regulation of a chloride current in heart. *Science* **244,** 983–985.
Harvey, R. D., and Hume, J. R. (1990). Histamine activates the chloride current in cardiac ventricular myocytes. *J. Cardiovasc. Electrophysiol.* **1,** 309–317.

Harvey, R. D., and Hume, J. R. (1991). Paradoxical agonist effects of the disulphonic stilbene derivatives SITS on the cardiac chloride current. *Biophys. J.* **59,** 90a.
Harvey, R. D., Clark, C. D., and Hume, J. R. (1990). Chloride current in mammalian cardiac myocytes. *J. Gen. Physiol.* **95,** 1077–1102.
Harvey, R. D., Jurevicius, J. A., and Hume, J. R. (1991). Intracellular Na^+ modulates the cAMP-dependent regulation of ion channels in the heart. *Proc. Natl. Acad. Sci. U.S.A.* **88,** 6946–6950.
Hescheler, J., Kameyama, M., and Trautwein, W. (1986). On the mechanism of muscarinic inhibition of the cardiac Ca current. *Pflüegers Arch.* **407,** 182–189.
Hilgemann, D. W. (1990). Regulation of cardiac Na^+-Ca^+exchange in giant excised sarcolemmal membrane patches. *Nature (London)* **344,** 242–245.
Horie, M., Hwang, T.-C., and Gadsby, D. C. (1992). Pipette GTP is essential for receptor-mediated regulation of Cl^- current in dialyzed myocytes from guinea-pig ventricle. *J. Physiol. (London)* **455,** 235–246.
Horowitz, B., Tsung, S. S., Hart, P., Levesque, P. C., and Hume, J. R. (1993). Alternative splicing of CFTR Cl^- channels in heart. *Am. J. Physiol.* **264,** H2214–H2220.
Hume, J. R., and Harvey, R. D. (1992). Chloride conductance pathways in heart. *Am. J. Physiol.* **261,** C399–C412.
Hutter, O. F., and Noble, S. M. (1961). Anion conductance of cardiac muscle. *J. Physiol. (London)* **157,** 335–350.
Hutter, O. F., and Padsha, D. (1959). Effect of nitrate and other anions on the membrane resistance of frog skeletal muscle. *J. Physiol. (London)* **146,** 117–132.
Hwang, T.-C., Horie, M., Nairn, A. C., and Gadsby, D. C. (1992a). Roles of GTP-binding proteins in the regulation of cardiac chloride conductance. *J. Gen. Physiol.* **99,** 465–489.
Hwang, T.-C., Horie, M., Dousmanis, A. G., and Gadsby, D. C. (1992b). Interactive modulation of PKA-regulated chloride current by forskolin, G_s, and G_i in guinea pig ventricular myocytes. *Biophys. J.* **61,** A395.
Hwang, T.-C., Horie, M., Dousmanis, A. G., and Gadsby, D. C. (1992c). Regulation of PKA-activated Cl conductance in guinea pig ventricular myocytes: Whole-cell studies. *J. Gen. Physiol.* **100,** 69a.
Hwang, T.-C., Horie, M., and Gadsby, D. C. (1993a). Functionally distinct phospho-forms underlie incremental activation of PKA-regulated Cl^- conductance in mammalian heart. *J. Gen. Physiol.* **101,** 629–650.
Hwang, T.-C., Nagel, G., Nairn, A. C., and Gadsby, D. C. (1993b). Dephosphorylation of cardiac Cl^- channels requires multiple protein phosphatases. *Biophys. J.* **64,** A343.
Jentsch, T. J., Steimeyer, K., and Schwarz, G. (1990). Promary structure of *Torpedo marmorata* chloride channel isolated by expression cloning in *Xenopus* oocytes. *Nature (London)* **348,** 510–514.
Kenyon, J. L., and Gibbons, W. R. (1977). Effects of low-chloride solutions on action potentials of sheep cardiac Purkinji fibers. *J. Gen. Physiol.* **70,** 635–660.
Kenyon, J. L., and Gibbons, W. R. (1979). 4-aminopyridine and the early outward current of sheep cardiac Purkinje fibers. *J. Gen. Physiol.* **73,** 139–157.
Levesque, P. C., Hart, P. J., Hume, J. R., Kenyon, J. L., Horowitz, B. (1992). Expression of cystic fibrosis transmembrane regulator Cl^- channels in heart. *Circ. Res.* **71,** 1002–1007.
Levesque, P. C., Clark, C. D., Zakarov, S. I., Rosenshtraukh, L. V., and Hume, J. R. (1993). Anion and cation modulation of the guinea-pig ventricular action potential during β-adrenergic stimulation. *Pflüegers Arch.* **424,** 54–62.
Matsuoka, S., Ehara, T., and Noma, A. (1990). Chloride-sensitive nature of the adrenaline-induced current in guinea-pig cardiac myocytes. *J. Physiol. (London)* **425,** 579–598.

Matsuura, H., and Ehara, T. (1992). Activation of chloride current by purinergic stimulation in guinea pig heart cells. *Circ. Res.* **70,** 851–855.
Moorman, J. R., Palmer, C. J., John, J. E., Durieux, M. E., and Jones, L. R. (1992). Phospholemman expression induces a hyperpolarization-activated chloride current in *Xenopus* oocytes. *J. Biol. Chem.* **267,** 14551–14554.
Morris, C. E. (1990). Mechanosensitive ion channels. *J. Membr. Biol.* **113,** 93–107.
Nagel, G. A., Hwang, T.-C., Nastiuk, K. L., Nairn, A. C., and Gadsby, D. C. (1992). The protein kinase A-regulated Cl channel resembles CFTR (Cystic Fibrosis Transmembrane conductance Regulator). *Nature (London)* **360,** 81–84.
Nakajima, T., Wu, S., Irisawa, H., and Giles, W. (1990). Mechanism of acetylcholine-induced inhibition of Ca current in bullfrog atrial myocytes. *J. Gen. Physiol.* **96,** 865–885.
Ono, K., and Trautwein, W. (1991). Potentiation by cyclic GMP of β-adrenergic effect on Ca^{2+} current in guinea-pig ventricular cells. *J. Physiol. (London)* **443,** 387–404.
Ono, K., Tareen, F. M., Yoshida, A., and Noma, A. (1992). Synergistic action of cyclic GMP on catecholamine-induced chloride current in guinea-pig ventricular cells. *J. Physiol. (London)* **453,** 647–661.
Overholt, J. L., and Harvey, R. D. (1992). Ionic selectivity of cAMP dependent chloride channels in isolated guinea pig ventricular myocytes. *Biophys. J.* **61,** A442.
Parson, T. D., Lagrutta, A., White, R. E., and Hartzell, H. C. (1991). Regulation of Ca^{+2} current in frog ventricular cardiomyocytes by 5′-guanylylimidodiphosphate and acetylcholine. *J. Physiol. (London)* **432,** 593–620.
Paulmichl, M., Li, Y., Wickman, K., Ackerman, M., Peralta, E., and Clapham, D. (1992). New mammalian chloride channel identified by expression cloning. *Nature (London)* **356,** 238–241.
Pfaffinger, P. J., Martin, J. M., Hunter, D. D., Nathanson, N. M., and Hille, B. (1985). GTP-binding proteins couple cardiac muscarinic receptors to a K channel. *Nature (London)* **317,** 536–538.
Picciotto, M., Cohn, J., Bertuzzi, G., Greengard, P., and Nairn, A. (1992). Phosphorylation of the cystic fibrosis transmembrane conductance regulator. *J. Biol. Chem.* **267,** 12742–12752.
Riordan, J. R., Rommens, J. M., Kerem, B.-S., Alon, N., Rozmahel, R., Grzelczak, Z., Zielenski, J., Lok, S., Plavsic, N., Chou, J.-L., Drumm, M. L., Iannuzzi, M. C., Collins, F. S., and Tsui, L.-C. (1989). Identification of the cystic fibrosis gene: Cloning and characterization of complementary DNA. *Science* **245,** 1066–1073.
Scamps, F., and Vassort, G. (1990). Mechanism of extracellular ATP-induced depolarization in rat isolated ventricular cardiomyocytes. *Pflüegers Arch.* **417,** 309–316.
Shenolikar, S., and Nairn, A. C. (1991). Protein phosphatases: recent progress. *Adv. Second Messenger Phosphoprotein Res.* **23,** 1–121.
Sorota, S. (1992). Swelling-induced chloride-sensitive current in canine atrial cells revealed by whole-cell patch-clamp method. *Circ. Res.* **70,** 679–687.
Sorota, S. (1993). Pharmacological inhibition of the swelling-induced chloride-sensitive current in dog atrial cells. *Biophys. J.* **64,** A210.
Sorota, S., Siegel, M. S., and Hoffman, B. F. (1991). The isoproterenol-induced chloride current and cardiac resting potential. *J. Mol. Cell. Cardiol.* **23,** 1191–1198.
Steinmeyer, K., Ortland, C., and Jentsch, T. J. (1991). Primary structure and functional expression of a developmentally regylated skeletal muscle chloride channel. *Nature (London)* **354,** 301–304.
Sullivan, M. M., Moss, R. B., Hindi, R. D., and Lewiston, N. J. (1986). Supraventricular tachycardia in patients with cystic fibrosis. *Chest* **90,** 239–242.

Tabcharani, J. A., Chang, X.-B., Riordan, J. R., and Hanrahan, J. W. (1992). The cystic fibrosis transmembrane conductance regulator chloride channel. Iodide block and permeation. *Biophys. J.* **62,** 1–4.

Takano, M., and Noma, A. (1992). Distribution of the isoprenaline-induced chloride current in rabbit heart. *Pflüegers Arch.* **420,** 223–226.

Tareen, F. M., Ono, K., Noma, A., and Ehara, T. (1991). β-adrenergic and muscarinic regulation of the chloride current in guinea-pig ventricular cells. *J. Physiol.* (*London*) **440,** 225–241.

Tareen, F. M., Yoshida, A., and Ono, K. (1992). Modulation of β-adrenergic responses of chloride and calcium currents by external cations in guinea-pig ventricular cells. *J. Physiol.* (*London*) **457,** 211–228.

Thiemann, A., Gründer, S., Pusch, S., and Jentsch, T. J. (1992). A chloride channel widely expressed in epithelial and non-epithelial cells. *Nature* (*London*) **356,** 57–60.

Tranum-Jensen, J., Janse, M. J., Fiolet, J. W. T., Krieger, W. J. G., D'Alnoncourt, C. N., and Durrer, D. (1981). Tissue osmolality, cell swelling, and reperfusion in acute regional myocardial ischemia in the isolated porcine heart. *Cir. Res.* **49,** 364–381.

Tseng, G.-Y. (1992). Cell swelling increases membrane conductance of canine cardiac cells: Evidence for a volume-sensitive Cl channel. *Am. J. Physiol.* **262,** C1056–C1068.

Vaughan-Jones, R. D. (1982). Chloride activity and its control in skeletal and cardiac muscle. *Philos. Trans. R. Soc. London, Ser. B* **299,** 537–548.

Vincent, G. M., Abildskov, J. A., and Burgess, M. J. (1974). Q-T interval syndrome. *Prog. Cardiov. Dis.* **6,** 523–530.

Walsh, K. B. (1991). Activation of a heart chloride current during stimulation of protein kinase C. *Mol. Pharmacol.* **40,** 342–346.

Walsh, K. B., and Long, K. J. (1992). Inhibition of heart calcium and chloride currents by sodium iodide. Specific attenuation in cAMP-dependent protein kinase-mediated regulation. *J. Gen. Physiol.* **100,** 847–865.

Weiss, M. S., Abele, U., Weckesser, J., Welte, W., Schiltz, E., and Schulz, G. E. (1991). Molecular architecture and electrostatic properties of a bacterial porin. *Science* **254,** 1627–1630.

Welsh, M. J., Anderson, M. P., Rich, D. P., Berger, H. A., Denning, G. M., Ostedgaard, L. S., Sheppard, D. N., Cheng, S. H., Gregory, R. J., and Smith, A. E. (1992). Cystic fibrosis transmembrane conductance regulator: A chloride channel with novel regulation. *Neuron* **8,** 821–829.

Yamawake, N., Hirono, Y., Sawanobori, T., and Hiraoka, M. (1992). Arrhythmogenic effects of isoproterenol-activated Cl^- current in guinea pig ventricular myocytes. *J. Mol. Cell. Cardiol.* **24,** 1047–1058.

Zakharov, S. I., and Harvey, R. D. (1993). Effect of tetramethylammonium on the isoproterenol activated chloride current in cardiac myocytes. *Biophys. J.* **64,** A389.

Zhang, J., Rasmusson, R. L., Hall, S. K., and Lieberman, M. (1992). Hypoosmotically activated Cl conductance in cultured chick heart cells. *J. Gen. Physiol.* **100,** 31a.

Zhang, K., and Ten Eick, R. E. (1993). A single type of chloride current (I_{Cl}) induced by activation of protein kinase A and C in feline ventricular myocytes. *Biophys. J.* **64,** A239.

Zygmunt, A. C., and Gibbons, W. R. (1991). Calcium-activated chloride current in rabbit ventricular myocytes. *Circ. Res.* **68,** 424–437.

Zygmunt, A. C., and Gibbons, W. R. (1992). Properties of the calcium-activated chloride current in heart. *J. Gen. Physiol.* **99,** 391–414.

Zygmunt, A. C., and Gibbons, W. R. (1993). Roles of the calcium-activated chloride current, $I_{Cl(Ca)}$, in the rabbit atrial action potential. *Biophys. J.* **64,** A239.

Index

C

D

E

Q

R

S

ISBN 0-12-153342-5

90018

9 780121 533427